Pollution of Lakes and Rivers

To my students, past and present

We possess nothing certainly except the past.
Evelyn Waugh, *Brideshead Revisited* (1945)

History . . . should be a vehicle rather than a terminus.
Modris Eksteins, *Globe and Mail* (1999)

Pollution of Lakes and Rivers
A Paleoenvironmental Perspective

Second Edition

John P. Smol

Blackwell Publishing

BLACKWELL PUBLISHING
350 Main Street, Malden, MA 02148-5020, USA
9600 Garsington Road, Oxford OX4 2DQ, UK
550 Swanston Street, Carlton, Victoria 3053, Australia

First edition published 2002 by Arnold
Second edition published 2008 by Blackwell Publishing Ltd

1   2008

Library of Congress Cataloging-in-Publication Data

Smol, J. P. (John. P.)
    Pollution of lakes and rivers : a paleoenvironmental perspective /
John P. Smol. – 2nd ed.
        p. cm.
    Includes bibliographical references and index.
    ISBN-13: 978-1-4051-5913-5 (pbk. : alk. paper)
    1. Lakes.   2. Rivers.   3. Paleolimnology.   4. Sediments (Geology)   I. Title.

    TD420.S63 2008
    363.739′4–dc22

                                                    2007016493

A catalogue record for this title is available from the British Library.

Set in 10.5/12.5pt Goudy
by Graphicraft Limited, Hong Kong
Printed and bound in Singapore
by Fabulous Printers Pte Ltd

For further information on
Blackwell Publishing, visit our website:
www.blackwellpublishing.com

Contents

Preface to the second edition

Look abroad through nature's range,
Nature's mighty law is change.
Robert Burns (1759–96)

We live in a constantly changing environment. Some changes are due to natural processes, but we now know that anthropogenic activities are responsible for many of the environmental problems we are currently facing. Humans have unintentionally initiated large, global "experiments," with processes such as titrating our lakes with acids from industrial emissions, to over-fertilizing our surface waters with sewage and agricultural runoff, to changing the physical and chemical properties of our atmosphere with greenhouse gas emissions, to synthesizing and releasing thousands of compounds into our biosphere. In most cases, we have little or no idea of the environmental repercussions of these actions until it is too late and the damage has been done. In almost all cases, we only start monitoring ecosystems *after* a problem is recognized.

We do not have a crystal ball to see where these environmental changes will take us, but we can learn many lessons from the past. Historical perspectives allow us to determine what an ecosystem was like before it was disturbed (providing realistic targets for mitigation) and to ascertain how the system changed as a result of human activities. Although very little long-term and pre-impact monitoring data are typically available for most ecosystems, much of this information is archived as proxy data in sedimentary records that can be interpreted by paleolimnologists in a manner that is useful to other scientists, environmental managers, politicians, and policy makers, as well as the public at large. As Confucius observed over two millennia ago, "Study the past to divine the future."

This book is about using the vast stores of information preserved in lake, river, and reservoir sediments to track past environmental changes. The first edition of this textbook was published in 2002, and I had no plans at that time to do a second edition in such a short time. However, the first edition sold out faster than I had initially anticipated, and the decision was then one of either simply reprinting the first edition or undertaking a revision. I chose the latter. The field of paleolimnology is rapidly evolving and improving, with new applications and approaches being developed at a frenetic pace. It is not possible to cover, even in a general sense, the myriad applications that paleolimnology offers in a book this size. I have therefore focused on problems dealing with the major forms of freshwater pollution. However, with the second edition, I have added a new chapter entitled "Greenhouse gas emissions and a changing atmosphere: tracking the effects of climatic change on water resources." The reasons for this new chapter were two-fold. First, it is clear that climate change is having myriad effects on water quality and other pollution issues, and so a short chapter describing some of the ways that paleolimnologists track past climatic changes appeared to be warranted. The second reason was more practical. Several professors had told me that they liked the first edition and used it in their upper year courses as a reference or text, but felt that, since paleoclimate research was only peripherally mentioned, it was not complete for a typical university course on paleoenvironmental research or global environmental change. I hope that this new chapter will at least partly fill this gap. In addition, a number of new sections and figures are now included throughout the book, as well as updates to many original entries from the first edition. About 250 new references are cited in this edition. Most of my examples continue to be from lakes, as the vast majority of published studies are from these systems. However, most paleolimnological approaches can also be applied to river and reservoir systems, provided that suitable sedimentary sequences are identified.

A frustrating aspect of writing this book was that I had to choose from so many good examples. Due to page limitations, I could only highlight a few studies, with my overall goal of showing a broad spectrum of applications. Almost all of these case studies are from Western Europe and North America, which is a reflection of where most of the applied paleolimnological work has been done to date. This, too, is changing, as scientists and managers become more aware that paleolimnological approaches can offer important and cost-effective answers to many serious water-quality questions, especially in developing countries. In fact, several of the new examples in this second edition are from these regions.

I have received excellent advice and assistance from many colleagues regarding the completion of this book. Foremost amongst these have been members of my laboratory, the Paleoecological Environmental Assessment and Research Lab (PEARL) at Queen's University in Kingston, Ontario, who read and commented on many sections of this book. Brian F. Cumming, William M. Last, and Andrew M. Paterson made especially thoughtful comments on overall structure and content. Many thanks are also due to Wiley-Blackwell Publishers, and especially my publishing editor Dr Ian Francis, for his assistance. Thanks also to my friend and colleague, Ray Bradley, for his encouragement to undertake this second edition. I also wish to acknowledge the original reviewers for their helpful comments.

As always, I owe a deep debt of gratitude to John Glew, who is my right hand (and often left hand as well) in the field and elsewhere, and whose immense artistic talents have enhanced the illustrations in this book. Jon Sweetman and Daniel Selbie were especially helpful in preparing the final photographic plates. I would also like to thank the many journal and book publishers, as well as individual scientists, who granted me permission to reproduce some of their figures.

The following individuals made important comments to various sections of the text and/or otherwise assisted with this volume: D. Antoniades, P. Appleby, J. Baron, R. Battarbee, B. Beierle, L. Benson, H. Birks, J. Birks, J. Blais, W. Blake, Jr, I. Boomer, D. Bos, R. Bradley, M. Brenner, J. Bright, K. Brodersen, R. Brugam, J. Casselman, M. Chaisson, D. Charles, S. Clerk, A. Cohen, S. Cooper, C. Dalton, F. Dapples, J. Dearing, P. DeDeckker, A. Derry, P. Dillon, A. Dixit, S. Dixit, M. Douglas, T. Edwards. A. Ek, S. Eisenreich, B. Finney, E. Fjeld, J. Flenley, R. Flower, R. Forester, F. Forrest, S. Fritz, A. Jeziorski, K. Gajewski, Z. Gan, R. Gilbert, B. Ginn, F. Ginn, J. Glew, I. Gregory-Eaves, C. Grooms, P. Guilizzoni, R. Hall, J. Havelock, M. Hermanson, D. Hodgson, E. Jeppesen, R. Jones, V. Jones, T. Karst, M. Kowalewski, W.C. Kerfoot, G. Kerson, K. Laird, S. Lamontagne, D. Landers, S. Lamoureux, D. Lean, P. Leavitt, M. Leira, L. Lockhart, A. Lotter, G. MacDonald, R. Macdonald, A. Mackay, H. MacIsaac, A. Martinez, M. McGillivray, R. McNeely, P. Meyers, K. Mills, N. Michelutti, B. Miskimmin, J. Morse, D. Muir, K. Neill, C. Nöel, P. O'Sullivan, A. Parker, A. Paterson, D. Pearson, R. Pienitz, M. Pisaric, S. Pla, R. Quinlan, E. Reavie, I. Renberg, S. Rognerud, N. Rose, K. Rühland, O. Sandman, C. Sayer, D. Selbie, C. Schelske, D. Schindler, R. Schmidt, W. Shotyk, F. Siegel, P. Siver, A. Smol, E. Stoermer, J.

Stone, A. Street-Perrott, J. Sweetman, E. Turner, D. Verschuren, W. Vincent, I. Walker, J. Webber, P. Werner, T. Whitmore, A. Wilkinson, A. Wolfe, B. Wolfe, N. Yan, and B. Zeeb. Any errors or omissions are, of course, my responsibility.

Funding for my research has come from a variety of sources, but primarily from the Natural Sciences and Engineering Research Council of Canada. I feel fortunate that I am able to pursue my research and teaching in what I believe is a remarkable country, at an excellent and supportive university, where I can participate in such an outstanding and rapidly growing field of science. I would also like to acknowledge the moral support supplied by my mother and my siblings, as well their families.

Paleolimnologists often tend to look backwards, and as I read over the final version of my text, I feel that I should acknowledge what will likely be obvious to any reader of this book – I think paleolimnology is wonderful! As such, I suspect some will find my views embarrassingly enthusiastic and perhaps hopelessly optimistic.

Certainly, as in all other branches of science, not all paleolimnological investigations are successful, nor can many conclusions be stated with sufficient confidence to warrant subsequent interventions and other actions. Nonetheless, I believe that there is now ample evidence that paleolimnological studies are consistently providing defendable answers to critically important questions. In many cases, due to the lack of monitoring data, proxy data provide the *only* answers attainable, and so even a partial answer is better than no answer at all.

I am excited and immensely proud of the accomplishments of my colleagues. The science of paleolimnology has moved rapidly and effectively to help solve many environmental problems, and has often blended itself seamlessly with many other disciplines. These are exciting times for paleolimnology. I hope this book captures some of this excitement.

John P. Smol

About the author

John P. Smol, FRSC is a professor in the Department of Biology at Queen's University (Kingston, Ontario, Canada), with a cross-appointment at the School of Environmental Studies. He is also co-director of the Paleoecological Environmental Assessment and Research Lab (PEARL), and holds the *Canada Research Chair in Environmental Change*. Professor Smol is Editor of the *Journal of Paleolimnology* (1987–2007), the journal *Environmental Reviews*, and the *Developments in Paleoenvironmental Research* book series. He has authored over 330 journal papers and book chapters, and has completed 15 books. Since 1990, he has won over 20 research awards and fellowships (including the 2004 *NSERC Herzberg Canada Gold Medal*, as Canada's top scientist or engineer), as well as five teaching awards.

1

There is no substitute for water

1.1 Our water planet

If extraterrestrial beings ever investigated this planet, it is doubtful they would name it *Earth*. They would almost certainly call it *Water*. Water covers approximately 75% of our world's surface, representing a volume of over one billion cubic kilometers. However, as noted later in this chapter, only a very small portion of this water is fresh and accessible. A second feature characterizing our planet is the incredible number and diversity of life forms. This truly is "the Living Planet" (Attenborough 1984). Water and life are intricately linked. Water makes up about 70% of our bodies. More than half of the world's species of plants and animals live in water, and even our terrestrial-derived food is totally dependent on and often largely composed of water. Civilizations have flourished and collapsed due to changing water supplies: Water can shape history; it can make or break a king (Ball 1999). It is not surprising that over 80% of Americans live within 8 km of surface waters (Naiman et al. 1995). Without water, there would be no life. Plain and simple.

The study of inland waters is called limnology. A common definition of a limnologist is someone who studies lakes and rivers, although the term includes those who work with other inland waters (e.g., ponds, swamps, saline lakes, wetlands). Limnology is a diverse science, which includes physics, chemistry, biology, geology, and geography, as well as a suite of other disciplines. It has strong links to applied fields as well, such as engineering and management. Many applications have a clear environmental focus.

Paleolimnology, which is the theme of this book, is the multidisciplinary science that uses the physical, chemical, and biological information preserved in sediment profiles to reconstruct past environmental conditions in inland aquatic systems. Paleolimnology has many applications. Although pollution studies make up a significant part of the literature, paleolimnological approaches are used to study a wide variety of basic and applied problems.

Most scientists seem to differentiate between limnology and paleolimnology, as if they are quite separate fields. As I have argued previously (Smol 1990a), this seems to me to be an artificial division. I prefer to use the word "neolimnologist" to designate scientists working with present-day aquatic systems, whilst "paleolimnologists" often address similar problems, but do so at much longer time scales, using sediments as their primary research material. Both are limnologists; it is primarily a distinction of time scales and temporal resolution. Whilst neolimnologists can typically use tools and approaches with higher resolution, paleolimnologists can extend many of these studies back in time, and back in time is where many of the critical answers to environmental problems are hidden. The two disciplines are tightly linked and complementary.

1.2 Water and aquatic ecosystems

Water (H_2O) is a peculiar compound and it is important to understand some of its characteristics (Box 1.1). In contrast to almost any other substance, water is less dense in its solid form (ice) than it is in its liquid form, and so ice floats. Water also has other interesting thermal and density properties in that it is most dense at about 4°C and less dense as it gets warmer, but it also becomes less dense as it gets colder, until it freezes at 0°C (these temperatures refer to pure water; water containing dissolved solutes, such as sea water, will freeze at lower temperatures). Because water layers at different temperatures will have different densities, three thermally (and hence density-) defined horizontal strata will often form in deeper lakes that have strong seasonal temperature differences (Fig. 1.1; the upper, warmer epilimnion; the middle portion of rapid temperature change, or the thermocline, which demarks the metalimnion; and the deep, colder layer called the hypolimnion, often at or near 4°C). In temperate regions, lakes are often stratified in this way throughout much of the summer but mix totally after ice-melt in the spring (spring overturn, which occurs when the lake is isothermal at or near 4°C) and again in autumn, as temperatures begin to cool and

thermal stratification weakens. Such lakes are called dimictic, because their water columns mix twice every year. In subtropical regions, only one mixing event in the cold season may occur (monomictic lakes), and other variations occur (a number of limnology textbooks deal with this topic in much more detail: Kalff 2001; Wetzel 2001). These stratification patterns have important ecological and environmental implications. For example, on a very windy, mid-summer day, a lake may appear to be well mixed, with large surface waves. But this is not really the case, as these waves are largely a surface phenomenon, and the deeper waters are still largely segregated into three layers. Thermal stratification is often weak or non-existent in shallow lakes or rivers, where wind mixing or currents are stronger than the density gradients set up by temperature differences, and so these systems mix frequently (polymictic).

In addition to vertical stratification, there are also marked horizontal differences in lakes and rivers (Fig. 1.1). Water depth (and associated light penetration) is often a major factor controlling horizontal zonation in water bodies, with of course shallower water occurring near the shore and deeper water typically farther from the shore (Box 1.2). The littoral zone is often defined as the part of the lake or river where rooted aquatic plants (macrophytes) can grow. This growth is primarily depth dependent, although it can be altered greatly by water clarity (for example, two lakes may have similar morphometries, but the lake with the clearer water will typically have the larger littoral zone). Shallower sites are often referred to as ponds; the distinction between lakes and ponds, though, is not standardized. For example, in some regions, a pond is defined as a water body where the entire bottom can support rooted aquatic macrophytes (i.e., the entire pond is technically a littoral zone). This definition, however, is difficult to apply universally. For example, some highly productive waters are quite shallow and so might intuitively be considered ponds, but because of the large amount of material in their water columns, the photic zone is much reduced and rooted macrophytes cannot grow. In polar

Box 1.1 The major properties of water

The water molecule consists of two hydrogen atoms bound to an oxygen atom, forming an isosceles triangle. Water molecules are attracted to each other, creating hydrogen bonds, which influence many of the physical as well as chemical properties of water. Pure water at sea level freezes at 0°C and boils at 100°C. At higher elevations, the boiling point of water decreases, due to the lower atmospheric pressure. If substances are dissolved in the water, the freezing point is lowered (this is why salt is often applied to some roads in winter; it deters ice from forming).

Like other liquids, the density of water is very closely related to temperature, as well as to the amount of solutes it contains. Perhaps the most striking feature of water is that it is less dense in its solid form (ice) than it is in its liquid form, and so ice floats. The highest density of pure water is reached just below 4°C, and so water of this temperature is often found in the deep waters of a lake. The addition of solutes also increases the density of water markedly (i.e., salty water will be denser than fresh water). Both these features influence thermal and chemical stratification patterns in lakes, with important environmental consequences.

Water has a very high specific heat, which is the amount of energy needed to warm or cool a substance. It has amongst the highest specific heats of any naturally occurring substance. This is why, for example, it is often cooler around an ocean or a large lake during summer, as the water remains cooler longer than land. Similarly, regions around large bodies of water are warmer in winter as the water is slower to cool. People who live close to large bodies of water are often said to enjoy a maritime climate, with reduced climatic extremes between the seasons. Farmers benefit as well, as crops are protected from nighttime freezing. In contrast, regions far inland are often said to have continental climates, with striking seasonal changes in temperature. In short, large bodies of water are the world's great heat reservoirs and heat exchangers. People exploit the high specific heat capacity of water when they bring hot water bottles into their beds on cold nights.

Water also has an extremely high surface tension, which is a measure of the strength of the water's surface film. Of the other liquids one is likely to encounter, only mercury (Chapter 10) has a higher surface tension. The epineuston is a community that takes advantage of this surface tension and lives on the surface of the water. Water striders (Gerridae) are insects commonly seen on the surfaces of lakes and ponds; they rely on this surface tension to walk on the water's surface. Another group of organisms, collectively referred to as the hyponeuston, live below the water line, but again attach themselves to the water's surface. This surface tension can also be a deadly trap for some organisms. For example, insects that touch the water may not be able to release themselves from this force.

One of the most important characteristics of water is that it is almost the universal solvent, with extraordinary abilities to dissolve other substances. Consequently, when water passes through soils or vegetation or a region of human activity (e.g., an agricultural field treated with fertilizers and insecticides, a mine tailings heap, a municipal or industrial landfill site, etc.) it changes its characteristics as it dissolves solutes. Even a drop of water falling as rain will dissolve atmospheric gases, and its properties will be altered (e.g., carbon dioxide dissolves readily in water, forming a weak acid, carbonic acid).

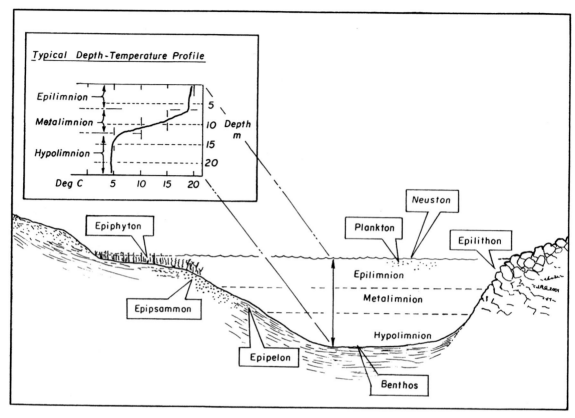

Fig. 1.1 A cross-section of a thermally stratifying lake, showing the density separation into three layers: the epilimnion, the metalimnion, and the hypolimnion. The lake cross-section also shows the major habitats of a lake, as described in Box 1.2.

regions, a pond is often defined as a water body that freezes completely to the bottom in winter.

Water is always on the move (Box 1.3). Using energy from the sun, over one thousand cubic kilometers of water evaporates from this planet every day. But then this water condenses and returns to Earth in precipitation, which may be in a liquid (rain), solid (snow), or gaseous (water vapor, fog) form. All of these phases can be affected by human activities. Once reaching the Earth, water may be deposited directly into lakes and rivers, but it is more likely to be intercepted by vegetation, or to become part of the surface runoff, which eventually makes its way to rivers and lakes and eventually to the oceans. Some

precipitation may land in polar or high alpine regions, and become frozen and then stored in ice caps and glaciers for millennia. Other water may percolate down through the soil and become part of the groundwater system, where it too may remain for a long period of time before resurfacing. About every 3000 years, this movement of water from the Earth to the sky and back again recycles an amount of water equivalent to the volume of the world's oceans.

Human activities alter the quality and quantity of water passing through all these systems by their activities, often with negative repercussions. This book will examine the repercussions of these activities.

Box 1.2 Horizontal zonation and the major habitats and communities in aquatic ecosystems

The littoral zone is typically defined as the shallow part of a lake where rooted aquatic macrophytes can grow. Although the littoral zones of some lakes are not very large, a significant, sometimes dominant, portion of the overall production occurs there. Many organisms, such as some fish, feed, find refuge, and reproduce in these areas. Since littoral zones are the closest to land (and to human influences), they are often critical areas in pollution studies.

Organisms, such as algae, living in the littoral zone are often attached to a substrate and are collectively called the periphyton. Substrates typically delineate this region (Fig. 1.1). For example, the epiphyton are organisms living attached to plants, whilst the epilithon are attached to rocks and stones, the epipsammon are attached to sand grains, and the epipelon live on the sediments.

The deeper, open-water region is often referred to as the pelagic region. Plankton live unattached to any substrate in the open-water system, and are largely at the mercy of water movements, although some have limited motility (e.g., with flagella). Plankton that are photosynthetic, such as algae, are called phytoplankton; those that are more animal-like, such as water fleas (Cladocera), are called zooplankton. Animals with strong locomotory capabilities, such as fish and large invertebrates, are called the nekton.

A smaller, poorly studied community, collectively called the neuston, lives closely associated with the surface tension of the water's surface. Organisms living in such a manner above the water line are called the epineuston; those below the water line are called the hyponeuston.

The profundal region typically refers to the deep waters in the middle of the system. Organisms living on and in the sediments are called the benthos.

Box 1.3 The hydrological cycle

Water is always on the move, cycling through the various compartments of our planet (e.g., the atmosphere, the lithosphere, the hydrosphere; Fig. 1.2). Driven by energy from the sun, water evaporates as water vapor into the atmosphere, where it can be transported long distances. During this gaseous phase, it can be transformed and contaminated by a variety of pollutants (e.g., acids). Eventually, the water vapor will condense and precipitate back to Earth in either a liquid (rain) or solid (snow) phase.

Some precipitation falls directly on water bodies, such as lakes and rivers, and so will not be directly influenced by watershed characteristics. However, catchments typically have much larger surface areas than the water bodies they drain into, and so much of the incoming water precipitates on land. Water quality is influenced by natural characteristics of the drainage basin (e.g., local geology, vegetation) as well as anthropogenic activities (e.g., industrial contaminants, agricultural fertilizers and pesticides, etc.). Because of water's ability to dissolve many substances (see Box 1.1), and the fact that moving water is often a very effective vector for eroding and

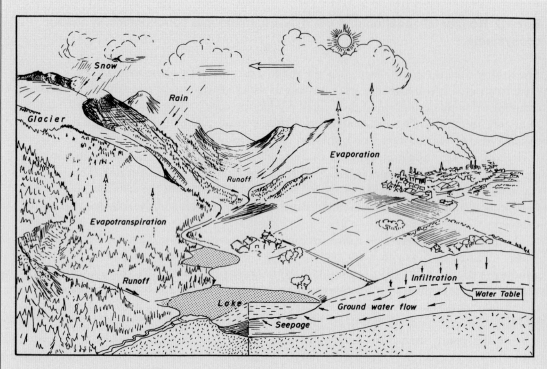

Fig. 1.2 A schematic diagram showing the major processes involved in the hydrological cycle.

transporting particulates, the water that eventually enters a lake or river can have dramatically different water-quality characteristics than the water droplets that had condensed in the atmosphere.

Moreover, not all the precipitation falling on a catchment will be flushed into the receiving water body. Some will be re-evaporated back into the atmosphere. Some of the water may percolate into the groundwater, where it will continue to move, albeit very slowly. It may stay underground for thousands of years before it resurfaces. The water quality may change dramatically while it is underground, as it is naturally filtered by passing through the cracks and pores in the soil and rocks. In addition, due to water's high solubility characteristics, it is constantly dissolving substances with which it is in contact. It may also be contaminated by human-produced compounds, such as pesticides.

Water may be intercepted by vegetation, taken up by plants as part of their physiological activities, and then released back into the atmosphere through the stomata in their leaves. For example, the roots of a typical birch tree may take up over 300 liters of water each day, but then the tree transpires much of it back into the atmosphere. This is called evapotranspiration. A good way to show how significant terrestrial vegetation is in evapotranspiring water is to remove it (e.g., logging), and then see how much more water is released from the catchment via overland flow (Likens 1985).

1.3 Water: a scarce resource, becoming more valuable daily

Although water characterizes this planet, much of it is not available to humans in a form that can readily be used as a source of drinking water or other uses (e.g., agriculture, industry), as the vast majority of it is salty (~ 95.1%) and contained in the world's oceans (Speidel & Agnew 1998). Of the small percentage that is fresh water[1] (~ 4.9%), most of this is inaccessible in groundwater or frozen in glaciers and polar ice. Once you account for this and other sources of water, only about 0.01% of the vast amounts of water on this planet is fresh and available as surface waters in a liquid form – equivalent to about a drop in a bucket. Of this small percentage, much of it is contained in lakes and rivers that are far from human populations. Hence, many of our water reserves are simply not readily accessible.

Certain regions are blessed with large amounts of fresh water. Wetzel (2001) estimates that the Earth contains approximately 100 million lakes that are greater than 1 ha in area, and approximately one million lakes that exceed 1 km^2. However, other regions, including areas of high population growth, have very serious water problems. Fresh water is a very valuable resource, and is getting more valuable every day. Each day, our limited water supplies have to be shared by a larger population. You can live without oil, but you can't live without water. As plainly stated by the 19th-century American author, Mark Twain: "Whiskey is for drinking; water is for fighting."

---

[1] There are a number of salinity boundaries used in the literature delineating fresh from saline waters. Most biologists define fresh waters as those with less than 1 g l^{-1} salt concentration, although 3 g l^{-1} is also commonly used as a boundary, especially amongst geologists.

1.4 Human influences on water quality

The Earth is about 4.5 billion years old and certainly many natural environmental changes have occurred on this planet during that time. Hominids only evolved a few million years ago, and our species, *Homo sapiens*, has only been around for about the past 150,000 years. If we put the 4.5 billion year history of the Earth on a 24-hour clock, the arrival of our species would only occur in the last 3 seconds before midnight! However, unlike any other species, our short visit on this planet has been very influential.

Throughout most of this approximately 150,000 years, humans made little impact on the environment. Their numbers were too small, too dispersed, and their technology was so primitive that they made little impact on the landscape except for footprints. Things began to change, though, about 10,000 years ago, when the experiment called "civilization" began. Starting at similar times in regions ranging from the Far East and the Near East to the west coast of Peru, people began clustering in larger groups with the development of agriculture. New technological advancements came shortly afterwards, as did food surpluses. Populations started to grow, slowly at first, but then faster – much faster. People first clustered in small groups, then towns, and eventually cities. The impact of humans on the environment grew in step with the population increases, but not on a linear scale. Human influences, due to continued technological developments, increased far more than their rising numbers might first predict. Fifty people with Stone Age technology made far less environmental impact than 50 people with our present-day technology. The human "footprint" became much larger and deeper very quickly, and the potential for a crisis began to loom. Anthropogenic impacts have become so profound that Crutzen (2002) has proposed the designation of a new geological epoch, called the Anthropocene, representing the recent human-dominated period of Earth's history.

The human population reached the 1 billion persons level around 200 years ago. Since that time, our population has grown to over 6 billion. Each day, about 200,000 more people are born than die. Humans need about 50 liters of water a day to stay healthy (e.g., for drinking, cooking, washing, sanitation). With increasing populations and increased technological growth, the ecosystems we depend on are under greater stress. The Earth's supply of accessible fresh waters is especially at risk. One out of three people in the developing world does not have access to safe drinking water. The United Nations Environment Programme (UNEP) estimated in 1999 that the present shortage of clean water will only get worse, to the point that by AD 2024 two-thirds of the world's population will not have adequate access to clean drinking water. In addition to municipal uses, water provides us with many other services, such as fisheries and other food resources, transportation, recreation, and other societal needs (Postel & Carpenter 1997). Trends outlined in the recent *Millennium Ecosystem Assessment* (2005) are even less encouraging. The quantity of water will not increase; it will stay the same. The quality of water will almost certainly decline further. Unlike resources such as coal, oil, or wheat, there is no substitute for water. We have a crisis on our hands, and we have to explore and develop new approaches to meet these challenges.

A major problem that faces water managers is that much of what happens in lakes and rivers is out of the sight of the public, below the water line. People tend to respond to obvious problems and symptoms, such as fish carcasses accumulating on a lakeshore or algal blooms fouling water intakes. However, many water problems begin with few obvious signs, but then steadily escalate to full-blown crises. It is often more difficult, or impossible, to correct the problem once it reaches these late stages. Extinction is forever. As Ricciardi and Rasmussen (1999) remind us, since 1900 some 123 freshwater animal species have become extinct in North America alone, and hundreds of other species are considered

imperiled. These alarming trends in loss of biodiversity appear to be accelerating each year (Millennium Ecosystem Assessment 2005). Human impacts are clearly implicated in almost every case. The numbers are staggering. For example, Ricciardi and Rasmussen (1999) estimated that species losses from temperate freshwater ecosystems are occurring as rapidly as those occurring in tropical forests. Surprisingly, these data from aquatic ecosystems receive little attention compared to the headline news of species diversity issues from tropical terrestrial ecosystems. We clearly need to use all the information we can garner to help managers and policy makers make responsible decisions to preserve these ecosystems.

1.5 Ecosystem approaches to water management

Lakes, rivers, and other water bodies cannot be treated as isolated systems. Although 19th-century naturalists at times referred to lakes as microcosms, often believed to be separate from the environment around them, we now know this is certainly not the case. Freshwater systems are intimately linked to their catchments or drainage systems (watersheds), as well as to the "airsheds" above them. They are also influenced by the quantity and quality of groundwater inflows. Lakes and rivers are downhill from human influences, and so much of what happens in a lake's catchment or watershed eventually has an effect on the aquatic system.

Environmental research requires this ecosystem perspective, involving trans-disciplinary studies explaining the processes and interactions among air, water, land, ice, and biota, including humans. Humans have currently transformed somewhere between one-third to one-half of the Earth's surface with their activities, such as agriculture and urbanization, and the rate of land transformation is increasing at an alarming rate. Furthermore, some industries have been treating the atmosphere as an aerial sewer, with the airways now transporting a suite of pollutants that can affect aquatic systems. The wind carries no passport; we are truly living in a global world of pollution transportation and deposition. Humans are capable of altering and affecting each part of the hydrological cycle (Fig. 1.2), whether via chemical, physical, or biological means. These repercussions, and some of the ways we can track and study them, are described in this book.

1.6 Pollution

Webster's dictionary defines the verb "pollute" as "to make foul." There is, however, a large range in the extent of pollution. As will be shown in the subsequent chapters, humans have now impacted all water bodies on this planet to some extent. Some impacts are minor and barely perceptible; others have degraded lakes and rivers to enormous degrees. This pollution is often chemical (e.g., contaminants, acid rain, nutrients) or physical (e.g., heat inflows), although "biological pollution," such as the introduction of exotic species, is also considered a form of pollution in its broadest sense.

Pollutants originate from many sources and are classified a number of ways. Pollutants are often discussed, in a general way at least, with respect to their source, and are divided into "point source" and "non-point source," or "diffuse source," pollutants (Fig. 1.3). A point source pollutant is simply one originating from a clearly defined source, such as a smoke stack from a smelter, an outflow pipe from a factory, or untreated sewage draining into a river. These point source pollutants are often easier to control, as they have a clearly defined origin, they can often be readily

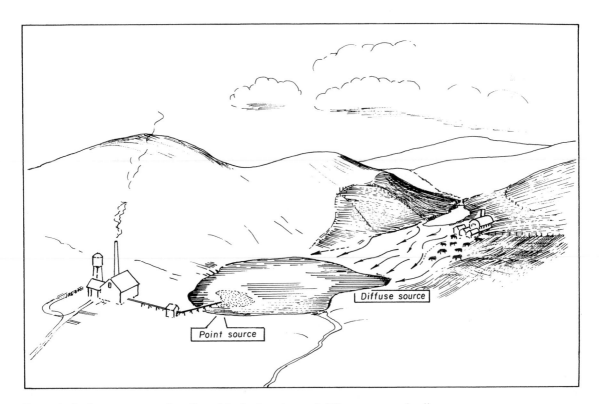

Fig. 1.3 Surface waters can be affected by both point and diffuse sources of pollutants.

measured and monitored, responsibility can be more easily assigned, and often technological fixes (e.g., a scrubber for a smoke stack or a treatment facility for an outflow) can be implemented more effectively and economically. Non-point or diffuse sources of pollution are often more challenging for managers and scientists. Pesticide or fertilizer runoff from agricultural fields in a lake's catchment would be examples of non-point sources.

1.7 The format of this book

This book is about the effects that humans have had on aquatic systems, and the ways in which we can study and assess these problems using long-term perspectives. Following this brief introduction, I discuss the need for long-term data (Chapter 2) and introduce the science of paleolimnology (Chapter 3). A brief overview of the basic techniques used by paleolimnologists and "thumbnail" sketches of some of the primary environmental indicators found in sediments are included in Chapters 4–6. I then explore some of the major pollution problems affecting freshwater ecosystems and show how paleolimnological approaches can be used to assist scientists, managers, and the general public. The first application deals with acidification (Chapter 7), a subject that received intense attention in the 1980s. Paleolimnology played an important role in these debates, and many of the paleolimnological tools that are used in applied studies were honed at this time. Discussion of other contaminants, such as metals and persistent organic pollutants, follow in the next chapters (Chapters 8–10). Collectively, these early chapters are loosely clumped together, as they often have an atmospheric transport component. A series of chapters more closely related to watershed and terrestrial changes follow (e.g., eutrophication, erosion; Chapters 11 and 12). The applications part of this book concludes with chapters that might best be described as dealing with relatively newer

environmental problems, such as exotic species invasions and extirpations (Chapter 13), greenhouse gas emissions and the resulting effects of climatic change on water resources (Chapter 14), ultraviolet radiation penetration (Chapter 15), taste and odor problems, and a suite of developing issues (Chapter 16). I conclude with some personal perspectives (Chapter 17) and a Glossary.

The paleolimnological literature is expanding very rapidly, and so many good examples could not be used. I acknowledge that I have tended to use studies with which I am most familiar, but I have attempted to provide examples that are applicable to other regions. My focus is on freshwater lakes, as almost all the pollution-related work to date has dealt with these systems. Consequently, I will often use the terms "lake" and "lake sediments" in this book, but in almost all cases these examples should be taken in a more collective sense, as most standard paleolimnological approaches, with some modifications and due caution, can be used in other freshwater systems, such as rivers, reservoirs, and wetlands. I have included such examples when possible. I have tried to emphasize ecosystem approaches, where individual symptoms are not treated in isolation, but examined and interpreted within a broader, more ecologically relevant framework.

1.8 Summary

Our planet is characterized by large volumes of water that cover about three-quarters of the Earth's surface. The vast majority (~ 95%) of this water is salty, and most of the fresh water is stored as groundwater or frozen in glaciers and polar ice. Only a tiny fraction (~ 0.01%) of our water resources is contained in freshwater lakes and rivers. The volume of water on our planet will not change, but its quality can be dramatically altered by human activities. With the establishment of civilizations, the development of technological advances, and a burgeoning human population, we have been polluting and altering our water supplies at an alarming rate. Degradation of lakes and rivers has taken many forms, ranging from the direct input of chemicals and other substances into receiving waters to more indirect sources of pollution, such as the emission of pollutants into the atmosphere that may then be transported long distances before they are redeposited back to Earth. Water is essential to life; we cannot live without it. Many researchers believe that we are fast approaching a crisis in our water supplies. This book explores the diverse ways in which paleolimnological approaches can be used to address some of these pressing environmental issues.

2

How long is long?

The farther backward you can look, the further forward you are likely to see.
Winston Churchill (1874–1965)

2.1 History as a vehicle for exploration and study

History is not just a retelling of the past. It provides context. It tells us where we were, and where we may be going. It reminds us of our successes and of our failures. The past can encourage us, but it also warns us. A central focus in this book is that history plays a fundamental role in environmental issues.

Observers of trends in society's attitudes have repeatedly noted that people tend to become more interested in history when they are worried about the future. There is much cause to worry about the future of our planet and the quality of our water supplies. It is not surprising, then, that interest in the historical development of environmental problems has increased in recent years.

As I (e.g., Smol 1990a,b, 1992, 1995a) and others (e.g., Anderson & Battarbee 1994) have argued in previous reviews and commentaries, historical perspectives often provide vital information for environmental assessments. Without long-term data, it is not possible to show how much a system has degraded (or recovered), is it possible to set realistic mitigation goals, nor determine what levels of disturbance have

elicited negative consequences in ecosystems, or determine the trajectories of environmental change. Each of these issues has a temporal component that requires long-term data.

But what kinds of long-term data sets are accessible to ecologists and environmental managers? The short answer is: "Not very good ones." I will use a few examples to emphasize this point.

2.2 How unusual are unusual events?

In 1986, Weatherhead asked the simple but powerful question: How unusual are unusual events? He wondered how often ecologists invoked "unusual events" to explain the outcome of their observations. He questioned whether these were truly unusual events or whether they were simply artifacts of the observational approach and perspective of the investigators.

Weatherhead explored these issues using a library search of 380 published ecological studies (averaging about 2.5 years in duration) and found that an astounding 11% of these studies invoked unusual events to explain the outcome of their observations. The first conclusion from this study was that supposed "unusual events" are not so unusual after all! This finding is especially relevant as none of these 380 papers was actually published to report specifically on the "unusual event."

A second finding of Weatherhead's analysis was that the duration of a study greatly influenced interpretations and conclusions. Put more plainly, the longer the study, the clearer were the interpretations. He showed that there was a relationship between the length of observations and the frequency with which researchers invoked unusual events to explain their observations. As one might expect, the frequency of reported "unusual events" increased with the length of studies, but interestingly this trend continued only until the length of study increased to about 6 years in duration. For longer-term studies (up to 15 years of observations), the reporting of unusual events actually decreased. When ecologists had access to longer-term data sets, "unusual events" became less common or, more correctly, events were no longer perceived as unusual, given the benefit of a longer-term perspective that sensitized investigators to the range of natural variability. As Weatherhead (1986, p. 154) concluded, ". . . we tend to overestimate the importance of some unusual events when we lack the benefit of the perspective provided by a longer study."

2.3 Managing aquatic ecosystems: a matter of time scales

Weatherhead's (1986) analysis warned us that the time frame we are using can influence our interpretations of observational data. But what types of data sets are we using as a basis for environmental decisions? Tilman (1989) undertook a library study to begin to answer this question. He documented the duration of field studies published in the internationally respected journal *Ecology* for a randomly selected subset of papers, encompassing a total of 623 observational and experimental studies. Of the observational studies, about 40% were under 1 year in duration (and

many were less than one field season), and 86% lasted less than three field seasons. Fewer than 2% of field investigations continued for more than 5 years. His analysis of experimental studies was even more biased toward short-term data sets.

Many of the papers published in *Ecology* are terrestrially based, and so I attempted a similar survey of 111 papers published in one randomly selected volume of *Limnology and Oceanography* (Smol 1995a). Over 70% of these aquatic studies were less than 1 year in duration. Environmental monitoring studies do not fare much better. A similar survey of 302 papers published in the international journal *Environmental Monitoring and Assessment* revealed that, again, over 70% of the monitoring and assessment studies were 1 year or less in duration (Fig. 2.1). By and large, ecologists, limnologists, and environmental scientists are using short-term data sets.

Before I continue to build the case for the need for long-term data, I acknowledge that short-term studies are often entirely appropriate for many types of investigations. For example, some of the aquatic papers cited above were physiological, and short-term experiments and observations may answer important questions. Moreover, it is not hard to see why most field-based ecological and environmental studies rarely exceed 3 years in length. This is the typical duration of a research grant, and it is the average length of time most universities would recommend their doctoral students devote to field data collection. But is it enough time to determine the range of natural variability? Can we distinguish reliably the rate and magnitude of environmental change? Can we assess ecosystems without knowing what they were like before human impacts? Can we distinguish between "environmental noise" and sustained environmental changes? Given the quality and short duration of most of the available data sets, it is not surprising that ecologists and other scientists are invoking "unusual events" to explain their observations (Weatherhead 1986).

Almost all ecological and environmental work is based on data gathered over a very short period (i.e., a few years of data, if lucky, but often

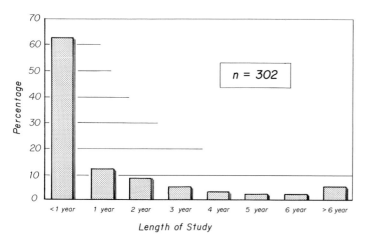

Fig. 2.1 The duration of studies published in the international journal *Environmental Monitoring and Assessment* between 1981 and 1993. A total of 302 papers were used in my analysis. Over 70% of papers published during this time period were based on observations of 1 year's duration or less. A survey of more recent papers (2006) indicated that most environmental monitoring is still based on relatively short-term data sets.

much shorter). Most environmental decisions are based on "the invisible present" (Magnuson 1990), with little understanding of how the system came to its current state. Much like someone taking a photograph of a train entering a station, we may have a good idea of what the train looks like now, but we have no idea where it came from or how it got to its present state. Managers are often presented with a system that is in some way altered, with little if any information on how it got to its present state. Almost all environmental assessments are done after-the-fact, after a problem has been identified. We have to find ways to separate important ecological and environmental "signals" from the "noise" that characterizes most ecosystems (Box 2.1).

Box 2.1 The importance of time scales in separating "noise" from "signal": a neolimnological example

An example from the Great Lakes illustrates how long-term monitoring data are often needed to put environmental changes into perspective. Nicholls (1997) explored how interpretations of data can be dramatically different when one is presented with, in his case, a relatively short-term (5–10 year) data set with longer-term data (30 years). Lake Erie (USA and Canada) has had two major interventions over the past 30 years, one planned (phosphorus reduction) and the second unintentional (the introduction of zebra mussels). Both would have had the effect of reducing algal standing crop. When Nicholls examined the short-term data (i.e., 5–10 years, which, for most monitoring programs, would actually be relatively long-term data!), one could easily have concluded that the invasion of the zebra mussels had dramatic consequences on algal concentrations, as a comparison of a few years of pre-invasion data with a few years of post-invasion data showed a reduction in green algal density of 85%. However, when viewed in the context of a 30-year data set, the 1988 arrival of the zebra mussels was now barely perceptible in the algal data, with much more significant declines in chlorophyte (green algae) densities beginning well before the arrival of this exotic bivalve. In fact, the annual chlorophyte density in the few years before the arrival of zebra mussels was only approximately 6% of their density during the 1970s (before phosphorus abatement programs were in place). The longer-term decline in chlorophyte abundance places the more recent effects of zebra mussels in their proper perspective, and underscores the need for long-term monitoring data. Short-term changes may simply be part of long-term noise.

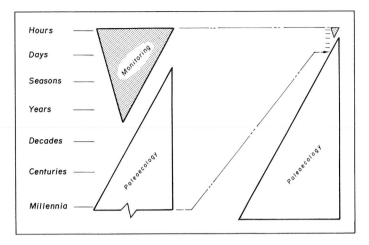

Fig. 2.2 Environmental monitoring – a matter of time scales. Ecosystem managers should ideally use a continuum of techniques, ranging from laboratory studies and bioassays (typically working on time scales of hours to days), to field studies (often representing seasonal or a few years of observations), to paleoenvironmental studies. With the tremendous advancements in techniques and the increased temporal resolution available to paleolimnologists, these approaches can now be used to dovetail with neolimnological approaches, and so allow managers to extend their sampling programs back in time. The time scales shown in the figure to the left are on a logarithmic scale; the figure to the right shows a more realistic representation of the relative amount of information potentially available from sedimentary deposits! Modified from Smol (1992).

Day-to-day environmental monitoring can be expensive and uninteresting. Both claims are, to some degree, true. But there are now alternative ways to obtain the long-term data sets for the variables we never bothered to monitor, or did not know how to monitor, or could not afford to monitor. In some cases, no one was even around to monitor some ecosystems. In this book, I hope to demonstrate that longer-term data are required and often attainable, at least in an indirect manner, from paleoenvironmental studies (Fig. 2.2). Aquatic sediments archive a tremendous library of information that can be interpreted by paleolimnologists, and then used by managers, other scientists, and the public at large.

2.4 A medical analogy to "ecosystem health"

A concept that has been espoused by some managers is "ecosystem health" and the melding of medical criteria with those used for environmental assessment. A medical analogy to aquatic ecosystems might be useful, but it does force some important comparisons (Smol 1992). As noted in a book review I read on medical history, the "retroscope" remains medical science's most accurate instrument. The same may be true of aquatic ecosystem health.

"What is a healthy ecosystem?" is a fundamental question for managers. I have argued that a healthy ecosystem is one that has similar characteristics to those that were present before significant human impacts occurred (Smol 1992). These data are important for determining whether a problem exists and for setting realistic mitigation goals or target conditions. To use the medical analogy, one of the first things a physician will do with a new patient is to obtain a medical history. In order to treat a problem, it is important to know the progress of any disease or condition. Although probably put more diplomatically by a medical doctor, it is also important to ask "Are you sure that you are sick?" The same is true of ecosystems. How can we treat an ecosystem if we do not know what it was like before it was polluted? For example, was the lake naturally acidic or naturally eutrophic, or did it naturally have large growths of aquatic macrophytes? Is any intervention justified? If so, to what conditions or attributes do we want to restore the system? (What was it like before it got "sick"?) Historical data, just as those supplied in a patient's medical history, are required to answer these questions.

In some cases, ecosystems may not yet be "sick" but are believed to be showing signs of "stress," similar to a patient who has been exhibiting some symptoms but has yet to have a diagnosed disease. The word "stress" has been used by ecologists in a number of ways, but, from an ecosystem perspective, few would argue with Schindler's (1990) definition: ". . . to denote any human agent that causes an ecosystem or community to respond in some way that is outside of its natural range of variation . . .". A medical definition of stress would likely be similar to the one noted above. But again, without an ecosystem or medical history, how do we know what is "outside of its natural range of variability"? Long-term data are needed.

Medical doctors know very well that not all patients are equally resistant to diseases. The same is true for water bodies. Based on their pre-disturbance (i.e., "healthy" characteristics), some lakes, for example, will be more susceptible to the effects of acid rain or to inputs of other contaminants. Establishing the susceptibility of ecosystems is an important management issue. Important issues include the determination of "critical loads" or thresholds (Box 2.2) for different types of ecosystems (see, for example, Chapter 7). Once again, long-term data are needed. By knowing when different types of systems began to show signs of impacts, these critical loads (and hence allowable pollution loads), and the resiliency of different ecosystems, can be estimated. With these data, the trajectory of ecosystem impairment (or disease) can be followed.

Just as it is important to study the development of disease, it is also important to monitor any recovery following treatment. Using the medical analogy, patients are advised to continue seeing their doctors following the prescription of a medicine or other intervention to determine if the treatment is working. However, just as there is very little monitoring data available documenting ecosystem degradation, there is often surprisingly little data gathered after implementation of mitigation efforts to determine if a system is responding. For example, are lakes increasing in

---

Box 2.2 Critical and target loads for pollutants

Determining the level of permissible pollution for an ecosystem forms a major part of environmental science and policy. Calculating the tolerable level of pollutants is often difficult. Critical loads (Fig. 2.3) are defined as the highest load of a pollutant that will not cause changes leading to long-term harmful effects on the most sensitive ecological systems (Nilsson 1986). As noted by the Swedish NGO Secretariat on Acid Rain (1995), this definition allows much room for interpretation.

Target loads are determined using similar criteria, but take other factors into consideration, such as national environmental objectives, tempered with financial and social considerations. As such, the target load may be lower (allowing for a margin of safety) or higher (reflecting a deliberate acceptance of some environmental damage or risk that is deemed acceptable) (Fig. 2.3).

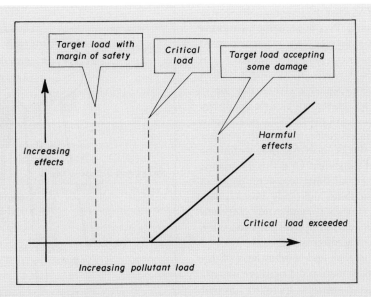

Fig. 2.3 A graph showing the effects of an increasing load of a hypothetical pollutant, showing the critical and target loads. Modified from the Swedish NGO Secretariat on Acid Rain (1995).

lakewater pH following declines in acidic emissions or after liming? Without monitoring data, these questions remain unanswered.

Most managers and politicians stress "looking forward." I would agree. But I feel we also have to "look backward" – for back in history is much of the information that we will need for effective management of our water bodies. Patients have memories and maintain medical records. Lakes and other water bodies accumulate sediments, which contain their records of environmental change. As Søren Kierkegaard (1813–55) astutely noted, "We live forward, we understand backward."

The proxy data contained in sediment profiles is not always clearly "written," nor is it always easy to "read" (Frey 1974). At times, it seems that some of the pages are missing or the lettering is smeared. But our abilities to read and interpret this information have increased at such a rapid rate that these approaches can now be applied to studies of water-quality degradation and recovery.

2.5 Requirements for effective ecosystem management

As discussed above, several important questions need to be addressed for effective ecosystem management, each of which has an important temporal component. These questions include the following:

1 What were the baseline (i.e., pre-impact, background or sometimes referred to as nominal or reference) conditions, prior to human disturbances? Without knowing what a system was like before a disturbance, how can we attempt to gauge the extent of a problem? Furthermore, how can we set realistic mitigation goals? We need to know what our restoration targets should be. For example, was the lake naturally acidic or eutrophic? Were deepwater oxygen depletions natural phenomena? Almost all environmental assessments are done after a problem has been identified, and so crucial

data are missing on what the system was like before human interference. People were not monitoring the system, but sediments were accumulating.

2   What is the range of natural variability? Ecosystems are often naturally "noisy" and it is often difficult to discern a signal of change from background variability. How then can we determine whether any trends that managers are now recording in monitoring programs over, say, the past few years have any significance? Are they within the range of natural variability? Or is this a sustained and significant trend? Only long-term data can answer these questions, and rarely are such data available from standard monitoring programs. The proxy data archived in sediments can often provide this missing information.

3   At what point in time and at what level of disturbance or contamination did negative impacts become apparent? Ecosystems are under constant stress from a variety of sources, some natural and others caused by humans. This will not change. What we need to know is the level of stress that a system can be subjected to before, for example, fish die or algal blooms develop. Long-term data are often needed for these assessments. What were the critical and target loads (Box 2.2) of pollutants for the ecosystem under study?

Historical perspectives put present and predicted changes into context. They can assess whether a change is occurring, as well as the direction, magnitude, and rate of change. They help us identify and understand the mechanisms and driving forces of change. We can understand nothing if it is without context. History provides this context.

2.6 Sources of information on long-term environmental change

Managers can potentially choose from four sources of information to address the management questions posed above. Each approach has advantages and limitations, and the strongest environmental assessments are made when all four approaches are used simultaneously (Fig. 2.4):

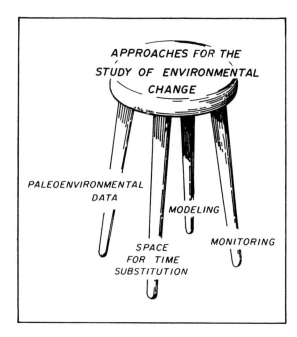

Fig. 2.4 The four major sources of data for environmental assessments. Each of these approaches has advantages and disadvantages but, like a four-legged stool, the most secure answer is attained when all approaches are used. Paleoenvironmental approaches provide critical data that cannot be directly attained from other approaches. For example, rarely were ecosystems studied before impacts had occurred; nor has the trajectory of change been recorded. We used a similar diagram of a stool as part of our final oral presentation to the National Acidic Precipitation Assessment Program (NAPAP) in 1989; I believe the analogy still holds true today.

1   Historical measurements. Instrumental and other observational measurements of environmental conditions and changes obviously represent the most direct approaches, and such data are very important, when available. However, as discussed above, such data sets are often very sparse and are rarely of sufficient quality or duration. Other sources of information must be used. Nonetheless, direct measurements are required for calibrating our other sources of data, and direct observational data are critical for understanding present-day processes, and for calibrating, evaluating, and interpreting other approaches.

2   Space-for-time substitutions (i.e., comparing conditions in affected lakes to those in similar but unaffected lakes). This approach has been used in a variety of environmental applications. For example, with acidification work, researchers have compared the present-day characteristics of water bodies from similar lake regions (e.g., similar climate, geology and vegetation), the only distinction being that one set of lakes was affected by acidic precipitation and the other set was not. The overall assumption is that the lakes in the low-deposition zones (i.e., relatively unaffected by acid rain) display the limnological characteristics that the affected lakes likely exhibited before they were subjected to acidic precipitation. With eutrophication assessments, managers may compare lakes with similar morphometric and catchment features, but with differing amounts of anthropogenic fertilization. For example, one lake may have a pristine, unaltered catchment, whilst a nearby lake may have similar features, except that agriculture has fertilized the lake with nutrients. The unaltered lake would be considered an analogue for the limnological conditions that existed in the latter lake before cultural eutrophication. This comparative approach makes several important assumptions, such as that the lakes are very similar in all respects, except for the intervention under study (e.g., acidic deposition, nutrient enrichment, etc.). One also assumes that other stressors (e.g., climatic change, etc.) have not altered the lakes used to estimate "pre-impact" conditions (i.e., the "unaffected lake" has been unchanged and is similar to what the affected lake was like before cultural disturbances).

3   Hindcasting past environmental conditions using empirical and dynamic models. The lack of direct monitoring data, coupled with the availability of high-speed computers, has prompted some environmental managers to exploit modeling approaches more fully. Based on our understanding of present-day conditions and processes, modelers have attempted to show how conditions have changed (and will change) with specific stressors (e.g., the amount of acidification one might expect with the additional deposition of 25 kg ha^{-1} yr^{-1} of sulfate on an acid-sensitive lake, or the expected eutrophication one should expect with the construction of 10 additional cottages and septic systems in a lake's drainage basin, and so forth).

Joint research collaborations between paleolimnologists and modelers are becoming increasingly common, as paleoenvironmental data can be used to evaluate many types of models (Box 2.3).

4   Paleoenvironmental reconstructions. These will be the focus of the remaining chapters in this book.

There is widespread evidence that paleolimnological approaches continue to be accepted and integrated into a large number of environmental assessments. For example, the European Water Framework Directive (WFD 2000/60/EC) is now the general policy document for European water management (European Commission 2000). The goal of the Directive, established in 2000, is that European water bodies should be returned to "good ecological statuses" by the years 2015–27. According to the WFD, this goal of good ecological status is characterized by "low levels of distortion resulting from human activity," and deviating "only slightly from those normally associated with the surface water body type under undisturbed conditions." Hence the key term "undisturbed conditions," and thus the identification of reference conditions, becomes a cornerstone of this European Commission initiative. A guidance document for the Directive (European Commission 2003) outlines the following methods that could be used to establish these reference conditions: (i) spatially based reference conditions using data from monitoring sites; (ii) reference conditions based on predictive modeling; (iii) temporally based reference conditions using either historical data or paleoreconstructions, or a combination of both; or (iv) a combination of the above approaches. As recognized in the fourth point, none of these approaches are mutually exclusive, and the most robust and defendable assessments will be attainable when data from a variety of sources are used (Fig. 2.4). Clearly, though, the WFD has recognized paleolimnology as a pivotal approach for defining reference conditions for a wide spectrum of European water bodies. Many new research opportunities are on the horizon (Bennion & Battarbee 2007).

Box 2.3 Model evaluation using paleolimnological hindcasts

A growing area of research is the use of paleoenvironmental data to evaluate model outputs. Many models are being developed to predict future environmental changes, based on set boundary conditions and anticipated changes in stressors (e.g., climate change, changes in contaminant emissions, etc.). Often, different models will provide different scenarios for future changes, and so it is difficult to determine which model is the best predictor of future change. One way to evaluate models is to simply wait and see which one was correct! This does little to help those who must make policy decisions. Instead of waiting for future changes, another way to evaluate models is to hindcast the models back in time. Because paleolimnolgical data can be used to reconstruct past conditions, it is possible to undertake a model-paleoenvironmental comparison. The models that are most successful at hindcasting environmental conditions correctly are likely the models we should use to forecast future environmental changes. Such comparisons are now being used commonly in assessing models of climatic change (so-called general circulation models, or GCMs), and are being used to evaluate the effects of many other types of stressors, such as the amount of acidic deposition, nutrient input, and so forth. Several of these applications will be described in subsequent chapters.

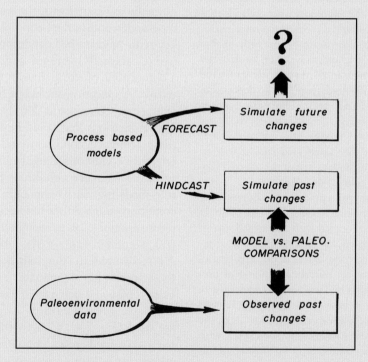

Fig. 2.5 Evaluating models with paleoenvironmental data. By hindcasting models back into time, they can be evaluated by comparison to paleoenvironmental data.

2.7 Summary

We live in a constantly changing environment; some of these changes are due to natural causes, whilst others are related to anthropogenic influences. Only short-term monitoring data (e.g., 3 years or less) are typically available for some aquatic ecosystems, whilst no monitoring data have been collected for most lakes and rivers. It is therefore difficult to determine whether a system has changed as a result of human influences, or whether any recorded environmental changes are within the boundaries of natural variability. Moreover, as most environmental assessments are done after a problem has been identified, knowledge of pre-disturbance background conditions was rarely recorded. Without these data it is difficult to determine if a system has changed, and if so, by how much. We cannot go back in time and sample lakes and rivers using neolimnological techniques, but in most cases we can reconstruct these missing data sets using the vast stores of proxy data archived in sediment deposits. Sediments, like other sources of paleoenvironmental data, are much like the flight recorders, or the so-called "black boxes," on aircraft – steadily recording information that can later be retrieved once problems arise and historical data are needed. Such paleolimnological data can be used to determine the trajectories of environmental changes, to help establish target and critical loads for pollutants, to set realistic mitigation targets, and to help evaluate computer model outputs.

3

Sediments: an ecosystem's memory

Coaxing history to conduct experiments.
Edward S. Deevey, Jr (1969)

3.1 Sediments and environmental change

Because lakes and their sediments collect and integrate regional and local environmental signals, they are often used as sentinel ecosystems (Carpenter & Cottingham 1997). Every day, 24 hours a day, sediments are accumulating. In many ways, sediments provide a history book of our environmental sins. This chapter provides an introduction to the overall approach that paleolimnologists use to read the record stored in sediments.

Many deep lake basins are natural traps for sediment, because the energy levels within them are seldom high enough to transport material out once it has settled to the lake bottom. This, however, is not always true for shallow lakes, where wind and other mixing can at times redistribute sediments, nor for flowing water systems, such as rivers, which are effective at moving sediments. Nonetheless, as will be shown by examples in this book, even these more problematic systems can be studied within a paleolimnological context if coring sites are carefully chosen and due care is given to interpreting the data.

As summarized by Håkanson and Jansson (1983), the distribution of sediments is a function of many factors, such as flow rates, topography, and climate. As a result of these processes, lake basins can typically be divided into three zones: an erosion zone, a transportation zone, and an accumulation zone. It is the latter zone that is of most interest to paleolimnologists, as it is least affected by wind-driven turbulence and other processes, allowing for the accumulation of even fine-grained sediments and resuspended material. In short, sediments in the accumulation zone should provide the most complete, continuous, and reliable record of past environmental change.

Accumulating sediments have many attributes that make them useful archives. The most fundamental of these is summarized by the *Law of Superposition*, which states that for any undisturbed sedimentary sequence, the deepest deposits are the oldest, since these are progressively overlain by younger material. Hence, a depth–time profile slowly accumulates. The job of the paleolimnologist is to interpret the information contained in these profiles. Paleolimnological approaches allow us to view present and predicted future changes against a historical backdrop.

As noted in other sections of this book, the sedimentary record can at times be blurred by mixing processes, such as bioturbation by benthic animals or physical mixing. However, using geochronological techniques (Chapter 4), such potential problems can often be assessed. Furthermore, some mixing does not preclude many interpretations, even though it may decrease the temporal resolution and precision of some studies. As with all scientific endeavors, a sense of proportion is required to deal with these issues. Moreover, paleolimnologists do not assume that sediments accumulating at the centre of a lake are solely originating from the water column above the coring site. Sediment focusing does occur, but this is an advantage, not a problem.

One reason paleolimnological approaches are so powerful is that they integrate information from many regions of the water body. A considerable volume of quality assurance and quality control data (e.g., multi-core studies, comparisons of inferred changes to long-term monitoring data, etc.) has been amassed over recent years to show that if sediment cores are collected and interpreted properly, they provide accurate reconstructions of past environmental conditions.

Sediment material is divided into two broad categories based on its sources: allochthonous and autochthonous (Fig. 3.1). Allochthonous refers to material that originated from outside the lake or water body, such as from its watershed or airshed. For example, clay and soil particles eroded into

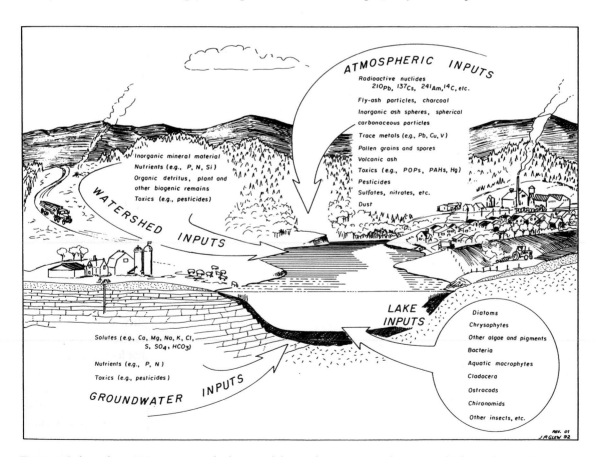

Fig. 3.1 Lake sediments are composed of material from a large variety of sources, which can broadly be divided into allochthonous (outside of lake, such as from the catchment or airshed) and autochthonous (from the lake itself) sources. From Smol et al. (2001a); used with permission.

a lake would be considered allochthonous, as would pollen grains from trees or contaminants from a smoke stack. Autochthonous material, such as dead algal material or the bodies of aquatic animals, remains of aquatic macrophyte plants, and chemical precipitates from processes occurring within the lake basin, originates from the water body itself. The environmental interpretation of this allochthonous and autochthonous material preserved in lake sediments is the focus of this book.

3.2 Sediment records from reservoirs, rivers, and other lacustrine deposits

The majority of paleolimnological studies have focused on lake sediments, as these records are the most prevalent and easiest to interpret stratigraphically. However, other types of water bodies may also archive paleoenvironmental information, although interpreting these proxy data is often more challenging.

Rivers, by definition, contain flowing waters. They are often of prime interest in environmental assessments, as many major cities have developed along rivers. However, being often high-energy systems, rivers typically have the opposite characteristics to those that paleolimnologists rely upon in lakes to preserve reliable sedimentary sequences (i.e., relatively still, low-energy regions, where sediments can accumulate in a well-ordered sequence). The fact that many rivers appear to be colored due to the high sediment loads that they are carrying attests to their potential for scouring and transporting sediment. In short, many river systems do not allow sediments to accumulate, but instead are removing them – exactly opposite to what is required for paleolimnological studies. A further complication is that water quality can change much more quickly in flowing waters than in lakes, as the latter have much longer residence times or flushing rates. As the pre-Socratic Greek philosopher Heraclitus (c. 535–475 BC) astutely noted, "You cannot step twice into the same river."

The situation, however, is not totally hopeless for understanding past riverine environments; paleolimnological and other historical perspectives are attainable for most river systems (Reid & Ogden 2006). In certain reaches of the river, the flow may be reduced so that sediments may accumulate in a similar manner to lakes. For example, fluvial lakes are regions where the river widens, water flow typically slows, and the river deposits sediments (Fig. 3.2). Other areas of sediment accumulation may include embayments or other regions that are outside of the direct flow of the river. In some cases, oxbow lakes or other cutoff features of old river systems contain sediments that can be used to track portions of a river's history. Paleolimnological analyses of deltaic sediments, where the river widens and slows as it enters the sea or another large body of water, may also provide a record of past river environments. In some cases, river sediments and the proxy data they contain may be carried out and deposited in the nearshore environment of the river's receiving waters, such as the ocean, where paleoenvironmental approaches can be used.

In regions where natural lakes do not exist, humans have created artificial reservoirs by damming rivers. Such reservoirs are needed to maintain reliable water supplies for municipal, industrial, and agricultural purposes, as well as to develop enough hydrological head to turn turbines, run mills, and provide hydroelectric power.

As summarized by Callender and Van Metre (1997), reservoir sediments may provide a reliable medium for detecting water quality trends if certain conditions are met. The first condition is that there should be continuous sedimentation at the sampling site since the time the reservoir was formed, or at least for the time period of interest. Reservoirs typically have three zones: (i) the riverine zone, which is most influenced by the river feeding the reservoir, and is typically characterized by complex sedimentation with periodic resuspension; (ii) a transition zone, as the riverine environment approaches a lake environment; and (iii) the lacustrine zone where, with increased water depth and slower currents, the water body more closely resembles a lake

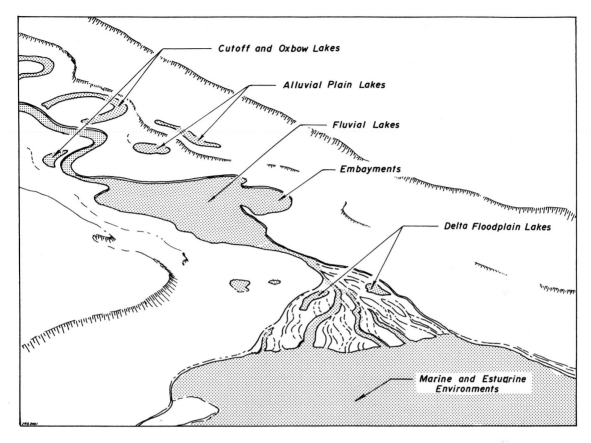

Fig. 3.2 A schematic diagram of a river system, showing areas where sediments may accumulate in a stratigraphic sequence. For example, fluvial lakes are regions where the river widens and flow is reduced. Embayments may also archive useful sedimentary profiles, as may delta or delta lake sediments, other alluvial floodplain lakes, or nearshore sediments in the river's receiving waters.

than a river, characterized by more steady and constant sedimentation (Fig. 3.3). Paleolimnological studies typically focus on the lacustrine zone.

The second condition is that the sediments and the information they contain should not be greatly altered by diagenetic processes. This is, of course, also true of lake studies. However, because sediment accumulation rates are typically higher in reservoirs than in nearby lake profiles, preservation may be better in some reservoirs,

Fig. 3.3 The three major zones in a river/reservoir system. Most paleolimnological studies of reservoirs focus on the lacustrine zone, where sedimentation patterns are often more easily interpretable. Modified from Thornton (1990).

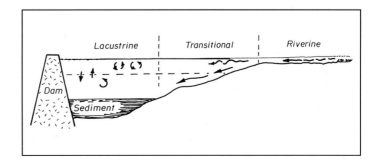

possibly because of faster burial (Callender & Van Metre 1997). The faster sedimentation rates may provide finer temporal resolution.

Other aquatic ecosystems, such as wetlands, bogs, and peatlands, have also been used in paleoenvironmental assessments. Although they are not the focus of this book, I have included a few examples in some chapters to illustrate how these natural archives can be used in applied studies.

3.3 The paleolimnological approach

The basic approaches used by paleolimnologists are not overly complex (Fig. 3.4), although some techniques and certainly taxonomic skills require

considerable patience and expertise. Significant errors and problems can be introduced at each step, and since many of these problems cannot be corrected after-the-fact, careful attention to details is a major requirement for a project's success. Short sketches of the eight basic steps are summarized below; they are of course discussed in much more detail in the subsequent chapters.

Step 1. Choice of study site(s)
The first step is the choice of study site. This selection will sometimes be based on a specific problem, such as contamination of a specific reservoir or problems associated with deepwater anoxia in a particular lake. In such cases, the choice is straightforward, and then one can proceed with the remaining steps. However, in some cases, the problem is not site-specific, as some

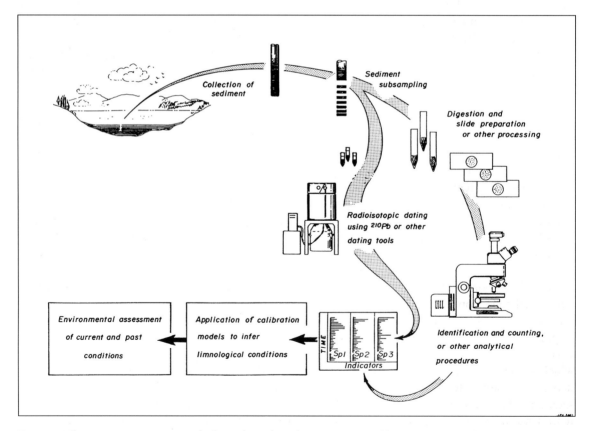

Fig. 3.4 The main steps in most paleolimnological studies, once a suitable coring site is chosen.

management questions are more general, such as "How much lake acidification has occurred in a particular region?" Careful thought must be given to what sites will be sampled. This requires a solid understanding of neolimnological processes that can affect or bias the study. For example: Has a representative range of lake sites been included in the sampling program? Are there any biases that can be addressed? Was the range of limnological variables sampled sufficiently broad? Was the range appropriate for the question(s) being addressed? – and so forth.

Step 2. Selection of coring site(s)

The second major step is to select a coring site in a lake, river, or reservoir. Because most paleolimnological analyses are time-consuming, often only a few or a single sediment core is analyzed in detail. It is therefore critical that this core be representative of overall limnological and other environmental changes. Variability studies have repeatedly shown the high reproducibility between multiple core analyses from the deep central regions of simple basin lakes (e.g., Charles et al. 1991), but the choice of coring site is critical. The overall goal is to retrieve a sediment profile from a region of the lake that integrates the most representative sample possible for the question(s) being posed. In most simple basin lakes, deep, flat, central areas have repeatedly been shown to archive representative profiles of overall lake changes. However, to some extent, the choice of coring site will be driven by the questions being posed. This step is discussed further in Chapter 4.

Step 3. Collection of sediment core(s)

The actual collection of the sediment core(s), the raw materials used by paleolimnologists, is the next step. Paleolimnologists have an enviable array of excellent coring equipment available, some of which are described in Chapter 4. The choice will depend, to some extent, on the location of the sampling site (Is it remote or easily accessible? Is it in a warm region or an area that might support a winter ice cover? – and so forth), the nature

of the sediments (e.g., organic, clay, etc.), and, very importantly, the length of the core and temporal resolution that is required.

Step 4. Sectioning the sediment core(s)

Equally important to the care taken in collecting the core is the sectioning of the sediment into appropriate temporal slices for the study at hand. As with most steps, compromises may have to be made. For example, as paleolimnological studies become more interdisciplinary, the requirements for sediment subsamples are increasing (e.g., sediment from the same slice may be needed for dating, for geochemical analyses, for isotope analyses, for biological examination, etc.). As most of these treatments are destructive (i.e., they destroy the sediment's usefulness for subsequent analyses), there may not be enough material for all analyses to be undertaken if the cores are sectioned at close intervals (i.e., thinner slices of the profile, resulting in higher temporal resolution but less available sediment). One option is to collect larger-diameter cores, but this causes sampling problems. Multiple cores can be taken, but this introduces the problem of correlation of temporal patterns amongst cores, and thus additional variability. As described in the subsequent chapter, a variety of techniques and equipment have been developed to section cores into the time intervals set by the investigator.

Step 5. Dating the sediment profiles

Establishing chronology is critical, for without a reliable depth–age profile, the timing of trends and events cannot be established. A wide spectrum of geochronological approaches is available, although most paleolimnological studies depend on ^{210}Pb, ^{137}Cs, and ^{14}C radiometric dating techniques (see Chapter 4).

Step 6. Gathering proxy data

Collecting proxy data is the biggest step. A plethora of environmental indicators can be retrieved from sediment cores, which includes physical, chemical, and biological information. Most of the effort in a paleolimnological study is devoted to this step. Summaries of the main indicators used in pollution-based studies are summarized in Chapter 5, and explored more fully in the subsequent chapters dealing with specific applications.

Step 7. Interpreting proxy data for environmental assessments

This important step has seen considerable progress and quantification over the past 25 years. As with the previous step, much of this book addresses the interpretation of paleolimnological data. In many recent studies, interpretation includes the construction of transfer functions, using surface-sediment calibration or training sets (Chapter 6).

Step 8. Presentation of data

Paleolimnological data may be complex. A challenge for paleolimnologists is to effectively present their data not only to their peers but also to other scientists, managers, politicians and decision makers, as well as the general public, which includes concerned citizens who are interested in the outcome of many applied paleolimnological studies. In my view, parallel development of both scientific and communication skills is important.

3.4 Recent advances in paleolimnology

Paleolimnology has enjoyed tremendous progress over the past two decades, with the implementation of many new techniques. It is therefore not surprising that this field has experienced considerable expansion and the development of new bridges to a diverse array of subdisciplines. Many recent advances have been instrumental in developing the applied aspects of paleolimnology, which is the focus of this book, but equally important advancements have been made in basic aspects of paleolimnological research.

About 15 years ago (Smol 1990b), I felt that there were three major areas of advances that have allowed paleolimnology to accelerate rapidly in its development, expand its range of potential applications, and demonstrate its credibility. Given the benefit of hindsight, I would still maintain that the major advances were the following:

1 Technological developments. Technology and infrastructure often drive progress, and this is also true for paleolimnology. Under this heading, I would include the ways we have learned to collect and section sediment cores (Chapter 4). This is especially important for applied aspects of paleolimnology, where high-resolution sampling of unconsolidated recent deposits is often important. Paleolimnology had been used for some time as a research tool; however, most of the earlier work focused on time scales of millennia and centuries, and not decades, years, and in some cases, sub-annual resolution. With new corers and sediment sectioners, some of which are discussed in the next chapter, as well as the availability of other laboratory equipment, many new and powerful applications have become possible. Major progress has also been made in the dating of sediments (geochronology), which is also discussed in Chapter 4.

2 The amount and quality of information we have learned to glean from the sediments. It was not long ago that if asked what information was preserved in sediments, many scientists would have said pollen grains, diatoms, and perhaps some geochemical information. Over the past 25 years, there has been an explosion of activity in the development of techniques and methods to glean a tremendous amount of information from sedimentary profiles. We now know that virtually every organism living in a lake (and many living outside the lake, in its catchment) leaves some form of morphological or biogeochemical marker (Chapter 5). Certainly, not all of these records are complete, nor are they of equal quality. Nonetheless, a surprisingly large library of information is archived in sediments. The job of the paleolimnologist is to find this information and analyze it in a scientifically sound and defensible manner.

3 Application of new procedures to interpret and use this information in a quantitative and robust manner. Most paleolimnological data are complex and multivariate. With increased computer

facilities, coupled with impressive advancements in statistical and data-handing techniques, paleolimnologists have taken advantage of these new developments and have made impressive progress in ways to "read" the information preserved in sediments (Chapter 6). These approaches are often highly quantitative and reproducible, although qualitative data are still very important and useful for many studies.

The above advances will be illustrated further with the examples used in this book. But first I will review some of the ways in which paleolimnologists collect sediment and date sediment cores (Chapter 4) and so provide a depth–time sequence, and briefly review some of the information preserved in these profiles (Chapter 5).

3.5 Summary

On the geological time scale, lakes are only temporary landscape features. Slowly – often imperceptibly at the time scales that we are usually accustomed to – lakes fill up with accumulating sediment until eventually they become land. Sedimenting materials can be broadly assigned into two categories:

■ autochthonous materials that are composed of materials produced in the lake or river (such as the remains of aquatic organisms, precipitates, etc.);
■ allochthonous materials that originate from outside the water body (such as eroded soils, terrestrial plant remains, etc.).

Provided that these materials are accumulating in a relatively undisturbed stratigraphic sequence (e.g., mixing processes at the sediment surface are not greater than the temporal resolution required by the study), then paleolimnologists can recover the information archived in these depth–time profiles and interpret these proxy data in a manner that is relevant to environmental managers, other scientists, and the public at large. Paleolimnology has witnessed major advances over the past 25 years or so, which allow these techniques to be used at the levels of accuracy and precision required to answer many lake and river management questions.

4

Retrieving the sedimentary archive and establishing the geochronological clock: collecting and dating sediment cores

The lake sets the time scale, not the investigator.
W.T. Edmondson, American Society of Limnology and
Oceanography conference, Boulder, Colorado (1987)

4.1 Collecting and dating sediment cores

After a research question is posed and a site is selected, the paleolimnologist's first task will be to collect a sedimentary sequence that is of the appropriate length, resolution, and quality for the study in question. The second task will be to establish an accurate depth–time profile, so that any environmental changes recorded in the sediments can be related to a chronological time scale. These two critical steps are summarized in this chapter.

4.2 Retrieving and sampling sediment profiles

Fieldwork can be difficult and at times uncomfortable. For some of us, it is the part of the job that we like the most! Nonetheless, collecting reliable and appropriate cores takes some planning

and skill. Sediment coring is arguably the most critical step in the paleolimnological process, as any errors or problems encountered at this stage can rarely be corrected after-the-fact.

The overall criteria of good core recovery have not changed since Hvorslev's (1949) seminal work on the subject over half a century ago. The three main goals are to obtain cores with (i) no disturbance of structure, (ii) no change in water content or void ratio, and (iii) no change in constituent or chemical composition. Although some changes and disturbances are inevitable, the overriding aim is to retrieve sedimentary profiles that represent, as closely as possible, the natural archive of sediment accumulation in the basin.

As noted in the introductory chapters, paleolimnologists are addressing a large number of diverse questions. As a result, they typically have to sample a wide range of water bodies that often differ markedly in their morphometric and other characteristics. Furthermore, sediments from different basins may vary strikingly in composition and texture. The rate of sediment accumulation will also vary. It is therefore not surprising that many corers and samplers have been developed to collect sedimentary sequences.

Papers, chapters, and entire books have been compiled to describe the methodologies and equipment that are available to retrieve cores from different limnological settings. Glew et al. (2001) summarize the major processes, challenges, and assumptions of coring, as well as describing the most commonly used equipment and approaches for collecting relatively short (e.g., from a few centimeters to up to a few meters in length) sediment cores. Leroy and Colman (2001) summarize the different types of corers that are typically used to retrieve much longer sequences, but the latter are more often used for paleolimnological questions that are beyond the scope of this book.

Only some of the most commonly used approaches to collect relatively short sediment cores (i.e., those most likely encompassing the period of human impacts) will be summarized here.

4.2.1 Choosing the coring site

Detailed paleolimnological analysis of a single sediment core often involves many days (or even years) to complete, and so typically only one or a few cores are analyzed per basin. This is not to say that core analyses are never replicated; often they have been, and typically show surprisingly good reproducibility in overall inferences if correct sampling approaches have been followed (e.g., Charles et al. 1991). Clearly, though, if an environmental history of a basin will be assessed from only one profile (or just a few), then the choice of coring site should not be taken lightly.

Ideally, the point of sediment collection should represent, as far as possible, an "average" accumulation of material (both in quantity and quality) for the entire basin. This is obviously a demanding and probably unrealistic goal, especially for larger bodies of water, but if chosen carefully, a strategically selected coring site can provide an integrated history of limnological conditions. Of course, the more cores that are analyzed, the better, and certainly multiple-core

analyses provide more complete and defendable histories of environmental change. Chapters in Last and Smol (2001a) summarize the major steps involved in basin analysis, and technological developments, such as subbottom acoustic data (Scholz 2001), that can increase the likelihood of choosing a more reliable coring site. Even if these more sophisticated techniques and equipment are not available, the paleolimnologist, at a minimum, should examine any available morphometric data and maps (and/or undertake transects of the lake with a depth sounder) to choose a coring site that would most likely contain a continuous, representative, and undisturbed sequence. There are no set rules for this, as conditions will vary from site to site. However, in general, coring sites that are located near steep morphometric gradients should be avoided, as slumping and other lateral movements of sediments are more likely to occur and blur interpretations. Deeper areas tend to be preferred over shallower sites, as the latter may be more prone to bioturbation or other mixing processes. Ideally, a flat, central basin in the deeper part of the water body would most likely archive the most complete and easiest to interpret sediment profile. However, such a coring site would not be appropriate for some studies (e.g., if the goal of the project is to document the history of a specific bay, then sampling sediment in the open-water region of the lake would not be appropriate). Furthermore, for some indicators (e.g., plant macrofossils; Birks 2001), much information can be gained by collecting cores from shallower, nearshore waters.

4.2.2 The coring platform

As sediment cores are typically taken through the water column (although some specialized paleolimnological studies may utilize exposed sediments, such as along an old river terraces), most cores are taken from a boat or some other floating platform. In colder regions, solid ice covers may supply a stable platform, which is an important consideration. The disadvantage of winter sampling is often the cold itself, which can make some aspects of coring uncomfortable and possibly hazardous, and can result in freezing up of mechanical parts of the corer and other equipment (and limbs!).

Many of the smaller types of corers, such as the gravity devices discussed below, can be effectively used over the side of small boats or canoes. This is especially important when sampling in remote locations, where road access is not possible. Such corers can also be used from the pontoons of a helicopter or float plane (Fig. 4.1).

With larger corers that retrieve longer sediment profiles, more elaborate platforms may be

Fig. 4.1 Collecting surface lake sediments using a small-diameter Glew (1991) gravity corer from the pontoons of a helicopter in Arctic Canada. The photograph, taken by Reinhard Pienitz, is of Ian Walker.

required. However, with careful planning, even these platforms can be made portable and carried to remote lakes. For example, two canoes or two inflatable boats can be tied together and anchored at each corner to make a relatively stable platform (Wright 1991).

4.2.3 Coring equipment

In paleolimnological work emphasizing human impacts, the focus is often on the more recent sediments (although, as will be shown with later examples, longer-term profiles are required for some management issues). In places such as North America, the main time period of interest is usually the past century or so, and in most lake regions this time window is contained in the surface 50 cm of sediment or less (although exceptions occur where faster sedimentation rates are encountered). In regions with longer histories of settlement, the time frame may be longer, but even here the uppermost sediments are often of most interest. These recent sediments may pose some challenges to collection and sampling, as they are often unconsolidated, with a very high water content. In many regions, this means that the paleolimnologist is trying to core and then subsection material that may exceed 95% water by weight. Although at first this may seem like a daunting task, a number of successful approaches have been developed to meet these challenges.

A variety of sediment corers have been constructed that can be used even by novices, with a little practice, to effectively sample the most watery sediments at high temporal resolution. Most of the equipment developed for sampling surface sediments falls into two broad categories: open-barrel gravity corers and freeze-crust samplers.

As the name suggests, open-barrel gravity corers are samplers that use primarily gravity to penetrate the sedimentary profile, and the sediment is then collected in an open barrel (typically a polycarbonate coring tube). In its simplest form, the corer consists of a tube (open at both top and bottom) that is driven vertically into the sediments. After the coring drive is accomplished, the top of the tube is closed (forming a seal), and the tube, now filled with sediment, is retrieved to the lake surface.

Gravity corers are further subdivided based on the method used to seal the core top. Some are close-on-contact types, which have self-triggering devices that seal themselves remotely once the coring tube is driven into the sediment. This is typically accomplished by relaxation of the line tension as the corer enters the sediment (e.g., Boyle 1995). Messenger-operated samplers are more common (and in my experience, much more reliable; Cumming et al. 1993). Most are sealed by a messenger, which is a weight captive on the corer recovery line that is released at the surface by the operator. The messenger is allowed to free fall down the line and trigger the sealing mechanism (Fig. 4.2). The earliest gravity corers were often referred to as Kajak–Brinkhurst (KB) corers (Kajak et al. 1965; Brinkhurst et al. 1969). Although the overall design has remained largely the same, significant modifications and changes have been made, such as in the Hongve type (Wright 1990), the HON–Kajak type (Renberg

1991), the Glew type (Glew 1989, 1991, 1995), the Limnos type (Kansanen et al. 1991), and many other designs (Glew et al. 2001).

Figure 4.2 shows the general operation of a messenger-operated gravity corer. Gravity corers are amongst the simplest samplers that a paleolimnologist can use, but they must be employed carefully to properly recover a sediment sequence. For example, it is critical for water to flow freely through the open coring tube as it is lowered through the water column, or else a pressure wave (often called a bow wave) can form below the lower end of the core tube as it descends. This pressure wave can literally "blow away" the surface sediments (Cumming et al. 1993). Using a properly designed corer, and then only lowering the device into the sediments, alleviates this problem (Glew et al. 2001). Given their portability and reliability, it is not surprising that gravity corers have become the main instruments used by paleolimnologists to retrieve recent sediment profiles.

Freeze-crust samplers represent another type of gravity type corer that was designed to sample the watery sediments characterizing the mud/water

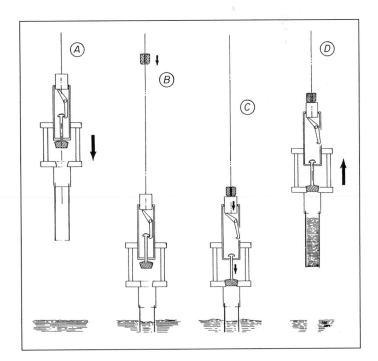

Fig. 4.2 The general operation of a messenger-triggered gravity corer. (A) The corer being lowered through the water column: water passes through the open-barrel core tube during descent. (B) The corer enters the sediment using its own submerged weight, and the messenger is released from the surface to close the corer. (C) The messenger strikes the corer, thus triggering the closure of the core tube. (D) The corer is recovered to the surface with the sediment sample.

interface (Renberg 1981). It is also the sampler of choice if one is dealing with sediments that contain gases that might disrupt the core profile during the coring process. In its simplest design, the freeze-crust sampler represents a weighted chamber (often wedge-shaped, but cylindrical and box-shaped samplers have also been used) that is filled with a coolant such as dry ice (i.e., solid carbon dioxide) immersed in alcohol, which is then lowered on a rope or cable into the sediment (Fig. 4.3). The supercooled apparatus is held in position by the sampling rope in the sediment for about 10 minutes (but this depends on the corer and the sediment characteristics) as it freezes, *in situ*, a crust of sediment on the surface of the sampler (Fig. 4.3). The device is then retrieved to the lake surface, and the frozen sediment crust is removed for subsequent subsampling. A chief advantage of this system is that the sediment profile is frozen *in situ*, so the integrity of the stratigraphy (as well as the mud/water interface) is retained. Depending on the dimensions of the freezing surface, such samplers may also collect a relatively large volume of sediment, which is becoming increasingly important with the trend towards multidisciplinary studies. Some of the disadvantages of the freeze-crust sampler include the extra logistical problems of carrying coolants into the field and then keeping the sediment frozen after collection,

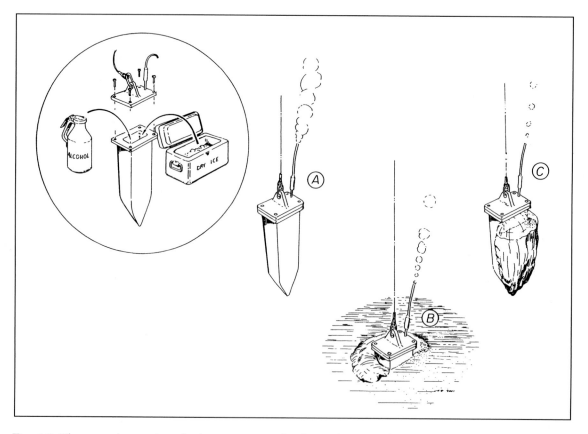

Fig. 4.3 The general operation of a freeze-crust sampler. Inset: the corer chamber is filled with dry ice and a coolant, such as alcohol. The corer top is secured. (A) The corer is lowered through the water column on the recovery line. (B) The corer is lowered into the sediment: it is held in a fixed position for a short period of time (i.e., until a sufficient amount of sediment has frozen to the corer freezing plate). (C) The sediment-encrusted corer is recovered to the surface.

as well as the understandable challenge of working with frozen materials in warm regions such as the Tropics. Several modifications have been proposed for the original designs (e.g., Verschuren 2000) to address some of these problems. Freeze-crust corers may also not be totally appropriate for some work assessing sediment contaminant inventories, as ice crystal formation may result in higher water contents of recovered sediments, thus lowering concentration inventories (Stephenson et al. 1996).

In some cases, longer sedimentary sequences are required. The corers described thus far will only retrieve relatively short sequences, typically less than 1 m in length and sometimes even much less (depending on the design). Many other coring devices are available. Rod-driven piston corers are frequently used to collect longer profiles. The operation of these corers is fairly straightforward, as summarized in Fig. 4.4. Most are based on the Livingstone (1955) design, although several modifications have been proposed (Glew et al. 2001). The typical piston corer consists of three components: the piston and cable assembly, the core tube, and the drive head and drive rods (Fig. 4.5). The core tube is assembled

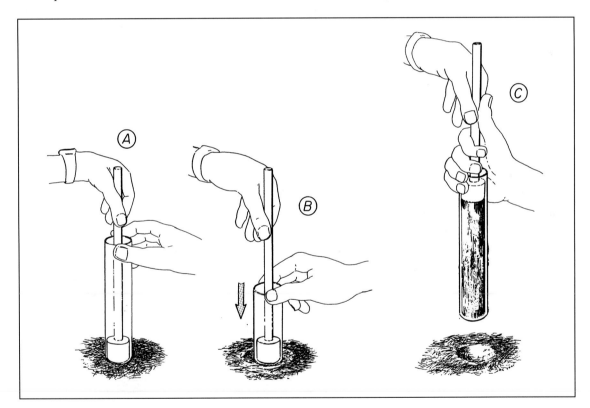

Fig. 4.4 A simplified diagram showing the basic principles used in piston coring. The closest analogy to how a piston corer performs is to the reverse operation of a medical syringe. (A) To recover an undisturbed core sample, the corer is positioned at the sediment surface (as in this example, or at a subsurface position to collect deeper sediment profiles), with the piston at the lower end of the core tube. (B) The piston is held stationary, while the core tube is pushed past it into the sediment using the coring rod. (C) The core section is recovered to the surface, with the core tube and piston locked together. The sealing of the piston in the core tube prevents any tendency for the sample to slide out or for the core material to be deformed. It is important to note that the blocking action of the piston as shown in (A) enables the corer to selectively sample well below the sediment surface.

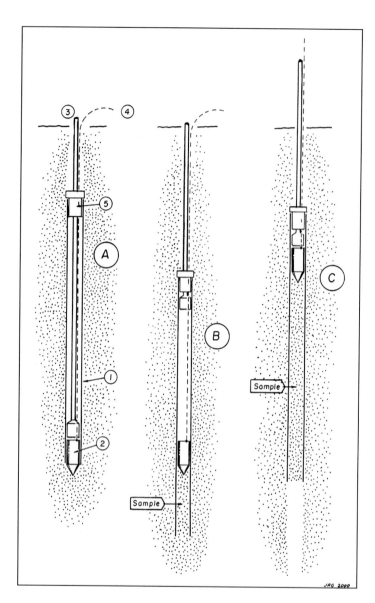

Fig. 4.5 The general operation of a cable-operated Livingstone-type piston corer, showing the lowering (A), sampling (B), and withdrawal (C) of a sediment sequence. The operator uses the drive rods to push the corer into the sediment to the desired depth. A cable, also held by the operator, keeps the piston in place. 1, Core tube; 2, piston; 3, drive rods; 4, piston cable; 5, locking drive head. From Glew et al. (2001); used with permission.

with a close-fitting piston that can move the length of the core sample. In operation, the corer enters the sediment with the piston at the bottom of the tube, effectively preventing sediment from entering the tube until the correct sampling depth is reached, at which point the piston is held stationary and the core tube is pushed past it into the sediment. Once the sample has been taken, the corer is recovered and the core sample is extruded from the core tube, or the core tube

containing the sediment sample is exchanged (with the piston) with a new, empty tube, and then coring of the next section can begin. One advantage offered by the piston is that coring can be carried out sequentially in sections to significant depths in such a manner that distortion of the sample is minimized. Retention of the sample in the tube is facilitated by the piston, because the pressure that is required to drive the core is accommodated by the piston seal and not the sediment sample. Disadvantages are mostly associated with control of the piston and its functioning. For lacustrine work, the piston corer is usually driven by extension rods, with the piston on a separate cable limiting its operating depth. In deeper systems, where coring rods are not practical due to the depth of the water column (e.g., > 40 m), cable-operated piston corers are often used, with internal weights being employed to drive the core tube in a single motion (e.g., Kullenburg 1947). Meanwhile, some designs replace push rods with compressed air, an example being the Mackereth (1958, 1969) corer, which is a pneumatically driven piston corer that can recover relatively long cores (e.g., 6 m) in an uninterrupted manner.

There are a large number of other corers that paleolimnologists can use, depending on the research questions being posed. Many of the most commonly used devices are illustrated and discussed in Glew et al. (2001). For example, chamber-type samplers (sometimes referred to as Russian peat corers) are becoming increasingly popular. With these devices, the sediment sample is "cut" and enclosed in a rotating chamber or half cylinder. Vibracorers, which use a vibrating device to impart high-frequency, low-amplitude standing waves along the length of a coring tube, are especially useful in collecting sediment sequences in material that may be difficult to penetrate using standard push rods, such as profiles containing sand lenses (Smith 1998; Fisher 2004).

I sometimes hear researchers, albeit almost always novices, making statements such as "we have developed a new corer and it is the only one you will ever need." My immediate response is "wait until you see the full spectrum of lake and sediment characteristics you will encounter, as well as the full range of paleolimnological problems you can actually investigate." Based on my experience, there is no "one corer" that will "do it all." There are some that will "almost do it all," but no single corer will be the ideal choice for all applications. The critical step of sample collection should not be made lightly; a lot more work is waiting for the paleolimnologist once the sediment is collected. Use the right equipment and make sure you have retrieved the best sediment sequence possible.

4.2.4 Sediment handling, sectioning, and subsampling

Once a core is collected, the next steps often include various logging and photographic techniques (e.g., Kemp et al. 2001; Saarinen and Petterson 2001; Zolitschka et al. 2001), followed by sectioning of the sediment profile. In many respects, the paleolimnologist can set the time scale of the investigation by determining at what temporal resolution (i.e., how finely) the core will be sectioned. Of course, this presupposes that the profile has not been significantly disturbed. For example there is little point in sectioning a core into 0.25 cm intervals if the sediments were mixed by benthic invertebrates within a 2 cm deep mixing zone. Nonetheless, the degree of mixing, if any, can often be assessed using independent methods (e.g., dating techniques, see below), and the integrity of the profile can be assessed. Although most sediments will have *some* mixing, a sense of proportion is required (Does the amount of mixing preclude or seriously compromise the study in question?).

Just as there were some challenges in coring the watery surficial sediments, care must be taken to section unconsolidated profiles without introducing any additional artifacts. I have often suspected that much of the "bio" in "bioturbation," with which some researchers seem to continually have problems, is human-caused mixing (as a result of improper coring or subsampling procedures). Appropriate sectioners and techniques are available to handle even the most watery sediments, but, like everything else, they must be used correctly. If practical, section the core at the lakeshore, or as soon as possible after collection, as transporting surface sediments, with their high water contents, can be problematic. A variety of portable, close-interval, vertical extruders are now available that can effectively section even

Fig. 4.6 Sectioning a surface, gravity sediment core at high resolution (e.g., 0.25 cm intervals), using a Glew (1988) vertical extruder. Photograph taken by M. Douglas.

the least consolidated sediments (e.g., Glew 1988; Glew et al. 2001; see also Fig. 4.6). Some gravity corers, such as the Limnos design (Kansanen et al. 1991), use segmented coring tubes that can be rotated horizontally to subsample the core, and so no additional sediment sectioner is required. Researchers have also developed a variety of resin embedding techniques that effectively turn the watery sediment into a solid form (Kemp et al. 2001; Lamoureux 2001), as well as microtomes that can section cores at very fine (e.g., 100 μm) resolution (Cocquyt & Israël 2004).

Frozen-crust samplers pose other challenges, but also opportunities. Frozen sediment is usually sectioned with a saw, knife, or hot wire, but because the sediment surface is exposed in a solid (i.e., frozen) form, sediment can be sectioned at finer scales. For example, individual laminae can be subsampled with a narrow cutting edge, such as a razor blade. Simola (1977), with later modifications by Davidson (1988), have introduced so-called "tape peel techniques," whereby pieces of transparent tape are placed directly onto the frozen sediment. When the tape is removed, adhered to it is a thin layer of sediment and associated proxy data (e.g., diatom valves, pollen grains). This sediment-coated tape can then be mounted on a microscope slide or on a scan-

ning electron microscope (SEM) stub, and examined at high magnification. The tape-adhered sediment contains an intact and undisturbed record of the sediment profile, and can be used, for example, to track even seasonal changes in lake biota in some cases (Simola et al. 1990).

4.3 Setting the time scale: geochronological methods

All the data collected by paleolimnologists have to be placed in a reliable temporal framework. Change implies time, and so a major challenge is to decipher the geochronometer that is preserved in a sedimentary profile. In short, we have to estimate how old a sediment slice is if we hope to use the proxy data it contains in an assessment of environmental change.

As with the other approaches described in this book, a wide array of tools are available to establish reliable depth–time profiles in sedimentary sequences (reviewed in Last and Smol 2001a). However, only a few of these techniques are used to establish the relatively short time scales typically required to address most management issues. The major approaches are summarized below.

4.3.1 Radioisotopic techniques

Many of the most commonly used dating techniques measure the decay of naturally occurring radioisotopes. For dating recent sediments (i.e., up to about the past ~ 150 years), the naturally occurring radioisotope of lead (^{210}Pb) in the uranium (^{238}U) decay series is by far the most commonly used. With a half-life of 22.26 years, ^{210}Pb is ideal for establishing depth–time scales over the past century or so (Fig. 4.7). Krisnaswami et al. (1971) originally applied the technique to

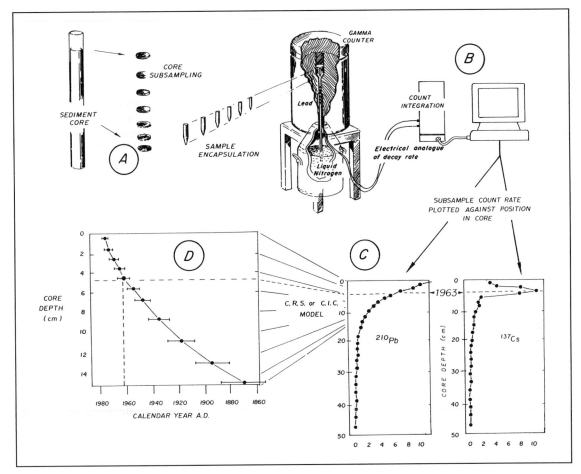

Fig. 4.7 A schematic diagram showing the major steps used in determining ^{210}Pb dates for sediment cores. (A) The core is sectioned into appropriately spaced intervals (e.g., 1 cm slices) for the study in question. (B) ^{210}Pb and other radioisotopes are counted using a low-background gamma counter (an alternate procedure is to use alpha spectrometry). (C) The activities of total ^{210}Pb and ^{137}Cs in this core are as follows: the level of supported ^{210}Pb activity reaches background concentrations at around 20 cm, while the peak in ^{137}Cs activity at around 5 cm is assumed to represent the height of atmospheric fallout in 1963 (see the text below for details). (D) The resulting age–depth relationship for the sediment core is based on calculations using a model such as the Constant Rate of Supply (CRS) or Constant Initial Concentration (CIC) model. Age determinations cannot be made accurately for depths below the level of detection for ^{210}Pb – in this example, for sediments below about 15 cm depth. Ages below this depth can only be extrapolated. The dotted line shows the placing of the ^{137}Cs peak on the dating curve, revealing (in this hypothetical example) a good match with the ^{210}Pb-estimated date of 1963. Error bars are standard deviations.

lake sediments and since then many thousands of cores have been dated using ^{210}Pb chronology. As summarized by Appleby (2001), the *total* ^{210}Pb activity in sediments has two components: supported ^{210}Pb, which is derived from the *in situ* decay of the parent radionuclide ^{226}Ra; and unsupported ^{210}Pb, which is derived from the atmospheric flux. Unsupported ^{210}Pb can be calculated by subtracting supported activity from the total activity.

The methodology developed in the original Krisnaswami et al. (1971) paper was based on the assumptions that: (i) the rate of deposition of unsupported ^{210}Pb from the atmosphere has been and is constant; (ii) the ^{210}Pb in water is quickly absorbed by suspended particulate matter, so that unsupported ^{210}Pb activity in sediments is essentially due to atmospheric fallout; (iii) post-depositional processes are not redistributing ^{210}Pb activity in the sediments; and (iv) ^{210}Pb decays exponentially with time, in accordance with the law of radioactive decay. In practice, things are a little more complicated: the supply of unsupported ^{210}Pb to the lake may include a small but significant fraction of ^{210}Pb deposited on the catchment and then transported to the lake (e.g., by soil erosion); a proportion of total inputs may be lost via the outflow; and/or deposits reaching the bed of the lake may be redistributed spatially by hydrologic processes and sediment focusing. Nonetheless, two simple models used to calculate sediment dates – the constant rate of ^{210}Pb supply (CRS) and the constant initial concentration (CIC) of ^{210}Pb (Appleby and Oldfield 1978; Robbins 1978) – have between them yielded reliable results in a great many cases. The CRS model has been the most successful and is most frequently used, but under certain conditions (usually where primary sedimentation rates have been constant and the core site has been impacted – e.g., by episodic slump events, or changes in the pattern of sediment focusing), the CIC model is more appropriate (Appleby 2001). Under some conditions, neither of the two simple models is adequate, and so various other models must be developed to provide a chronology.

Useful primers on ^{210}Pb dating include Oldfield and Appleby (1984) and Appleby (1993), and Appleby (2001) has published a detailed discussion on recent methodological developments. Without a doubt, ^{210}Pb dating has become the technique of choice for providing age estimates of recent sediments.

^{210}Pb is a naturally occurring radioisotope, constantly being produced by natural processes on our planet and constantly decaying. However, paleolimnologists can also use other isotopes that are not natural, but have been manufactured by the nuclear industry. These can be used to pinpoint certain time periods in sedimentary sequences. The most commonly used isotope is cesium-137 (^{137}Cs), which was released for the first time in 1945 with the dawning of the nuclear age (Pennington et al. 1973). Stratospheric testing of atomic weapons has been a major source of ^{137}Cs; releases accelerated with the initiation of the nuclear arms race and atmospheric weapons testing, which began in earnest with the high-yield thermonuclear tests beginning in November 1952. The radioactive debris injected into the stratosphere was transported around the world, but eventually redeposited as fallout. In the Northern Hemisphere, fallout reached significant levels by 1954, and then increased steadily until 1958, when there was a short-lived moratorium on testing, but then increased again in 1961 with a sharp peak by 1962–3. The Test Ban Treaty was signed in 1963, and thereafter concentrations dropped steadily. Figure 4.8 shows the ideal sediment profile that one might expect if the paleolimnological record directly tracked ^{137}Cs fallout. Although some profiles approach this idealized pattern, many are less clear, due to post-depositional ^{137}Cs mobility. This is a significant problem in many sediments, especially those with high organic matter content (Davis et al. 1984) and at least some saline systems (e.g., Foster et al. 2006). Although it frequently results in a "tail" of ^{137}Cs extending to depths substantially preceding 1954, in many cases the position of the peak identifying 1963 does remain fixed.

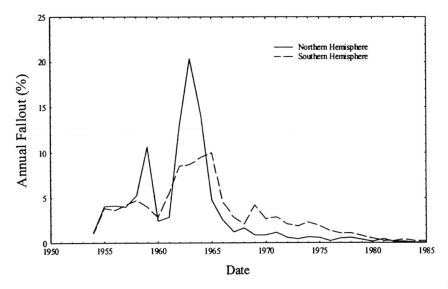

Fig. 4.8 ^{137}Cs fallout in the Northern (solid line) and Southern (dashed line) Hemispheres from atmospheric testing of nuclear weapons up until 1985. Under ideal conditions, identical patterns in ^{137}Cs would be observed in lake sediments, which could then be matched to the fallout curve for the region and used as a geochronological tool. In most cases, the sedimentary ^{137}Cs profile is not identical to the curves shown here, but it may still be used to supplement and verify ages obtained from other sources, such as ^{210}Pb. Additional ^{137}Cs was released during the 1986 Chernobyl accident, which affected some lakes. From Appleby (2001); used with permission.

^{137}Cs was also released in the 1986 Chernobyl nuclear power plant accident, but the plume was mainly confined to parts of the former Soviet Union, Europe, and Turkey (see pp. 300–1). Other radioisotopes from the nuclear industry, such as americium (^{241}Am), which appears to be less mobile in sediments than ^{137}Cs, also have considerable potential as dating tools (Appleby et al. 1991).

For longer sedimentary sequences, the dating method of choice is almost always radiocarbon (^{14}C) dating (Björck & Wohlfarth 2001). ^{14}C is an isotope of carbon, which forms in the atmosphere as a result of cosmic ray bombardment of ^{14}N. As ^{14}C is isotopically unstable, it decays back to the stable element of ^{14}N at an estimated decay rate half-life of 5730 ± 40 years. However, the Libby half-life (named after Willard Libby, who first determined this value) of 5568 years (also referred to as the "conventional half-life") has been used since the 1950s, and so by convention it is still used for age determinations to keep dates consistent. During the life of an organism (e.g., an alga or a plant), ^{14}C is taken up as ^{14}CO$_2$ until an equilibrium is reached (i.e., by photosynthesis), and then animals can consume this organic matter containing ^{14}C. Once an organism dies, carbon replenishment stops and ^{14}C begins to decline at the decay rate. In short, the "carbon clock has started." ^{14}C is the ideal choice for dating organic deposits that are from about 500 to 40,000 years old.

As with all techniques, there are several assumptions and potential problems with ^{14}C dating. Contamination is always a potential problem, either by reworking of older or younger organic material into the sample being dated, contamination through rootlets, fungal growth and infiltration by younger humic acids, or by incorporation of "old carbon." The latter is especially a serious problem in areas with calcareous bedrock or soils, and results from aquatic plants and algae incorporating not only "recent" carbon during photosynthesis, but also some of the old, "^{14}C-dead" carbon (e.g., millions of years old) leached or eroded from the surrounding bedrock. The resulting "hard water effect" will yield radiocarbon ages that may be considerably older than their correct ages (MacDonald et al. 1991). For these reasons, paleolimnologists try to date material that is unaffected by this contamination, such as terrestrial plant remains (e.g., seeds, other macrofossils), as terrestrial organisms are in equilibrium with the atmospheric reservoir of carbon.

Until the 1980s, most radiocarbon determinations were based on so-called conventional or bulk dates, which measured the amount of ^{14}C in a sample usually by employing gas proportional or liquid scintillation counters. This required a relatively large volume of organic matter that would yield a minimum of about one gram of pure carbon. This limitation resulted in ages that were less precise than many geochronological problems required. Also, as these were bulk samples (e.g., sections of an organic sediment sequence), typically the sample contained a mix of many different types of organic matter. This could have included, for example, carbonates such as shells, which may have incorporated "old carbon" into their carapaces, and so would provide erroneously old dates.

Many of the shortcomings of bulk ^{14}C dating were alleviated with the development of accelerator mass spectrometry (AMS) and its application to radiocarbon dating in the 1980s. AMS dating requires a much smaller sample (only a few mg), such as a seed or a conifer needle, to provide reliable ages. This immediately had two major advantages. First, paleolimnologists would now know what they were dating. As opposed to bulk dating, when one typically submitted an organic section of sediment that could contain both allochthonous and autochthonous organic matter, one could now submit a specific organic sample, such as a maple seed or a spruce needle. Such a sample would provide a radiocarbon date for a specific time period (i.e., an estimate of the

year the seed or needle was shed), and not the much broader time period that might be contained in organic matter making up, for example, a 5 cm section of sediment. Second, because a specific sample (e.g., a terrestrial seed) was now being dated, many of the problems with the "old carbon effect" could be circumvented (i.e., only dating terrestrial-based samples that would not be taking up old carbon as part of their photosynthesis). One of the most celebrated applications of AMS ^{14}C dating was the dating of a small strip from the Shroud of Turin, believed by some to have been the burial cloth of Christ. AMS dating revealed that the shroud was actually medieval in age (Damon et al. 1989).

Once an analyst has obtained a date in "radiocarbon years," this number is not yet equivalent to calendar years. The ^{14}C age has to be converted to "calendar years" or "years before present (BP)" using correction factors determined from ^{14}C dating of the annual growth rings of German trees covering the last 12 millennia, in combination with ^{14}C-dated laminated marine sediments and ^{14}C/U-Th dated corals (Stuiver et al. 1998). This correction step is required because of natural variations in ^{14}C production. By convention, the "present" is set at AD 1950.

4.3.2 Pollen chronologies

Pollen grains and spores are a common component of most sedimentary sequences and have wide-ranging applications in paleoenvironmental work (see pp. 55–6). Because the introduction, proliferation, or demise of certain plant taxa can be linked to known time periods, the stratigraphic distribution of pollen grains from these taxa can be used to provide additional geochronological control. The best-known example in eastern North America is ragweed (*Ambrosia*) pollen. This taxon often proliferates after native forests are cut, and the soil layers are disturbed by plowing and other activities for the implementation of European style agriculture. As a result, the so-called "ragweed rise" has become a standard stratigraphic marker in many pollen diagrams, and can be used as an approximate dating horizon for the time of European settlement. Other pollen assemblage changes can be used, such as increases in other weedy species or cereal grains with agriculture (Bennett & Willis 2001). In eastern North America, the chestnut blight fungus, introduced in New York City in 1904, resulted in the elimination of this tree throughout its range – a decline that can be tracked in sedimentary pollen assemblages (Anderson 1974).

4.3.3 Episodic events

In certain settings, past episodic events (specific, short-term events) that can be definitely related to a known time period can be used to help date sediment cores. For example, different volcanoes expel different types of ash (tephras). If an ash layer can be identified in a sediment profile and linked to a specific volcanic eruption of known age, then this tephra layer can be dated with a high degree of certainty (Turney & Lowe 2001). Similarly, if a forest fire is known to have occurred in a certain year, a charcoal layer in the paleolimnological record can be used to identify this time slice (Whitlock & Larsen 2001). A silt layer of erosive material (e.g., clays and silts) can sometimes be linked to a specific watershed event, such as the building of a highway (Smol & Dickman 1981), and so forth. Although a single episodic event will not provide enough information to establish a reliable depth-time sequence, identifying such "benchmarks" are important markers for evaluating chronological sequences established using other approaches, such as ^{210}Pb.

4.3.4 Other anthropogenic time markers

In addition to some of the radioisotopes that are related to human activities (e.g., ^{241}Am, ^{137}Cs), there are a variety of other anthropogenic markers preserved in sediments that can be used as time markers (e.g., Kilby & Batley 1993). Many of these include chemicals related to various industrial applications, such as lead and other metals (Chapters 8 and 10), and the chemical signatures of a variety of organic compounds, such as insecticides and other organochlorides (Chapter 10). Such analyses can provide additional geochronological control. For example, DDT began to be applied as an insecticide in 1939. Hence, any sediment containing this organochloride would be younger than 1939 (assuming no post-depositional mobility). The increase in lead and mercury in sediments is often associated with the middle 19th century, as this period coincided with accelerated industrialization that released these metals to the atmosphere (e.g., Blais et al. 1995). Black carbon (soot) particles that are produced by industries can similarly be used as a geochronological tool (Renberg & Wik 1984; Rose 2001; Rose and Appleby 2005).

4.3.5 Annually laminated sediments (varves)

Under certain limnological conditions (e.g., little or no bioturbation in basins where sediment deposition is linked to a strong seasonal cycle), sediments may be laid down in annual couplets, called varves (Fig. 4.9). Relatively few lake basins contain such annually laminated profiles, but certainly when they are available they present the paleolimnologist with outstanding opportunities for high-resolution, temporally controlled studies (Lamoureux 2001). The resolution available from varved sequences is similar to those exploited by dendroecologists using tree rings to track past climatic and other ecological changes. However, finding a sedimentary profile with laminae does not necessarily imply an annually varved sequence, as many laminae are not necessarily annual couplets. False varves and missing laminae are common, and the quality and reliability of the profile must be assessed, usually involving other geochronological tools. If the sequence is indeed varved, then this outstanding temporal resolution allows the paleolimnologist to pose a large number of new and powerful questions. In some cases, even seasonal resolution of limnological change is attainable if sedimentation rates are sufficiently high (Simola et al. 1990).

Fig. 4.9 Annual couplets (varves) of sediment from Nicolay Lake, Nunavut, Arctic Canada. Photograph taken by S. Lamoureux.

4.4 Correlating multiple cores from the same basin

For certain types of paleolimnological studies, it is important to sample and analyze a large number of cores from a single basin. The resource allocations required to date many cores using ^{210}Pb or other procedures are often prohibitive. In such cases, it may only be possible to date one "master core" and then use core correlations of sedimentary features that are common to all the cores so as to assign approximate dates to the undated cores. To use this approach, changes in some easily measured parameters, such as

sediment texture, composition, pollen markers (e.g., *Ambrosia* rise), and/or magnetic properties must be recorded in all cores. The measurement of magnetic susceptibility variations (see p. 51) has proven to be a very effective technique for between-core correlation in some limnological settings.

4.5 The "top/bottom approach": snapshots of environmental change

Detailed paleolimnological analyses (e.g. cm by cm analyses of sediments) can be very informative, but they are also often quite time-consuming. In some regional cases where many lakes have to be sampled, such detailed analyses are not practical or possible due to resource and personnel limitations. Moreover, for many management questions, such detailed analyses are desirable, but not always totally necessary, as a "before and after" type of environmental assessment can provide many important management answers. For example, with questions such as "Are lakes currently more acidic than they were before the 1850s?" or "Have 20th-century agricultural practices affected water quality in a suite of lakes?", an assessment of what limnological conditions were like before and after the putative impacts may suffice. By inferring the pHs for a suite of lakes before the 1850s (i.e., before acid rain was a problem in eastern North America) and comparing these estimates to present-day pHs, then the first question can be answered. Similarly, if we can infer lakewater nutrient levels that existed in a lake before European-style agriculture was initiated, and compare these hindcasts to post-agricultural water quality, a general answer to the second question can be attained. The data required for such broad questions are from two points in time – before and after the impacts being study. Detailed centimeter-by-centimeter paleolimnological analyses are not required for this type of "before and after" assessment. Paleolimnologists typically use a simpler approach, developed primarily during the acid rain work (Cumming et al. 1992b; S.S. Dixit et al. 1992), which has come to be colloquially called the "top/bottom approach" (Box 4.1).

The major advantage of the "top/bottom" approach is that it is very time and cost effective: typically only two slices of the core profile are analyzed for each lake (not including material used for replicate and variability studies), as opposed to the 20 or more sediment slices that are often analyzed in detailed paleolimnological studies of recent environmental changes. The main disadvantage of this approach is that it only provides two "snapshots" of the lake's history. It does not provide any detailed information on overall trends or specifics regarding the timing of any changes; it simply provides inferences of conditions "before" and "after" a suspected stressor. Nonetheless, this information often comprises some of the most important limnological data that are required for effective management decisions. In order to address some environmental problems, it is an inescapable fact that it may be more

instructive to have, for example, only two time-slices analyzed from 100 different lakes than to have 100 time-slices analyzed from only two lakes. By knowing background (i.e., pre-impact) conditions, the magnitudes of changes can be assessed and more realistic mitigation goals can be set. For example, if a lake was naturally at a pH of 5.6 before acid rain became a problem, then it is not realistic, nor desirable, to "restore" the lake to a pH of 7.0. There are many other applications of these types of data sets, as described later in this book.

Like all scientific procedures, the "top/bottom approach" has some assumptions. The surface sediments (e.g., the top 1 cm, as is often used) quite clearly represent recent conditions, but it is a time-integrated sample that will vary between lakes. For example, the surface centimeter of sediment in Lake A may contain about 3 years of accumulation, whilst it may be about 4 years in Lake B, and may potentially be even less than 1 year in Lake C. It is difficult to precisely determine the accumulation rates of the topmost

---

Box 4.1 The "top/bottom," or "before and after," paleolimnological approach

Detailed centimeter-by-centimeter paleolimnological analyses provide much information on limnological trends over long time frames, but they can rarely be used in large-scale studies on a regional basis (e.g., hundreds of lakes at a time), as they are labor intensive. Broad-scale questions can sometimes be answered by simply using "top/bottom" sediment sampling methods (Fig. 4.10). This approach is very straightforward. Surface-sediment cores are collected in the same manner as they would be for detailed paleolimnological studies, but only two time slices are typically analyzed. Indicators are analyzed from the surface centimeter of sediment, which represents an integrated sample of recent limnological conditions (i.e., the past few years). Then, the so-called "bottom" sediment sample is analyzed in an identical manner. This bottom sample is a sediment slice from the core that represents pre-impact lake conditions. For example, in acid rain work in many North American lake regions, sediments deposited below about 25–30 cm would represent pre-1850 lake conditions (this time estimate is based on a large number of ^{210}Pb-dated cores from the region). Inferences based on the bottom samples provide estimates of background (sometimes referred to natural, pre-impact, or nominal lake conditions) and can be compared to the inferences from the surface samples. This provides a before-and-after type of analysis. Reproducibility should be evaluated by analyzing replicate cores from several lakes (Cumming et al. 1992b).

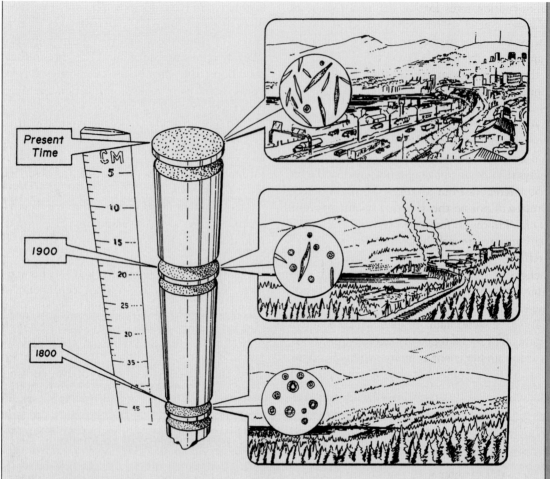

Fig. 4.10 Using the "top/bottom" approach, typically only the surface (the "top") sample of sediment, representing present-day limnological conditions, and one deeper sediment sample (the "bottom"), representing conditions that existed prior to marked anthropogenic influences, are analyzed. In some cases, additional sediments are analyzed, representing other time slices (e.g., diatoms from c. 1900 in this example).

sediments. Nonetheless, like most paleolimnological analyses, we are working with "running averages" and these sediment slices have repeatedly been shown to capture accurately recent limnological conditions.

If the core is not dated (due to resource limitations), then another assumption is that the "bottom sample" was in fact deposited prior to the time that the stressors were introduced

(e.g., before acid rain). To use the North America acid rain example mentioned above, the bottom sample must date from before c. 1850 to capture pre-acid rain conditions. Ideally, paleolimnologists have some independent measure to ensure that sediments predate the disturbance (e.g., they are working in a region where a large number of cores have previously been collected and dated, and so they have an estimate of the average sedimentation rates for the region, or they use an independent stratigraphic marker, such as a low percentage of a pollen grain – for example, ragweed in eastern North America). ^{210}Pb dating of the top and bottom sediments using a gamma counter will also provide an estimate of the minimum age of the bottom sediment sample.

Another assumption is that overall background conditions, estimated from the one bottom sample, have not fluctuated greatly. There is always the danger when using only point sampling that inferences from a sample at, for example, 30 cm will not be the same as from a sample at 32 cm. However, for many regions where the top/bottom approach has been used and evaluated with replicates from a number of lakes, background conditions for the time periods of interest have been effectively captured by these one-point bottom samples. If these assumptions are clearly stated and investigated as much as practical, then the top/bottom approach is a very powerful and efficient management tool.

4.6 The paleolimnologist's option of setting the most appropriate time scale

I began this chapter with a quote from the late Professor W.T. Edmondson, who stated in a keynote address that "the lake sets the time scale, not the investigator." Few would argue with this statement; ecosystems change and respond to stressors at rates that are not dictated by schedules imposed by managers, scientists, or politicians. As much environmental and ecological work is based on time frames of less than a few years, much of the critical information required for sound management decisions is not captured in these short time windows. Paleolimnologists are fortunate in that they have considerable latitude in setting the time scale for their assessments. It is this major advantage – the ability to push the monitoring record back into time – that will be the focus of the remainder of this book.

4.7 Summary

Collecting and dating sediment cores represent critical steps in any paleolimnological study. A variety of different types of coring equipment are now available to retrieve sedimentary sequences from almost any basin, and sampling procedures are available to provide the temporal resolution required for most management questions. Equally important have been advances in geochronological techniques used to date sedimentary sequences. For paleolimnological studies focusing on recent (e.g., the past century or so) environmental changes, radiometric techniques such as ^{210}Pb and ^{137}Cs are most frequently used. For longer time frames (e.g., thousands of years), radiocarbon (^{14}C) dating is often employed. The temporal resolution achievable in some sedimentary sequences, such as those found in annually varved lakes, may approach sub-annual (e.g., seasonal) resolution. Detailed analyses of sediment cores at close sampling intervals, however, are often very time-consuming. In some cases, much coarser sampling resolution can be used, especially if regional-based questions are posed. In such cases, a "top/bottom" sediment approach may be adequate, where typically only two sediment samples are analyzed from each core (excluding samples used for quality assurance and quality control considerations). The "top" sample is from the surface sediments (usually the top centimeter), representing recent (the past few years) environmental conditions. The "bottom" sediment slice represents material deposited before the purported stressor may have affected the lake (e.g., in North American acid rain work, sediments deposited before the 1850s were considered to represent pre-impact conditions). The main advantage of this approach is that it allows one to analyze a large number of lakes on a regional scale; however, because only two sediment slices are analyzed, it does not provide detailed information on the timing and trajectories of changes.

5

Reading the records stored in sediments: the present is a key to the past

Limnology is the science of trying to make sense out of nonsense.
Robert H. Peters, American Society of Limnology and
Oceanography conference (1985)

5.1 Environmental proxy data in sediments

When I am doing an initial examination of paleolimnological data, I sometimes have the same uneasy feeling I get when I am browsing the Internet: I am thirsting for knowledge but drowning in information! Physical, chemical, and biological data contained in sediments are often complex, and it may be difficult at times to see patterns in this "blizzard of information." I have often heard the expression that "the data speak for themselves." Frankly, I have often sat at a desk full of data and have heard nothing. Nonetheless, despite the complexity and multivariate nature of many data sets, paleolimnologists have developed ways to interpret this information to an enviable level of precision, and improvements are constantly being sought and implemented.

Once sediment cores have been collected using appropriate sampling techniques and the depth–time profile is established (Chapter 4), the next job is to begin searching for the paleo-environmental information contained in the sedimentary sequence. This is often the most

time-consuming and important part of a paleolimnological study as it provides the data that will be used in subsequent interpretations. No amount of statistical finesse will compensate for a poor data set: Garbage in → garbage out.

In a book of this size, it is not possible to summarize in any detail the myriad techniques available to paleolimnologists. I will only provide "thumbnail sketches" of some of the common methods and indicators, and expand on these techniques in the subsequent chapters, where examples are discussed in more detail. Several reference volumes exist on methodology (e.g., Berglund 1986), and since this is such a rapidly expanding field, protocols are always being refined. For example, a new book series is dedicated to this topic with volumes on basin analysis, coring, and chronological techniques (Last & Smol 2001a), physical and geochemical methods (Last & Smol 2001b), as well as biological techniques, such as terrestrial, algal, siliceous, and zoological indicators (Smol et al. 2001a,b). A subsequent volume on data handling and statistical methods is currently being written (Birks et al., in preparation). These handbooks outline in considerable detail the advantages, but also the caveats, associated with a diverse array of paleolimnological approaches. Brief summaries of the main sources of proxy data are discussed below.

5.2 Visual inspection of sediments and logging techniques

Once a core is collected, a considerable amount of information may be gained from a simple visual inspection of the sequence (e.g., color, texture, presence of laminations). For example, the input of allochthonous material can be strongly influenced by drainage basin morphology and the type and distribution of vegetation. If the vegetation is removed, erosion typically increases and may result in the influx of silts and clays, and even larger particles such as sand (see Chapter 12). Organic input is usually strongly dependent on biological productivity (from both within and outside the lake), and the degree to which the lakebed environment can preserve this material (e.g., oxygen levels). Many of these and other changes in a sediment profile are visible to the naked eye, and so one of the first steps for the paleolimnologist is simply to note changes in colour, texture and bedding, and so on (Troels-Smith 1955; Aaby & Berglund 1986; Schnurrenberger et al. 2003). A variety of logging and recording techniques are available, especially with the advent of digital technology (e.g., Saarinen and Petterson 2001; Zolitschka et al. 2001). Moreover, radiography and X-ray imaging (Kemp et al. 2001) and other approaches (chapters in Last & Smol 2001b; Francus 2004) can often reveal bedding features and other useful information that is not readily visible with the naked eye.

5.3 Determining the relative proportions of organic matter, carbonates, and clastic matter in the sediment matrix

After the core has been visually examined, photographed, and logged, one of the first procedures typically undertaken by paleolimnologists is to characterize the sediment into a few commonly used components. This is routinely done using weight-loss techniques (Dean 1974; Heiri et al. 2001). For example, the "percentage water" of the sediment is easily calculated by weighing a known amount of wet sediment, and drying it in an oven or freeze drier. By weighing the sample again after drying, the "% water" can easily be calculated by subtraction and division. In the scientific literature, "% water" has been expressed both as a percentage of the dry weight and a percentage of the wet weight of sediment, with most analysts choosing the former. This information is useful as it provides some indication of the amount of compaction the sediment has been subjected to, as well as clues to the paleo-environmental history of the system. Typically, "% water" can be very high in surface sediments, often exceeding 90%.

The resulting dry sediment can be further characterized into three commonly defined components. By placing a known amount of the dry sample in a muffle furnace set typically at 550°C (although other temperatures are sometimes recommended) and ashing it for one to several hours (Heiri et al. 2001), any organic matter in the sediment should be combusted (Dean 1974). (The so-called "self-cleaning ovens" found in some of our kitchens use a similar process.) By subsequently cooling this ashed sample in a desiccator, and then re-weighing it, the "% organic matter," sometimes referred to as the "% loss on ignition" (or "% LOI"), can again be easily calculated by subtraction and division (Dean 1974).

To determine the carbonate content of the sample, the ashed sample is returned to the muffle furnace, and now heated to between 925°C and 1000°C. The carbonates will be converted to CO_2 during this process, and so only the non-carbonate fraction (typically comprising of clay minerals, feldspar minerals, quartz, etc.) will be left in the sample. Once again, the "% carbonates" are calculated by subtraction and division. What is left in the sample is "% siliciclastic material," but this is sometimes incorrectly referred to as the

"% clastics" or the "% mineral matter." Heiri et al. (2001) summarize some additional protocols that paleolimnologists should observe when using these approaches.

The above analyses are easy to perform, but the data are not always so easy to interpret on their own. Intuitively, one might expect that an increase in "% organic matter" in sediments would immediately signal an increase in lake productivity. This is not necessarily so, for a number of reasons. First, the amount of organic matter preserved in the sediments is also related to the limnological conditions in the water column and sediments themselves (e.g., oxygen levels). Higher water temperatures may enhance lake productivity and so increase the amount of organic matter deposited in the sediments, but warmer waters also typically have higher decomposition rates of organic matter. For these reasons, the sediments of some Arctic and sub-Arctic lakes have higher organic matter contents than some tropical lakes. Perhaps more importantly, though, the source of the organic matter should be identified if possible. For example, is it from the terrestrial (i.e., allochthonous) or the aquatic (autochthonous) system? Rapid delivery of terrestrial-based organic matter does not necessarily mean higher within-lake productivity.

Finally, it is critical to remember that organic matter and carbonate and the non-carbonate fractions are all typically expressed as percentage data, and so are "closed data" (i.e., an increase in one component will immediately be reflected by a proportional decrease in another, as the total cannot exceed 100%). For example, with land-use changes, such as deforestation and plowing of fields in a lake's catchment, one might expect two limnological repercussions. One expectation might be an accelerated influx of nutrients to the lake from the catchment, thus resulting in higher lake productivity and increased organic matter production. A corresponding increase in % organic matter in the sediments may occur. However, it is possible that, with increased erosion, a considerable amount of clay material was eroded into the lake, and so the paleolimnologist may actually record a relative decrease in % organic matter, as the clay fraction increased disproportionately higher. So even though lake productivity increased with disturbances, the % organic matter may not reflect this. If sedimentation rates can be reliably estimated, these data can be expressed as an accumulation rate. However, other approaches, such as those summarized in the remainder of this chapter, should be used in conjunction with these data to disentangle these relationships.

5.4 Particle size analysis

The inorganic, siliciclastic portion of the sediment matrix can be characterized further with respect to size (Last 2001a). On a broad scale, particles are divided into clay (< 0.002 mm in diameter in most European work, but < 0.004 mm in most other studies), silt (from the clay upper limit boundary up to 0.06 mm in diameter), sand (0.06–2.00 mm in diameter), and, in some cases, gravel (over 2.0 mm in diameter) fractions. Different approaches can be used to determine particle sizes, including older hydrometric techniques to more sophisticated laser counters (Last 2001a).

Particle size analyses are employed in a variety of applications, but have been mostly used by sedimentologists to help determine the process, and to some extent the source, of detrital sedimentation. For example, a high concentration of clay particles is often interpreted as an erosion signal. An increased frequency of larger-sized particles, such as sand grains, would suggest delivery of this material in a high-energy environment, such as from river flow. The location of the coring site in the lake basin must be carefully considered when interpreting these data. In studies with an environmental focus, these approaches have mostly been applied to the study of erosion patterns and sediment infilling (Chapter 12).

5.5 Magnetic properties of sediments

Sediments also archive a magnetic signal, reflecting the magnetic properties of the sediment matrix. Magnetic grains are dominated mainly by iron oxides and sulfides. Details regarding methodology are presented in Nowaczyk (2001) and Sandgren and Snowball (2001); the techniques available are relatively rapid, inexpensive, and non-destructive. Maher and Thompson (1999) provide many chapters on the applications of magnetism to paleoenvironmental issues that are relevant to the overall theme of this book. Therefore, only a few introductory comments will be mentioned here, as examples dealing with environmental magnetic techniques are discussed primarily in Chapter 12 on erosion, although magnetic properties are also used to track and monitor heavy metal and other pollution sources (Petrovský & Ellwood 1999).

Maher et al. (1999) provide a photomicrograph atlas of some of the natural and anthropogenic magnetic grains found in sediments. Although sediment samples contain several magnetic properties, paleolimnologists are primarily concerned with three magnetic units: magnetic field, magnetization, and magnetic susceptibility. Magnetic fields are set up by the motion of electric charges. Magnetic induction (B, measured in Teslas or T), the response of any materials in this space, comes from both the magnetic field (H) and the magnetization (M) of the material. The magnetization (M, measured in ampere per meter, or A m^{-1}) is the response of a material to a magnetic field passing through it. Magnetic susceptibility (κ, and has no units) is a measure of how much magnetism sediment retains after being exposed to a magnetic field, and is related to M and H by the equation $M = \kappa H$. Using a variety of ratios (Maher et al. 1999, and some examples in Chapter 12), paleolimnologists can glean important information on sedimentary history, especially as it relates to past erosion sequences. It is also useful for correlating multiple cores taken from the same basin (King & Peck 2001) and

for locating tephra layers in sediment profiles (Turney & Lowe 2001).

5.6 Geochemical methods

Entire volumes have been written on geochemical techniques (Last & Smol 2001b), and so only some salient features will be mentioned here. Moreover, geochemical data are the focus of several chapters in this book (e.g., Chapters 8–10).

Geochemical analyses of sediments can yield much information but, like most paleolimnological data, inferences have to be considered carefully, as post-depositional changes in geochemical signatures can occur and obscure the record (Boyle 2001). For example, if managers wanted to reconstruct past nutrient levels in a lake, one might initially think that simply analyzing the sediments for total phosphorus and nitrogen concentrations would provide a record of past nutrient loading. As has often been shown, this is not as simple as it may sound. For example, phosphorus can be easily remobilized from sediments under anoxic conditions, which itself may be more common during periods of highest nutrient loading (see p. 185). So phosphorus release (not accumulation) from sediments may be highest under periods of highest nutrient loading.

Nonetheless, with an understanding of geochemical processes, much information can be gleaned from the sedimentary record about processes occurring within the water body, as well as in the surrounding drainage area (Boudreau 1999). Following the pioneering work of Mackereth (1966) in the English Lake District, a large number of geochemical studies have been completed on sediments. These data have been used to interpret changes in inputs and processes within lakes and rivers, to clarify nutrient fluxes (Chapter 11), as well as the accumulation of pollutants such as metals (Chapters 8 and 10) and persistent organic contaminants (Chapter 9).

5.7 Isotope analysis

Some isotope methods have already been mentioned in Chapter 4, as they are often the prime tools used by paleolimnologists to date sediment cores (e.g., ^{210}Pb, ^{137}Cs, ^{14}C). However, these radioactive isotopes are just a small part of the paleolimnologist's tool kit, as many stable isotopes can be used to track, for example, the source of pollutants and materials, as well as a variety of other limnological processes (e.g., Ito, 2001; Talbot 2001; B.B. Wolfe et al. 2001). For example, sulfur isotopes have been used in work on acidic precipitation (see p. 93), and carbon isotopes have been employed in lake eutrophication studies to reconstruct past patterns in aquatic primary production (see pp. 190–1). As shown in Chapters 8 and 10, the isotopes of various metals, such as lead and mercury, can also track pollution sources. Isotopes of nitrogen have been used to estimate past salmon and seabird populations (see pp. 290–5). The potential for using compound-specific isotope ratios from lipids and other biomarkers is now also being explored using sediments (e.g., Huang et al. 2004), which could lead to important applications for water quality and related issues.

The use of isotopes in paleolimnology is a rapidly expanding field, with some recent approaches summarized in a volume edited by Leng (2006). Many of these applications are discussed in more detail in subsequent sections of this book, and surely many new applications will soon be available as these techniques become more fully incorporated into ongoing paleolimnological studies.

5.8 Fly-ash particles

A rapidly developing area of applied paleolimnology is the use of different types of fly-ash particles, which can be used to track past industrial activities related to fossil fuel combustion (Wik & Renberg 1996; Rose 2001). When fossil fuels such as coal and oil are burned at high temperatures to generate electricity or heat, or used for some other industrial application, two types of particles are produced that are now being used increasingly by paleolimnologists. Porous spheroids of primarily elemental carbon (Goldberg 1985), now often referred to as spheroidal carbonaceous particles (SCPs), are one of the products of this burning process (Rose 2001). The second type of particles includes those formed from the mineral component of the fuel (Raask 1984), called inorganic ash spheres (IASs) (Rose 2001). Collectively, SCPs and IASs are referred to as fly-ash particles (Figs. 5.1 & 5.2).

Rose (2001) provides a review on developments in the methodology and uses of fly-ash particles in sediments, where they have been used primarily for three types of applications. First, stratigraphic changes in particles have been interpreted in conjunction with other inferred changes in sediment cores, such as decreases in diatom-inferred lakewater pH associated with lake acidification (Chapter 7). As the industries producing the particles (e.g., coal combustion) are also often producing other environmental stressors (e.g., acid rain), including fly-ash particles in multi-proxy paleolimnological studies can be very useful in correlating the timing and potential causes of change. For example, if the timing of acidification inferred from diatoms (see below) matches the increases in fly-ash particles, the industry producing the particles may be implicated in the acidification process.

Second, particles can be used to provide additional geochronological control in sedimentary profiles. SCPs are not produced by any natural mechanisms, and so their first appearance in sediment cores can be used to identify the start of industrial burning of fossil fuels. For example, in the United Kingdom, the SCP record begins at about AD 1850–1860 (Rose 2001). The use of SCPs as a dating tool is primarily done via calibration of temporal concentration profiles of SCPs from, ideally, a number of sediment cores from a region to reliable and independently

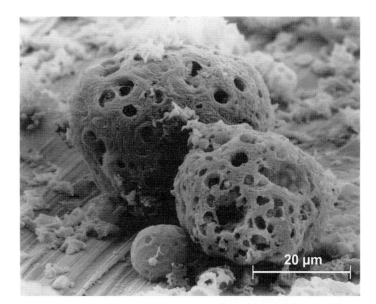

Fig. 5.1 A scanning electron micrograph of spheroidal carbonaceous particles (SCPs) from oil combustion. From Rose (2001); used with permission.

Fig. 5.2 A scanning electron micrograph of inorganic ash spheres (IASs) from the combustion of coal. From Rose (2001); used with permission.

derived dates (e.g., from ^{210}Pb and varve counting) for the same sediment cores. Once reasonably reliable dating schemes and associated SCP profiles are established for a region, then SCP concentrations from undated cores can be used as a geochronological tool (e.g., Rose & Appleby 2005). Furthermore, if the history of industrial fuel combustion is documented, then the stratigraphic changes in sedimentary fly-ash particles can be correlated to known dates. For example, the increase in particles should coincide with the increases in the industry's fossil fuel burning. Similarly, if mitigation efforts have been initiated and the release of particles has declined, this decrease should also be recorded in the sediments and can be used as another time horizon.

Finally, particles can be used as surrogates for other pollutants, whose concentrations are correlated with fly-ash particle numbers. For example, fly-ash particles can often be correlated with trace metals, sulfur, and polycyclic aromatic hydrocarbons (PAHs).

Fly-ash particles are a relatively new group of indicators, but work is progressing rapidly. For example, one active area of research is on particle source apportionment – determining which industry produced specific SCPs. As summarized by Rose (2001), this has been attempted using morphological data on particles, as revealed through high-resolution microscopy using thin-sectioning and size characteristics, and more recently via multi-element chemical apportionment. The morphological classification of combustion particles, as documented for Arctic regions by Doubleday and Smol (2005), may also prove to be a useful approach.

5.9 Biological indicators

One of the biggest advances paleolimnology has experienced over the past few decades has been in the description and application of a suite of new biological indicators. A central theme in ecology has always been the attempt to classify ecosystems by the flora and fauna they contain. In fact, we make these associations every day. For example, presenting someone with a picture of cacti and camels will immediately suggest an arid and hot climate. A picture of a polar bear, a musk ox, and barren tundra vegetation will indicate cold, Arctic climates. Sea urchins and seaweeds will indicate a marine environment, whilst certain species of snails and fish will only survive in fresh waters, and so on. To a large extent, the same types of ecological associations are used by paleolimnologists, although they typically use microscopic indicators of past biota and deal with many species simultaneously. Biological approaches offer the additional advantage in that organisms integrate changes in limnological variables over longer-term time frames than can typically be captured in periodic

water chemistry sampling programs. As Rosenberg (1998) remarked, "Chemical measurements are like taking snapshots of the ecosystem, whereas biological measurements are like making a videotape."

Fortunately, a large number of biological indicators are represented as morphological fossils, such as the hard parts of organisms (e.g., diatom frustules, resting stages or the exoskeletons of invertebrates, or pollen grains and spores). But equally rapid advancements have occurred in biogeochemical fields, where progress has been made in the identification and interpretation of "chemical fossils" (e.g., fossil pigments, organic geochemistry).

In order for an organism to be a useful paleoindicator, it must meet two important requirements. First, it must be unequivocally identifiable in the sedimentary record as some sort of morphological or biogeochemical remain of a particular taxon. Fortunately, as will be described below, a surprisingly large spectrum of organisms leave reliable and interpretable fossil records. Problems certainly do occur with both physical (e.g., breakage) and chemical (e.g., dissolution) degradation under certain conditions, as well as other taphonomic processes. These problems, however, can often be assessed and the quality of the record evaluated. In fact, if the processes responsible for these problems are well understood, paleoenvironmental information can, in some cases, be surmised from the preservation status of indicators (e.g., if the conditions required for poor preservation of a specific indicator are understood, and the paleolimnologist is faced with these conditions in a sediment core, then he or she can surmise that these environmental conditions had existed in the system under study).

The second overall requirement of paleoindicators is that the ecological characteristics of the organism should be known to at least some level of certainty (for without some understanding of its ecological preferences, one cannot use it to infer past conditions). Some species are generalists, but others are specialists and are quite restrictive in their distributions with respect to physical and chemical variables. Powerful tools and statistical treatments are available to explore and quantify these relationships (Chapter 6). Ideal paleoindicators should also be abundant and relatively easy to identify and quantify.

For almost all biological indicators, some form of sediment preparation is required before they can be easily identified and analysed, and almost all morphological fossils require microscopy (sometimes with very high magnification and resolution) for identification (Berglund 1986; Smol et al. 2001a,b). In general, the goal is to remove (clean) as much extraneous material as possible, which leaves the indicators relatively unobscured for further examination. For some larger indicators, the sediment is first sieved through screens of different mesh sizes. Most sediments require treatment with various strong acids and other reagents to complete the separation and cleaning process. As different indicators are composed of different materials (e.g., silica, chitin, etc.), there are many ways that sediment samples are prepared (Berglund 1986; Smol et al. 2001a,b).

Below are brief overviews of some of the most important bioindicators used by paleolimnologists; additional details on these groups are contained in later sections of this book, where these indicators are used to address a variety of environmental problems. The list of organisms available to paleolimnologists is steadily increasing as new approaches and techniques continue to be developed (Smol et al. 2001a,b).

5.9.1 Pollen and spores

Plants produce either pollen grains (e.g., from angiosperms and gymnosperms) or spores (e.g., from so-called "lower plants," such as mosses, ferns, and fungi; e.g., van Geel & Aptroot 2006) as part of their reproductive strategies. Collectively, these microfossils are referred to as palynomorphs, but that term also refers to, for example, the vegetative remains of algal cells (e.g., *Pediastrum* colonies: Jankovská & Komárek 2000; Komárek & Jankovská 2001) or their resting stages (e.g., van Geel 2001). Other indicators often

observed on pollen slides include the stomatal cells of leaves (MacDonald 2001).

Pollen analysis is a large field, with many applications to topics such as ecology, climatology, archeology, forensic science, and various medical fields (e.g., allergies). Although palynological investigations are often focused on reconstructing terrestrial vegetation (Bennett & Willis 2001), they often have strong links to many questions posed by paleolimnologists, such as inferring changes occurring in the catchment of a lake.

Plants may release pollen in extremely high numbers (some trees produce several billion grains each year). Anyone who suffers from pollen allergies, such as hay fever, will likely already know this! Some plants rely on wind to transport their grains to other plants for fertilization (anemophilous pollen), and these tend to be the high pollen producers. Others rely on insects and other animals, as well as water currents and other vectors, to transport their pollen to other plants.

The outer layer (exine) of grains contains the polymer sporopollenin (also spelled sporopollinen), which is one of the most inert organic substances known. Hence, it is not surprising that pollen grains and spores are amongst the most common microfossils preserved in sediments. Differences in the exine structure and sculpture, as well as other morphological features, typically allow palynologists to identify grains to the generic level and sometimes even to the species level, although some taxa can only be identified to the class or family level. A glossary of the terminology used by palynologists has recently been published (Punt et al. 2007). By identifying and enumerating grains and spores in sedimentary profiles, palynologists can reconstruct past trends in terrestrial vegetation succession, as well as past shifts in aquatic macrophytes and nearshore vegetation. A large number of books and several journals are dedicated to this field, as palynology has many applications quite separate from those that might be used by paleolimnologists. As vegetation is often closely linked to

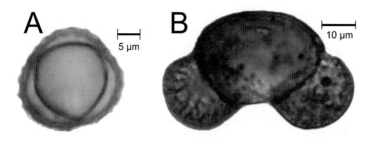

Fig. 5.3 Light micrographs of pollen grains. (A) ragweed (*Ambrosia*); (B) pine (*Pinus*).

climatic variables, palynology is often used to reconstruct past climatic trends (Chapter 14).

An important application of palynology is to use specific anthropogenic indicator pollen to reconstruct how catchments have been affected by human impacts, such as the removal of forests, replacement by weeds and cereal crops, and so forth. As noted in the previous chapter, palynology can be used to help establish (or confirm) depth–time profiles. For example, one of the "signatures" of the arrival of European settlers to the northeastern United States and parts of eastern Canada has been the proliferation of ragweed (*Ambrosia*) pollen (Fig. 5.3), as this weed does very well in disturbed soils following forest clearance. The "ragweed rise" in pollen diagrams (i.e., an increase in ragweed pollen percentages) is often used as a chronostratigraphic marker. In other regions, cereal grains, grasses (the family Poaceae, referred to as the Gramineae in some of the older literature), or other taxa are used. Pollen grains can be categorized as either arboreal pollen (AP) or non-arboreal pollen (NAP), and expressed as a ratio or percentage, with NAP typically increasing with decreasing forest cover, as would occur with land clearance. As noted above, some pollen grains are transported by wind (sometimes hundreds and even thousands of kilometers). This has both advantages and disadvantages; pollen can be used to track regional changes, but at times it is difficult to reconstruct changes occurring within, for example, a single watershed. Plant macrofossils (see the next section) are often more helpful at delineating local, watershed patterns of vegetation change.

5.9.2 Plant macrofossils

Plant macrofossils are defined as any identifiable plant remains that can be seen with the naked eye, even though some form of magnification is often needed for identification. Typical macrofossils include seeds, fruits, cones, needles, bark, pieces of twigs, and moss fragments. Some algal macrofossils would also be included, such as reproductive structures (oospores) of charophytes (e.g., García 1994; Haas 1999). Because macrofossils are larger and heavier than pollen grains and spores, macrofossils are not transported long distances, and so are especially useful for describing *local* as opposed to *regional* changes. Conversely, macrofossils are found in far lower numbers than pollen, and so recovering representative sample numbers may be difficult in some cases.

Within the focus of this book, plant macrofossils have primarily been used to reconstruct past aquatic macrophyte and nearshore vegetation (e.g., Birks 2001). For example, tracking macrofossils associated with aquatic macrophytes can be used to document the development of littoral zones. Furthermore, macrofossils directly associated with anthropogenic activities, such as wood chips from lumber mills, can provide important time markers for paleolimnological studies (e.g., Reavie et al. 1998).

5.9.3 Charcoal

Associated with plant remains, and often studied using microscope slides prepared for pollen analysis or some modification to standard palynological techniques (Whitlock & Larsen 2001), are charcoal particles – the burned residues of past fires. Similar to pollen grains, small charcoal pieces (e.g., < 100 μm in size) can be transported long distances by wind currents, and therefore provide information on regional fire frequencies. Larger charcoal particles more likely reflect local wood burning. Although more often used in long-term studies of ecosystem development, such as tracking past vegetation responses, and linking fire frequency, to past climatic changes, charcoal analyses have been used in applied paleolimnological studies to track different types of human activities, such as industries that depended on wood burning (e.g., some earlier mining operations). Charcoal profiles have also been used as geochronological tools to help determine the onset of, for example, slash-and-burn agriculture, and to study post-fire succession.

5.9.4 Cyanobacteria (blue-green algae)

Although the cyanobacteria (more commonly referred to as blue-green algae) are prokaryotic and should probably be considered under a separate heading, I will deal with them in the subsequent section dealing with algae. Because blue-green algae have chlorophyll *a* and other pigments, they photosynthesize much like other algae and therefore, in the context of lake management issues, are functionally and ecologically more similar to algae.

5.9.5 Algae

Primary production in water bodies comes principally from two major groups: algae (including cyanobacteria or blue-green algae) and aquatic macrophytes. With the exception of some shallow-water bodies that have extensive littoral zones and large growths of macrophytes, algae are the main photosynthesizers in most deep aquatic systems.

The word "algae" has no formal taxonomic status, as it represents a large group of diverse organisms from different phylogenetic groups, representing many taxonomic divisions. In general, algae can be referred to as plant-like organisms that are usually photosynthetic and aquatic, but lack true roots, stems, leaves, or vascular tissue, and have simple reproductive structures. Most are microscopic, but some are certainly larger, such as several marine seaweeds that can exceed 50 m in length, as well as certain freshwater taxa that, at first glance, may appear to share morphological affinities with higher plants (e.g., charophytes). The study of algae, called phycology or algology, is a major discipline of biology, with many textbooks, journals, and university courses dedicated to this field.

As most of the primary production in freshwater systems comes from algae, and given that there are many thousands of algal taxa, it is not surprising that paleophycology has played a leading role in paleolimnology. All algal groups leave some form of fossil record in lake sediments, whether it be a morphological (e.g., diatom valve, chrysophyte scale) or biogeochemical (e.g., algal pigments) marker, although many of these indicators are still rarely used and/or not fully explored. Below is a brief overview of some of the indicators most frequently employed by paleolimnologists.

5.9.5.1 Diatoms

Until relatively recently, if a paleolimnologist referred to algal indicators in his or her sediment cores, most other paleolimnologists would intuitively think that he or she simply meant diatoms (Bacillariophyta). The reason for this is that diatoms, with few exceptions, were the only algal group used in most studies until the past 20 years or so. It is now recognized that many other algal groups can provide important proxy data; nonetheless, diatom valves still often form the mainstay of many paleolimnological investigations. There are many good reasons for this, as few indicators can rival the important qualities diatoms share that make them such powerful biomonitors of environmental change (Stoermer & Smol 1999). For example, diatoms are often the dominant algal group in most freshwater

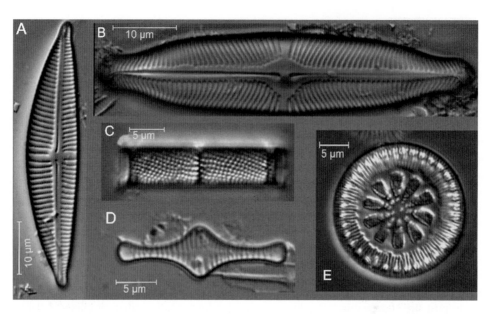

Fig. 5.4 Light micrographs of diatom valves. (A) *Cymbella hebridica*; (B) *Pinnularia microstauron*; (C) *Aulacoseira ambigua*; (D) *Tabellaria flocculosa* str. IV; (E) *Cyclotella antiqua*. Photographs taken by Kathleen Laird and Brian Cumming.

systems, often contributing more than half of the overall primary production. In addition, there are thousands of species of diatoms, whose taxonomy is based primarily on the size, shape, and sculpturing of their siliceous cell walls (Fig. 5.4), called frustules (two valves make up one frustule). Different taxa have different environmental optima and tolerances, and so analyses of fossil species assemblages can be used to reconstruct a suite of environmental variables (e.g., pH, nutrient levels, etc.). Diatoms have very fast migration rates, and so can colonize new habitats quickly. Hence, assemblage changes closely track environmental shifts (i.e., lag times are minor). Equally important, from a paleoecological perspective, silica (SiO_2) is generally resistant to bacterial decomposition, chemical dissolution, and physical breakage, and so diatom valves are often very well preserved in sedimentary deposits. Battarbee et al. (2001) provided methodological details related to fossil diatom analyses, as well as summarizing their historical development in paleolimnological studies.

Diatom assemblages are excellent microscopic indicators of a large number of important environmental variables. The impressive number and diversity of applications span such diverse fields as paleolimnology and paleoceanography, forensic science, and archeology (Stoermer & Smol 1999). Because there are so many taxa (often estimated in the range of 10^4 taxa, but many believe that this is likely an underestimate by perhaps one order of magnitude), and because taxa have different ecological optima and tolerances, species assemblages can be effectively used to infer past environmental changes. For example, certain taxa are adapted to colonizing specific habitats of a lake or river, such as attached substrates in the littoral zone (i.e., the periphyton) or the open-water community (i.e., the plankton). As is shown in the coming chapters, different taxa are also reliable indicators of a wide spectrum of limnological conditions, such as pH, nutrient concentrations, and salinity. New paleoenvironmental applications for diatoms are constantly being sought and found.

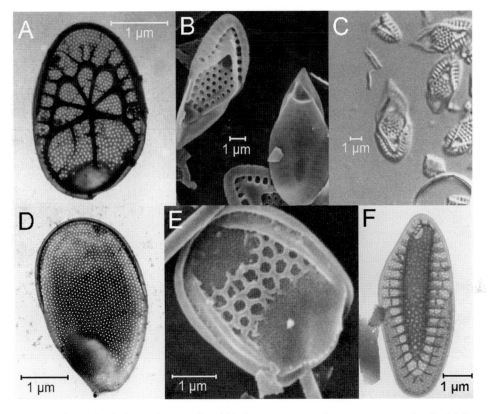

Fig. 5.5 Micrographs of fossil chrysophyte scales. (A) A transmission electron micrograph of a *Mallomonas hindonii* scale. (B) A scanning electron micrograph of M. *pseuodocoronata* scales. (C) A light micrograph of M. *pseuodocoronata* scales. (D) A transmission electron micrograph of a M. *hamata* scale. (E) A scanning electron micrograph of M. *punctifera* scale. (F) Transmission electron micrograph of a *Synura petersenii* scale.

5.9.5.2 Chrysophyte scales and cysts

Another group of siliceous algal microfossils that is being used more often are the scales (Fig. 5.5) and resting cysts (Fig. 5.6) of the classes Chrysophyceae and Synurophyceae (hereafter referred to as chrysophytes). Over 1000 species have been described thus far, from about 120 genera. However, like the diatoms, these numbers are considered to underestimate their true diversity, as many new taxa likely remain to be described. Most chrysophytes have flagella and are especially common in the plankton of oligotrophic lakes, although benthic and eutrophic taxa also exist. Sandgren et al. (1995) collated a series

of review chapters on different aspects of the taxonomy, biology, and ecology of chrysophytes, including one focusing on their paleolimnological applications (Smol 1995b). Zeeb and Smol (2001) summarized methodological aspects relevant to microfossil work.

Although only about 15% of chrysophyte species have siliceous scales and bristles, and all chrysophytes form resting stages, to date most paleolimnological studies have focused on the scales. The reason for this is mainly taxonomic – the often intricately shaped and sculptured scales of taxa in the class Synurophyceae (Fig. 5.5) are species specific (Siver 1991), whilst the taxonomy of cysts (Fig. 5.6) is still in development

Fig. 5.6 Scanning electron micrographs of fossil chrysophyte stomatocysts. (A, B) Light and scanning electron micrographs of Stomatocyst 204 (*Uroglena volvox*). (C) A scanning electron micrograph of Stomatocyst 97 (biological affinity not yet determined). (D, E) Scanning electron and light micrographs of Stomatocyst 41 (*Dinobryon cylindricum*). (F) A scanning electron micrograph of Stomatocyst 35 (biological affinity not yet determined). Modified from Duff et al. (1995).

(Duff et al. 1995; Pla 2001; Wilkinson et al. 2001). Even though considerable work has been completed on the taxonomy of chrysophyte scales (Siver 1991), their use in paleolimnological research only began in the 1980s, with the almost simultaneous publication of three independent studies (Battarbee et al. 1980; Munch 1980; Smol 1980).

Most scales identified in lake sediments are from two major and diverse chrysophyte genera, *Mallomonas* and *Synura*, but other genera (e.g., *Chrysosphaerella*, *Paraphysomonas*, and *Spiniferomonas*) are also used. Scanning or transmission electron microscopy is needed to study the ultrastructure of scales, but (with few exceptions) most scales can be identified using high-resolution light microscopy, once an adequate proportion of the flora has been investigated (Smol 1986). Similar to the diatoms, scales possess many characteristics that make them important paleoindicators (e.g., well-preserved due to their siliceous structure, reliable indicators of a suite of limnological conditions, etc.).

Unlike scales, which characterize only some genera, all chrysophyte taxa produce an endogenously formed, siliceous resting stage (Fig. 5.6), known as a stomatocyst (or statospore in some of the older literature). Because of their siliceous nature, these microscopic cysts are well preserved in most deposits. Moreover, cyst morphotypes (e.g., the cyst shape, surface sculpturing and ornamentation, as well as pore and collar morphology) appear to be species specific, although only about 15% of morphotypes have, as yet, been linked to the taxa that produce them (Duff et al. 1995; Wilkinson et al. 2001). This presents the paleolimnologist with some interesting challenges: Contained in most sediments are often diverse assemblages of stomatocysts, which can be differentiated and – using surface-sediment calibration sets (Chapter 6) – can be related to important environmental gradients (e.g., pH, nutrients), but cannot yet be linked to known extant taxa. From a paleoecological perspective, where the goal is mainly to infer limnological changes, differentiating cyst morphotypes is critical, but actual species identifications are not critical (although desirable). As a result, a well-defined system of morphotype descriptions has developed, beginning with the guidelines of the International Statospore Working Group (ISWG), first summarized in Cronberg and Sandgren (1986), and later modified by Duff et al. (1995) and then by Wilkinson et al. (2001), with the publication of two atlases of cyst morphotypes. With these standardized guidelines and the numbering system, it has been possible to identify many of the morphotypes present in Northern Hemisphere temperate lakes, although there has been much less study of tropical and Southern Hemisphere floras. The long-term goal of paleolimnologists working with cysts is to eventually abandon these temporary nomenclature systems and link morphotypes to known species, but progress in this area has been slow. Nonetheless, chrysophyte stomatocysts have been used in a number of important paleolimnological applications (Zeeb & Smol 2001).

5.9.5.3 Other siliceous microfossils

Other siliceous microfossils are often preserved on diatom microscope slides, including ebridians (Korhola & Smol 2001), invertebrate remains such as freshwater sponge spicules (Frost 2001) and siliceous protozoan plates (Douglas & Smol 2001), and phytoliths (Piperno 2001) from grasses and other plants. Each of these can be used in applied investigations, although their full potentials have yet to be explored in pollution-based studies.

5.9.5.4 Biogenic silica

Identifying and counting microscopic indicators, such as diatom valves and chrysophyte scales and cysts, is often time-consuming and requires trained analysts. Nonetheless, the amount of information one can gather from species assemblage data is often worth the effort. If, however, a paleolimnologist simply wants to estimate the past biomass of diatom populations and other siliceous organisms, he or she can use a chemical technique to determine the amount of biogenic silica (BiSi) in sediments (Conley & Schelske 2001). BiSi estimates have primarily been used to track past diatom populations in eutrophication studies (Chapter 11), but also in other environmental applications.

5.9.5.5 Other algal morphological indicators

Siliceous indicators, and especially diatom valves, have dominated paleophycological work, but other algal remains may also be preserved as morphological fossils. These include vegetative structures such as colonies of green algae, such as *Pediastrum* (Jankovská & Komárek 2001; Komárek & Jankovská 2001), or filaments, lorica, and various resting stages of other algae and cyanobacteria (van Geel 2001). For example, dinoflagellate cysts may be quite common in some profiles (Norris & McAndrews 1970). Typically, these non-pollen palynomorphs are studied alongside pollen grains and spores on microscope slides. Although still rarely used, several examples have highlighted their considerable potential in applied studies. For example, van Geel et al. (1994) used fossil akinetes (resting stages) of the blue-green algae *Aphanizomenon* and *Anabaena* in a Polish lake to indicate phosphate eutrophication dating back to medieval times.

5.9.5.6 Fossil pigments

Although the list of known morphological indicators of algae and cyanobacteria increases steadily, there are still many taxa that do not leave reliable morphological fossils, and so their populations can only be reconstructed using biogeochemical indicators. These generally take the form of fossil pigments or other forms of organic geochemistry. For example, long after many morphological remains of algae and bacteria are lost due to various degradation processes, sedimentary carotenoids (carotenes and xanthophylls), chlorophylls, their derivatives, and other lipid-soluble pigments can be used to track past populations (Leavitt & Hodgson 2001). This, of course, assumes that these chemical bioindicators can be isolated from sediments, and can be identified and linked to the organisms that produced them. With the advent of new technological developments (such as high performance liquid chromatography, or HPLC, and mass spectrometry), considerable progress has been made in this regard, and many fossil pigments have been used in different applications where the abundance, production, and composition of past phototrophic communities are important response variables (Leavitt & Hodgson 2001). Some pigments, such as β-carotene, are found in all algae and some phototrophic bacteria, whilst other pigments are much more specific (for example, alloxanthin is specific to the Cryptophyta, and peridinin is specific to the Dinophyta). Pigments such as myxoxanthophyll, oscillaxanthin, and aphanizophyll are specific to various blue-green algal taxa, and have been used to track nuisance populations of these cyanobacteria in lake eutrophication work. Leavitt and Hodgson (2001) provide a table listing the commonly recovered pigments from lake sediments and their taxonomic affinities.

5.9.6 Other forms of organic geochemistry

Fossil pigments are just one form of organic biochemical fossils found in sediments. Although most organic compounds are lost from the paleolimnological record by diagenetic processes, lake and river deposits often contain a complex mixture of lipids, carbohydrates, proteins, and other organic molecules from organisms that lived in the lake or river and in its catchment. Most organic matter is from plants, which can be divided into two major groups: (i) nonvascular plants (such as algae) that contain little or no carbon-rich cellulose and lignin; and (ii) vascular plants (such as trees, grasses, aquatic macrophytes) that contain much larger proportions of these fibrous tissues (Meyers & Lallier-Vergès 1999). Cranwell (1984) summarized some of the earlier paleolimnological work done using organic geochemistry; more recent applications are reviewed by Meyers and Teranes (2001).

One approach that has been used in paleolimnology is to measure the ratio of carbon to nitrogen (C/N) in sediments as a proxy of past algal versus higher plant production. This approach is based on the observation that algae have low C/N ratios, of about 4 to 10. Meanwhile, vascular land plants, which have high cellulose content and therefore high C content, but are relatively low in proteins and hence N, generally have much higher C/N ratios of about 20 or greater (Meyers & Lallier-Vergès 1999; Meyers & Teranes 2001). These types of analyses can also be linked to isotope studies, whereby the source material for the organic matter can further be traced to, for example, C_3 or C_4 land plants (Fig. 5.7).

Other approaches include Rock-Eval™ pyrolysis, originally developed in the petroleum industry, but recently applied to lake sediments. By studying the ratios of hydrogen to carbon and oxygen to carbon, organic matter can be linked to its most likely source material, such as from microbial biomass, algae, land plants, and so forth (Meyers & Lallier-Vergès 1999; Meyers & Teranes 2001).

Stable isotope analysis of different organic fractions, determining the concentrations and ratios of specific compounds (e.g., Cranwell 1984; Bourbonniere & Meyers 1996), and microscopical examination of the sediments (Meyers & Lallier-Vergès 1999) have all been used effectively to better document paleoenvironmental changes. Only a few of these studies have been applied to management issues (most have focused on climatic

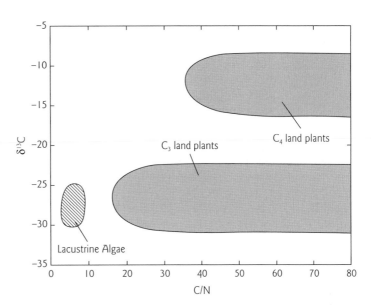

Fig. 5.7 The proportion of sedimentary organic matter that originated from non-vascular aquatic (algal) versus terrestrial (vascular land plant) sources can be estimated by the elemental ratio of carbon to nitrogen (C/N). By also using stable isotope methods (e.g., $\delta^{13}C$), material sourced to C_3 land plants can be distinguished from C_4 land plants. From Meyers and Lallier-Vergès (1999); used with permission of Springer Science and Business Media.

change), but the potential is certainly there to adapt these approaches to many of the problems discussed in this book.

5.9.7 Bacteria

In some respects, bacterial paleolimnology has been covered to a limited extent in the above treatment of fossil pigments, as the cyanobacteria or the so-called blue-green algae are often discussed alongside the algae. Although far less used, other prokaryotic organisms can also be traced in the sedimentary record. Fossil pigments have taken the lead, and Brown (1968) summarized some of the pioneering work in this area, which has been updated in the review by Leavitt and Hodgson (2001). Because different species of photosynthetic bacteria in lakes have different ecological requirements for variables such as light, nutrients (e.g., sulfur), and oxygen levels, past changes in their populations can be used to infer several important limnological conditions. For example, Brown et al. (1984) used the fossil pigments of photosynthetic bacteria from a meromictic lake to track past stratified populations of a number of different bacterial groups. Chlorobiaceae, or the brown-colored varieties of the green sulfur bacteria

(inferred from the pigment β-isorenieratene), live in deepwater strata, and would have thrived during the most oligotrophic (highest light penetrating) periods of the lake's recent history. Chromatiaceae, or the purple sulfur bacteria, represented by the pigment okenone, live in higher strata of the water column, and are at a competitive advantage relative to the green sulfur bacteria when light penetration decreases (e.g., with eutrophication). Brown et al. (1984) concluded that these organisms can be used as "historical Secchi discs," reflecting the trophic status of the waters above them.

Some studies have attempted to directly count bacterial remains. For example, Granberg (1983) counted in a Finnish lake sediment core the number of *Closteridium perfringens* colonies, a non-motile bacterium that lives in the human digestive system, to track the effects of human effluents on this system. Nilsson and Renberg (1990) and Renberg and Nilsson (1991) showed how paleolimnologists can use sterile microbiological techniques to culture viable bacteria found in sedimentary profiles, and thereby establish past population trends. For example, they estimated past annual accumulation rates of viable endospores of *Thermoactinomycetes vulgaris*, a terrestrial bacterium known to be common in decaying vegetation, composts, hay, and manure. Because it is a thermophilic bacterium, once it is washed into a lake and deposited in the sediments, it does not reproduce, and so its abundance in sedimentary profiles can be used as a proxy of its past abundance on land. By using sterile microbiological techniques, such as culture media and plating sediment samples on agar plates, Renberg and Nilsson (1991) showed that *T. vulgaris* endospores were very rare in Swedish lake sediments deposited before agriculture began (c. AD 1100), but increased greatly with the expansion of farming and other land-use activities. With the advent of new biochemical and DNA techniques, it is likely that paleobacterial studies will become more common in the future.

5.9.8 Invertebrates

In addition to the primary producers discussed earlier, representatives from higher trophic levels are also typically preserved in sediments. Invertebrate fossils are, by far, the most common zoological indicators recorded, with Cladocera, larval insects, and ostracodes being used most often in applied studies. From an ecological perspective, these invertebrates often form important intermediate positions in food webs (e.g., they graze algae, but are then themselves preyed upon by other invertebrates and vertebrates such as fish).

5.9.8.1 Cladocera

The invertebrate class Branchiopoda is divided into four orders: Cladocera (water fleas), Anostraca (fairy shrimp), Notostraca (tadpole shrimp), and Conchostraca (clam shrimps). Each of these groups leave some morphological remains in sediments (Frey 1964), but certainly the chitinized body parts of Cladocera are more common and have been the most widely used (Korhola & Rautio 2001), with many applications related to studying past human impacts (Jeppesen et al.

2001). Today, Cladocera have a global distribution, with approximately 400 described species from 80 genera.

Cladoceran species are rarely larger than 1 mm in length, but still play pivotal roles in the functioning of many freshwater systems. Taxa are common in both littoral zone (e.g., Chydoridae) and open-water (e.g., *Daphnia* spp.) habitats. Depending on ambient conditions, they can exert intense grazing pressures, and so can influence the abundance and composition of algal communities. Equally, the abundance, species composition, and even morphology (Kerfoot 1974) of cladocerans can be influenced by predation from other animals, such as larger invertebrates and fish (Carpenter et al. 1985).

Cladocera are preserved in sediments primarily as three chitinized body parts: head-shields, shell or carapace, and postabdominal claws (Figs. 5.8 & 5.9). Although Cladocera typically reproduce via parthenogenesis, when conditions become unfavorable, they can switch to sexual reproduction and produce resting stages that can remain dormant for many decades until conditions become favorable again. These ephippia (singular, ephippium; Fig. 5.9D) are also preserved in sediments and, for at least some taxa, are sufficiently morphologically distinct that they can be linked to the species that produced them (Mergeay et al. 2005). Ephippia are now being used in a variety of other applications, including DNA analyses (i.e., paleo-genetics; see p. 70) as well as the developing field of "resurrection ecology" (Kerfoot et al. 1999; Kerfoot & Weider 2004), where diapausing eggs are recovered from surface sediments (e.g., Vandekerkhove et al. 2005) or sediment profiles (see pp. 142–4) and hatched in the laboratory.

Korhola and Rautio (2001) summarize the current techniques used to study Cladocera in paleolimnological research, and Jeppesen et al. (2001) review some applications of fossil Cladocera to the reconstruction of anthropogenic impacts. Most studies have been related to trophic dynamics and eutrophication issues, but Cladocera have also been used in assessments of lake acidification,

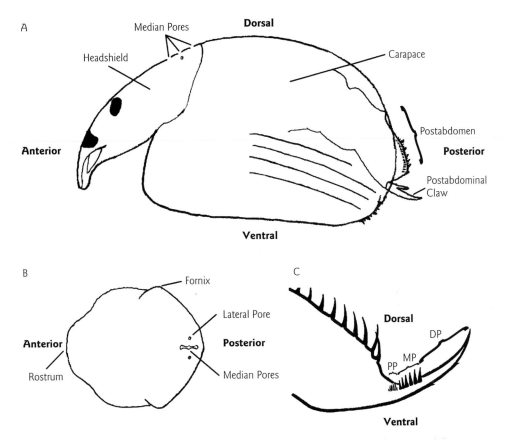

Fig. 5.8 (A) A schematic drawing of an *Alona* cladoceran, showing the main chitinized body parts that are useful for taxonomic identifications in sediments. (B) Detail of the head shield. (C) Detail of the postabdominal claw. PP, proximal pecten; MP, medial pecten; DP, distal pecten. Drawing by Darren Bos.

metal contamination, salinity changes, shifts in water levels, exotic species invasions, as well as other applications.

5.9.8.2 Chironomidae and related Diptera

Although we tend to think of Diptera (true flies) as primarily terrestrial organisms, often flying or crawling around us, many species spend large portions of their life cycles in fresh water. Several dipteran groups are preserved as various morphological indicators in sediments, but certainly the most commonly used insect remains are the chitinized head capsules of the Chironomidae (non-biting midges).

Chironomids are the most widely distributed and often the most common insects in freshwater systems. Only a few taxa may surpass them in global distributions, as they are found from Ellesmere Island in the Canadian High Arctic to Antarctica. It is likely that at least 10,000 species exist (Armitage et al. 1995). The adult stage of many chironomids may only last a day or two, whilst the various larval and pupal stages may last several years (Walker 2001). The head capsules shed by the third- and fourth-instar larval stages during ecdysis (or moulting) represent the bulk of chironomid fossils (Fig. 5.10), which can often be identified to the genus level, and sometimes even to the species level. First- and

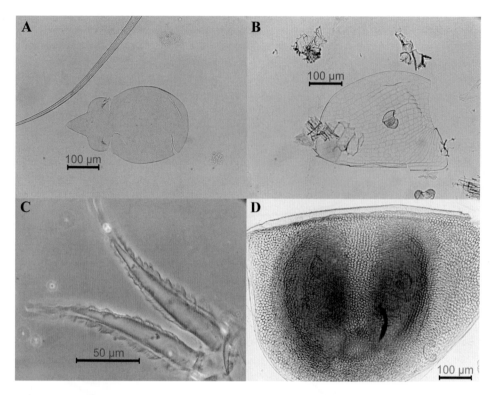

Fig. 5.9 Light micrographs of cladoceran body parts. (A) *Chydorus piger* head shield; (B) *Graptoleberis testudinaria* carapace; (C) *Daphnia dentifera* postabdominal claw; (D) ephippium of a *Daphnia*. Photographs taken by Darren Bos.

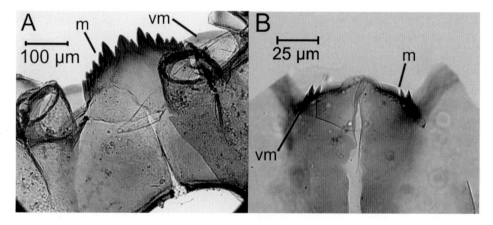

Fig. 5.10 Light micrographs of fossil chironomid head capsules. (A) *Chironomus*; (B) *Protanypus*. m, mentum; vm, ventromental plate. Photographs taken by Saloni Clerk.

second-instar stages are less heavily chitinized, and so are rarely preserved.

The possibility of using chironomids as environmental indicators dates back to the early part of the 20th century, when Thienemann (1921) proposed a chironomid-based lake classification system for German lakes (e.g., oligotrophic *Tanytarsus* lakes, eutrophic *Chironomus* lakes, etc.). Considerable progress has been made since that time, with chironomid head capsules now included in many different types of applications. For example, chironomid indicators have been used extensively in paleoclimate work, as their distributions appear to be correlated with important climatic variables (Walker 2001). From a lake or reservoir management perspective, though, their most important applications may be to infer deepwater oxygen levels (e.g., Quinlan et al. 1998; Little et al. 2000; Quinlan & Smol 2001; Walker 2001; Brodersen & Quinlan 2006). Whereas most of the indicators discussed so far live either in the littoral zones or in the open-water, upper-layer strata of lakes, many chironomid larvae are adapted to living in profundal regions. A key factor in determining which taxa will be most successful is deepwater oxygen concentration, as different taxa have strikingly different tolerances to hypoxia and anoxia. For example, some taxa – the so-called "blood worms" in the genus *Chironomus* – have hemoglobin in their hemolymph and can withstand anoxia for prolonged periods by binding oxygen with hemoglobin. Other physiological and behavioral adaptations also vary amongst taxa, such as the capacity to maintain osmotic and ionic regulation of their hemolymphs with the build-up of metabolic end products during periods of low or no oxygen. Certain taxa possess physiological adaptations to low oxygen levels, such as ventilating and re-oxygenating their surrounding micro-environment by undulating body movements. As a result, changes in benthic chironomid assemblages can often be related to hypolimnetic oxygen levels, which can be estimated using a variety of measures related to the duration and intensity of deepwater hypoxia and anoxia

(Quinlan et al. 1998; Quinlan & Smol 2001). Such issues are of prime concern to many limnological assessments, as deepwater anoxia is an important symptom of eutrophication (Chapter 11).

Not all chironomids live in the deepwater regions, as many species are adapted to life in the littoral zone, where they can provide additional information on, for example, macrophyte development (Brodersen et al. 2001). Moreover, in some cases, deepwater anoxia may be so severe that even the most tolerant benthic chironomids are extirpated, and only littoral zone taxa survive (Little et al. 2000). Although almost all paleolimnological studies have focused on lakes and reservoirs, Klink (1989) and Gandouin et al. (2006) showed how chironomids and other insect remains can be used to track riverine changes.

Other insect larval parts also preserve in sediments, but have been used far less than chironomids. For example, the chitinized mandibles of the Chaoboridae (phantom midges) are well preserved, although typically much less common than chironomids. Because certain taxa are very sensitive to fish predation (e.g., *Chaoborus americanus*; Fig. 5.11), their presence can be used to track historical fish populations (Uutala 1990; Sweetman & Smol 2006). This approach was effectively used in determining if past fish extirpations were related to lake acidification (see pp. 105–7), as well as the study of other types of fish kills, such as from chemical additions (see pp. 153–5). Uutala (1990) provides photomicrographs and a key to identify the fossil mandibles of North American species.

The chitinized remains of the Ceratopogonidae (the biting midges or "no-see-ums") and the Simuliidae (the black flies) also have some potential applications (Walker 2001). The Simuliidae, in particular, whose larvae are found exclusively in streams, may provide information on past inflows and paleohydrology (Currie & Walker 1992). Although rarely used thus far in pollution-based studies, other insect fossils, including Coleoptera (beetle) and Trichoptera (caddisfly) remains (Elias 2001; Greenwood et al. 2006), can be used by paleolimnologists.

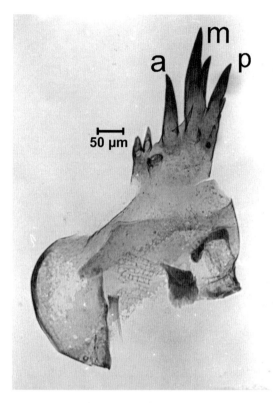

Fig. 5.11 Light micrograph of a fossil *Chaoborus americanus* mandible (a, anterior tooth; m, medial tooth; p, posterior tooth). The presence of this taxon is a good indication of fishless conditions, as it is very susceptible to predation. Modified from Uutala (1990).

5.9.8.3 Ostracoda

The ostracodes (also spelled ostracods) have a long history of paleolimnological applications, but mostly in the context of paleoclimatic and paleohydrological studies (DeDeckker 1988; DeDeckker et al. 1988; Holmes 2001; Chivas & Holmes 2003). Like the Cladocera discussed earlier, the subclass Ostracoda is classified under the Class Crustacea, but ostracodes differ from cladocerans in their external morphology by being characterized by two calcitic valves or shells, hinged dorsally to form a carapace (Fig. 5.12), which partially encloses the body parts and the seven or eight pairs of appendages. The carapace, which in most non-marine taxa ranges in size from about 0.5 to 2.5 mm, is often species specific. Because the carapace is calcareous, ostracode carapaces are typically only preserved in neutral to alkaline environments.

Until recently, most studies have focused on using ostracode species assemblage data to reconstruct past environmental conditions, in much the same way that paleolimnologists use other biotic indicators. This has been a powerful technique, as many ostracode taxa have specific environmental optima and tolerances, and can be related to, for example, habitat, pH, salinity, hydrology, and climate-related variables such as temperature (e.g., Curry 1998). Paleolimnological

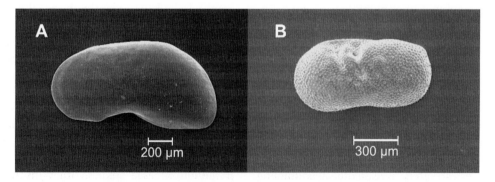

Fig. 5.12 Scanning electron micrographs of ostracode valves. (A) A male, left valve of a *Candona* species, believed to be endemic to Bear Lake, Utah. (B) Left valve of *Ilyocypris bradyi*, a cosmopolitan taxon. Photographs by Jordon Bright.

studies have focused on lakes, but there is also potential for using ostracods in tracking changes in rivers, springs, and wetlands (e.g., Mezquita et al. 1999; Smith et al. 2003).

New approaches have extended the potential applications of ostracods to paleoenvironmental reconstructions. The trace element chemistry and stable isotope composition of the ostracod external shell can further refine paleoenvironmental reconstructions (Holmes 1996, 2001; Ito 2001; Wansard & Mezquita 2001). Because the ostracode shell is secreted over a very short time, a "memory" of past limnological conditions is preserved in the isotopic and geochemical composition of the ostracode shell (Holmes 1996). Such studies have thus far focused on paleoclimatic questions, but likely many other applications exist for studying past water-quality problems.

5.9.8.4 Other invertebrates

The above summary focused on the most commonly used invertebrate indicators for water-quality issues. Many other morphological remains are found in lake sediments, such as protozoa (Patterson & Kumar 2000; Beyens & Meisterfeld 2001), rotifers (Swadling et al. 2001; Turton & McAndrews 2006), bryozoans (Francis 1997, 2001), oribatid mites (Solhøy 2001), beetles and other insects (Elias 2001), mollusks (Miller & Tevesz 2001), some copepod remains (Knapp et al. 2001; Cromer et al. 2005), and other indicators described in various chapters in Smol et al. (2001a,b). Although most have been applied to some water-quality issues, these groups are still largely untapped resources for these types of studies.

5.9.9 Fish

Most lake and river managers, as well as the general public, are concerned with fish. Unfortunately, fish have often represented the "missing link" in many paleolimnological studies, as they rarely leave a good fossil record, although a variety of scales, skeletal bones, and otoliths are recoverable from most sedimentary profiles (Fig. 5.13). Fish scales have been used, in a limited context (e.g., Daniels & Peteet 1998; Patterson & Smith 2001; Davidson et al. 2003), and some identification guides for freshwater taxa have been produced (Daniels 1996). However, in most lacustrine systems, fish scales are not sufficiently abundant to reconstruct past fish abundances in a quantitative manner.

The calcareous inner ear bones (otoliths) of several fish taxa are species specific and are being used increasingly in paleolimnological studies, although rarely, as of yet, in dealing with management issues (Patterson & Smith 2001; Panfili et al. 2002). Because otoliths have accretionary growth, internal zonation often reveals annual (and sometimes even daily) growth rings. Microsampling these growth rings and analyzing the bands for stable isotopes and other variables can reveal important information on past environments, at fine temporal resolution. This is part of the developing field of sclerochronology, the science that uses calcified structures, such as fish otoliths, mollusk shells, and corals, to reconstruct environmental conditions. To date, otoliths have been used primarily in paleoclimatic studies (Patterson & Smith 2001), although there is considerable potential for other applications (e.g., Campana & Thorrold 2001; Panfili et al. 2002; Campana 2005).

The use of fish bones, scales, or other fish debris can be directly related to fish populations, but rarely are the concentrations of these fossils sufficiently abundant in lacustrine sediments to provide defendable estimates of past fish populations. To date, most inferences have been based on more indirect techniques, using the species composition, size and/or morphology of invertebrate indicators influenced by the presence of fish (Palm et al. 2005). For example, as discussed above with the Chaoboridae, the presence of *Chaoborus americanus* mandibles in sedimentary profiles indicates fishless conditions, as this taxon is especially sensitive to fish predation (Uutala 1990; Sweetman & Smol 2006). Jeppesen et al. (1996) developed a transfer function based on

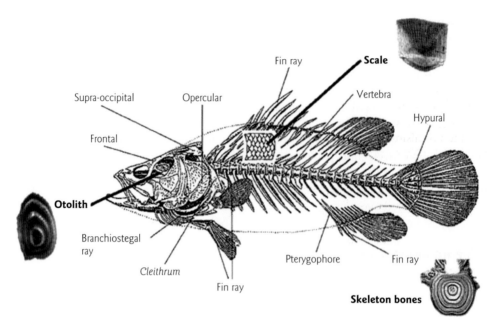

Fig. 5.13 The various calcified structures from fish that can be used in paleolimnological studies. From Pontual et al. (2002); used with permission of IRD Editions.

cladoceran remains to infer past planktivorous fish populations, and cladoceran indicators have been used in a variety of limnological settings to track changes in fish status (e.g., Amsinck et al. 2003; Jeppesen et al. 2003). Likely other morphological indicators can be used as well, and new approaches are always being explored. For example, stable isotopes of nitrogen in lake sediments can be used to track past populations of Pacific sockeye salmon in nursery lakes (see pp. 290–5; Finney et al. 2000).

5.9.10 Paleo-genetics and DNA analyses

Molecular techniques are rapidly being incorporated into many ecological sciences, and paleolimnology is no exception. This exploratory work has, thus far, centered mainly on using DNA from resting eggs preserved in lake sediments, such as the ephippia of Cladocera (see pp. 252–3). If the eggs can be hatched, it is possible to trace long-term changes in genetic composition as well as shifts in physiological phenotypes; if eggs cannot be hatched, DNA can often still be extracted to reconstruct ecological and evolutionary dynamics (Hairston et al. 2005). The use of ancient DNA and paleo-genetics may become especially valuable for paleolimnologists in reconstructing populations that are difficult to track using morphological fossils, such as copepods (e.g., Bissett et al. 2005) or perhaps rotifers (e.g., Gomez & Carvalho 2000). Interesting work has also been initiated on the extraction of ancient DNA from algae (Coolen et al. 2004) and pollen grains (Bennett & Parducci 2006). Meanwhile, Willerslev et al. (2003) have shown that permafrost and temperate sediments may preserve diverse plant and animal genetic records, including those of megafauna such as mammoths.

5.10 Other approaches

The indicators and methods described above are only a subsample of the potential information relevant to lake and river managers that is pre-served in sedimentary profiles. Certainly many other sources of proxy data are available, and new methodologies are continually being developed. For example, near infrared spectroscopy (NIRS), although widely used in some industries, has only recently been applied to paleolimnological studies, with some promising results (Korsman et al. 2001; Wolfe et al. 2006). Paleolimnology as a field is moving very quickly, and I fully expect that my above list of major indicators will continue to increase in the years to come.

5.11 The importance and challenges of multi-proxy inferences

There is little doubt that major advancements have been made in paleolimnology due to the increased attention to the added power of using multiple lines of evidence. There are many advantages to these multi-proxy approaches, not least of which is that they provide much more holistic overviews of ecosystem development. It has been repeatedly shown that different types of indicators do not always simply reproduce or reinforce information gleaned from another type of indicator, but very often *additional* information is provided. A better, more complete, under-standing of environmental change is possible. In some cases, however, the new information is simply corroborative, but this again is important, as it greatly strengthens interpretations. If preservation or other taphonomic issues are a problem, using multiple indicators can also avoid having hiatuses in our profiles. For example, if chemical conditions are such that siliceous indicators are poorly preserved, paleoenvironmental information could still be gathered from the remaining calcareous indicators and/or fossil pigments, and so forth.

As summarized by Birks and Birks (2006), large multi-proxy data sets also pose significant challenges. They are often very time-consuming and demand significant resources, typically involving many specialists. With large amounts of data, there is the problem of data storage and handling. Certainly, interpretations can be challenging to accommodate the many available lines of evidence, and one must avoid falling into the trap of the "reinforcement syndrome," where interpretations are slanted to reinforce each other, rather than to be used to fully explore the possible range of interpretations. Paleolimnologists must be vigilant to objectively use multi-proxy data to show converging validity in interpretations.

Not surprisingly, these complex studies are a major challenge to synthesize and eventually write up. For example, a detailed, multi-proxy, paleolimnological study that was completed on one core from Kräkenes Lake, Norway, by Hilary Birks and colleagues (Birks et al. 2000) yielded about 250,000 data points! Most would agree, though, that the huge scientific effort was worth the result.

5.12 Presenting paleolimnological data

Paleolimnologists typically present their data in more-or-less standardized formats, using a few basic approaches. The depths of the sediment slices (and hence the time scale) are almost always presented along the vertical y-axis, with the percentages (or other estimates of abundance) along the horizontal x-axis. Relative abundances (percentages) are used most often in taxonomic-based biological work. These data have the advantage of not being closely influenced by changing sedimentation rates (see below), but have the disadvantages inherent with closed data sets (i.e., an increase in one taxon is mirrored by a relative decrease in other taxa, as the net sum cannot exceed 100%).

Some data are expressed as concentrations, usually calculated relative to the dry or wet weight of sediment (e.g., diatoms per mg dry weight sediment), but sometimes relative to other sediment components, such as the amount of organic matter (e.g., some fossil pigment work).

These data may appear to be more quantitative than relative percentages, but they introduce other elements of uncertainty, as changing sedimentation rates will influence concentration data. For example, if a sediment core is collected from a lake that experienced a doubling in the abundance of diatoms in the water column, one would expect to see a similar doubling of the diatom concentration in the sediment profile during that time period. This may well happen, provided that the accumulation of other components of the sediment matrix has also remained constant (i.e., sedimentation rates have not changed). But suppose that the influx of siliciclastics from the catchment quadrupled (say, from increased erosion) at the same time that the diatom population doubled. This extra allochthonous material would dilute the diatom concentrations in the sediments, and the net result is that the diatom concentration data would actually drop by half due to the diluting effect of the extra sediment matrix.

Paleolimnologists have tried to compensate for changing sedimentation rates by calculating accumulation rates (sometimes referred to as influx data). Here, the concentration data are multiplied by the sedimentation rates. Using the diatom example cited earlier, an accumulation rate would report the number of diatoms accumulating on a known area of lakebed per year (e.g., diatoms per cm^2 per year). If examined uncritically, these data appear to be very quantitative, but one must realize that their accuracy is tightly linked to how reliably one can estimate sedimentation rates, and so significant (and often major) errors can be introduced. It is perhaps not too surprising that most biological data are simply expressed as relative percentage data. Furthermore, percentage data are also typically used in the development of transfer functions using surface-sediment calibration sets (Chapter 6), again primarily because it is notoriously difficult to calculate sedimentation rates for surface sediments.

Other types of paleolimnological data are typically expressed as ratios. Isotope data in particular are almost always presented this way, but ratios can be calculated for other types of proxy data. For example, biologists sometimes determine the ratio of planktonic to littoral (P/L) taxa in order to estimate past shifts in the relative importance of these two lake regions (e.g., Korhola & Rautio 2001). The ratios of two taxonomic groups, such as the number of chrysophycean cysts to diatom frustules, have also been used in several applications (Smol 1985), and other examples will be highlighted later. Similarly, a variety of other community metrics can be calculated, such as species diversity, but these too must be interpreted critically, as they can be strongly influenced by changes in sedimentation rates (Smol 1981).

5.13 Summary

Sediments typically archive a vast array of physical, chemical, and biological information that can be used to reconstruct limnological conditions. A simple visual inspection of the sediment matrix can aid interpretations, which are now often complemented by more sophisticated logging and recording techniques (e.g., digital and X-ray technology). The physical and chemical properties of sediments have been used in a large number of applications, such as tracking past erosion events, the source and magnitude of metal pollution, and other types of contaminants. Sediments also preserve a broad spectrum of biological indicators (most of them microscopic) that can be used to track past changes in both terrestrial (e.g., pollen) and aquatic (e.g., algal and invertebrate remains) communities. Morphological remains (e.g., diatom valves) are most often employed; however, considerable progress has been made in the development of biogeochemical techniques (e.g., fossil pigments, organic geochemistry) that can be used to reconstruct populations that do not leave reliable morphological remains (e.g., cyanobacteria). Not all indicators are well preserved, and some taphonomic problems (e.g., dissolution, breakage) occur under certain conditions, thus limiting some interpretations. Nonetheless, the ever-increasing repertoire of proxy data available to paleolimnologists has been an important factor in the successful application of this science to lake and river management problems. Multi proxy approaches offer the most robust assessments of environmental change.

6

The paleolimnologist's Rosetta Stone: calibrating indicators to environmental variables using surface-sediment training sets

No rational man would choose to become an ecologist or limnologist.
Ralph O. Brinkhurst (1974)

6.1 Using biology to infer environmental conditions

A central tenet in ecology is that the distributions and abundances of organisms are largely controlled by physical, chemical, and biological constraints imposed by the environment. It therefore follows that if the present-day distributions and abundances of biota can be related to environmental variables, then fossil communities can be used to infer past environmental conditions. This approach makes some assumptions, such as that the ecological characteristics of indicators have not changed over time, but assumptions such as these can be addressed (see pp. 84–6).

Biota have long been known to be reliable indicators of environmental conditions. For example, if presented with pictures of a polar bear, a camel, a palm tree, and a mountain goat, most people would immediately associate Arctic, desert, tropical, and alpine environments with these images. The study of indicator species and assemblages has a long history, as a central theme in ecology has

been the attempt to relate the distributions and abundances of species to the environmental conditions in which they live. Some taxa are quite specific indicators (i.e., specialists), whereas others have less well-defined environmental optima and tolerances (i.e., generalists). Paleolimnologists are often presented with an abundance of complex biological data, sometimes involving several hundred species, as well as other paleoenvironmental information. Unlike polar bears and palm trees, the ecological characteristics of many of these microscopic taxa are as yet poorly known, and so an active area of research involves the empirical estimation of their environmental optima and tolerances.

Given the large number of environmental variables that can influence species distributions, as well as the large number of taxa paleolimnologists are often dealing with, it may first appear to be a daunting, if not impossible, task to "calibrate" thousands of taxa to a multitude of environmental variables. Several approaches are available. Some ecological information is typically available from previous surveys or laboratory experiments, and all sources of data should be evaluated and used. However, surface-sediment calibration or training sets are the most powerful ways to relate the distributions and abundances of taxa to environmental variables. Over 15 years ago, I had referred to training sets and the associated statistical techniques as our new "Rosetta Stone" – our key to deciphering the complex information preserved in sedimentary profiles (Smol 1990a). Although statistical and methodological techniques are constantly improving, the same overall approaches continue to be employed today.

6.2 Surface-sediment calibration or training sets

Before an indicator can be used to infer quantitatively past conditions, its environmental optima and tolerances must be estimated. This is one of the most daunting challenges facing paleolimnologists, as often they are working with hundreds or thousands of indicator taxa, which can be influenced by many important environmental variables (e.g., temperature, pH, nutrients, and other water chemistry, as well as additional physical and biological variables). As Brinkhurst (1974) pondered: Why would any rational person even try to decipher such complex relationships?

Although some ecological information can be gleaned from various sources (e.g., limnological surveys, laboratory experiments, etc.), the most widely used approach for quantitative environmental reconstructions involves "surface-sediment calibration sets," or "training sets." As discussed below, these calibration sets involve sampling a range of lakes (or other sites) for a suite of environmental variables, which are then related to the indicators preserved in the surface sediments using statistical techniques. These data are critical, as the ultimate goal is to use the known present-day ecology of organisms to infer past conditions. If the characteristics of taxa can be quantified to a high enough degree of certainty, quantitative transfer functions can be developed,

whereby species distributions can be used to infer specific environmental variables, such as lakewater pH, nutrient concentrations, benthic oxygen levels, and so forth. Paleolimnologists (such as John Birks and Steve Juggins), working with statisticians (particularly Cajo ter Braak), have developed a number of relatively simple but highly effective techniques to interpret this complex biological information, and these advances have certainly accelerated progress in making paleolimnology more applicable and acceptable to applied fields, such as those concerned with water management issues.

Surface-sediment calibration or training sets are now routinely used to develop objective criteria of the indicator status of various taxa. The approach is fairly straightforward. A paleolimnologist may wish to develop a transfer function to infer lakewater pH from chrysophyte species composition in an acid-sensitive lake district, to develop inference equations from chironomid head capsules to infer benthic oxygen levels, to use diatoms to infer nutrient levels, and so on.

I will use an actual example of developing a transfer function to infer lakewater acidity from the Adirondack Park region of New York in the northeastern United States (Dixit et al. 1993); this is part of a larger project described more fully in Chapter 7. The major steps used to develop a transfer function are illustrated in Fig. 6.1.

The first step is to choose a suite of reference (calibration) lakes that span the environmental variables of interest, and the range of environmental conditions that one is likely to encounter in the limnological history that one wishes to reconstruct. Most importantly, the taxa that will eventually be encountered in cores must be well represented in the modern, surface-sediment samples. Careful consideration of this criterion is useful when determining how many lakes should be included in a calibration study. For each of the calibration lakes, relevant limnological data must be available, and preferably for several years, as these present-day environmental data will be used to estimate the environmental optima and tolerances of the taxa. Charles (1990) provides

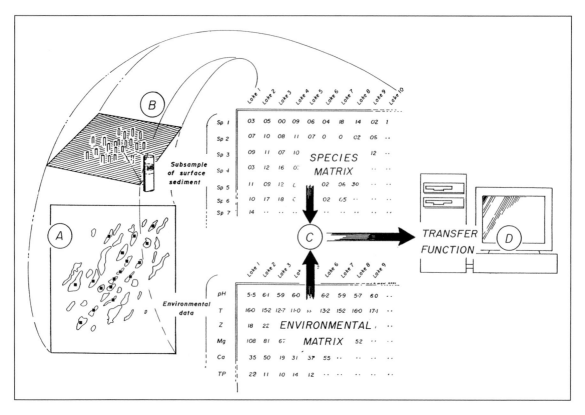

Fig. 6.1 A schematic diagram showing the main steps used in constructing a transfer function using a surface-sediment calibration set or training set. A suite of appropriate calibration lakes (A) is chosen for which limnological data are available (or will be collected); these environmental data provide the first data matrix used in the development of the transfer function (C). Surface sediments (often the top 1 cm of sediment) are collected from the same set of lakes (B), and indicators (e.g., diatom valves, chironomid head capsules, etc.) are identified and enumerated from these samples, thus producing the second matrix used in the training set (C). By using a variety of statistical techniques discussed later in this chapter, environmental variables that are highly related to the species abundance data are used in the development of transfer functions that include estimates of species parameters (e.g., optima, tolerances) based on this training set (D).

a useful checklist for some of the environmental variables that should be included in a calibration set. Typically, at least 40 or so reference lakes are chosen, but generally the more the better. For example, as part of the PIRLA-II acidification project (Chapter 7), we chose 71 reference lakes that spanned a pH gradient of 4.4 to 7.8 (Dixit et al. 1993). The variable that the paleolimnologist is most interested in (with this example, pH) is often emphasized, but it is useful and important to include other data, as different variables will also influence the composition of assemblages. This

step produces the first data matrix required: the calibration lakes and the relevant limnological data.

The next step is to construct a second data matrix of the same calibration lakes with the associated biological data. But how does one estimate, for example, the diatom assemblages that have lived in a particular calibration lake over the past few years?

The most effective and efficient way to collect an annually integrated estimate is to sample the surface sediments (e.g., the top 1 cm of sediment, which in many systems represents about the last 2 or 3 years of sediment accumulation) from a representative basin of each calibration lake (for most simple basin lakes, this would be near the middle and often the deepest part of the basin). The surface sediments integrate biota, such as diatom valves, 24 hours a day, every day of the year, from the entire water column, as well as integrating material from the littoral zone and other habitats of the basin under study. Moreover, surface-sediment assemblages are a more direct comparison to the assemblages that will be preserved in a sediment core than are point samples from different habitats in the basin, although the latter information is also useful from a paleolimnological perspective. This is an empirical approach: paleolimnologists "calibrate" the assemblages that are deposited on the surface sediments with the overlying limnological characteristics.

From each calibration lake, typically the surface centimeter of sediment is collected, often using some type of gravity corer, such as a Glew (1991) mini-corer. These samples are then digested and prepared for microscopic examination, identification, and enumeration. Using the acidification example we began with, this would result in the construction of the second data matrix: the percentages of diatom taxa found in the surface sediments of the same lakes included in the first data matrix (i.e., environmental data).

In some cases, such as in river environments, surface sediments cannot be used effectively, since they are not deposited in a continuous fashion. Modifications of the above approaches, such as using organisms attached to periphytic habitats in different stretches of a river, can be used in lieu of sediment samples (see pp. 195–6).

The next job is to combine these two matrices and develop a transfer function, so that the taxa (e.g., diatom species assemblages in this example) can be used to infer environmental variables from fossil assemblages. Considerable progress has been made in the development of these quantitative techniques. Some of the commonly used approaches are summarized in the next section.

6.3 Assessing the influence of environmental variables on species distributions and the development of quantitative transfer functions

The final step in the development of a quantitative transfer function is the statistical analysis of the data. At this stage, the paleolimnologist has two data matrices: One of environmental data for the calibration sites, typically different lakes or sections of a river; and a second matrix of the taxa (often as percentages) or other response variables from each of the training-set sites (Fig. 6.1). The next step is to determine which environmental variables are influencing the taxa (i.e., the response variables), and how robustly this variable or these variables can be reconstructed. Many of the statistical treatments that paleolimnologists typically use are summarized in Jongman et al. (1995), and two review articles by Birks (1995, 1998), specifically directed at paleolimnologists, provide readable introductions to these statistical issues. Below, I summarize some of the salient points.

Most organisms respond to environmental gradients in a unimodal fashion, such as the Gaussian curve shown in Fig. 6.2. That is to say, if we track the distribution of a species across a broad environmental gradient, such as a temperature gradient of −50°C to +50°C, there will be a temperature at which that particular species will be most abundant (i.e., best adapted). This optimal temperature would be the optimum (u) in Fig. 6.2. Different species will have different optima. For example, a polar bear will have a low optimum for temperature, and a camel will have a higher optimum. A second important parameter is the tolerance (t) in Fig. 6.2, which can be estimated by the standard deviation of the unimodal curve. Some species will be generalists, and thus have broad unimodal curves for specific environmental variables (i.e., a large tolerance); others will be quite specific, such as stenothermal species such as polar bears and camels, which would have relatively narrow unimodal curves

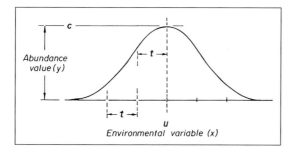

Fig. 6.2 A Gaussian unimodal curve for a hypothetical species, showing the optimum (u) and tolerance (t) to an environmental variable. Modified from Jongman et al. (1995).

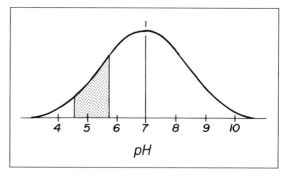

Fig. 6.3 The same Gaussian unimodal curve as shown in Fig. 6.2, except that in this case only a small part (shaded area) of the environmental gradient (i.e., only lakes that had a pH between 4.5 and 5.8) was used in the training set. In this example, the abundance of this hypothetical taxon would appear to be linearly related to lakewater pH between 4.5 and 5.8, and so linear statistical techniques would be most appropriate for this analysis. However, if the entire pH gradient was sampled, then this taxon would exhibit a unimodal response to pH, and unimodal techniques would be most appropriate.

for temperature. However, even a eurythermal species such as *Homo sapiens* would still show a unimodal curve if the temperature gradient is long enough. Although the distribution of humans with respect to temperature would be broad (confirmed by our colonization of this planet from the High Arctic to the equatorial zones), we would, as a species, still have an optimum temperature (perhaps about 25°C). If the temperature was raised too high, eventually we would be less successful and die from over-heating. Similarly, eventually we would freeze to death at the other end of the gradient.

The same types of relationships can be modeled with the distribution of diatoms, chironomids, and other paleolimnological indicators. Using the Adirondack Park study referred to earlier, some diatoms or chrysophytes were shown to have restricted pH distributions in this data set. For example, the scales of the chrysophyte *Mallomonas hindonii* were only common in very low pH waters, whilst *Synura curtispina* may have a similarly narrow (low tolerance) unimodal curve at the other end of the pH gradient, reflecting its restricted distribution to high pH waters. Meanwhile, a more generalist chrysophyte taxon, such as *M. crassisquama*, has an optimum closer to a neutral pH and a broadly shaped unimodal curve (large tolerance) for pH.

Not all taxa will show unimodal responses to environmental gradients, and paleolimnologists can use several different curves to model the distribution of taxa to environmental variables (Birks 1995). For example, if the sampled environmental gradient is not long enough, certain taxa may appear to have linear responses to changing conditions (Fig. 6.3), and in this case, linear statistical methods would be appropriate (Jongman et al. 1995).

The major questions remain: Which measured environmental variables are actually exerting the greatest influence on the distribution of taxa? Are these species-environmental relationships strong enough to develop robust transfer functions? In order to use organisms to reconstruct a specific environmental variable, their distributions have to be influenced by (or at least linearly related to) that variable. With the large number of taxa in most calibration sets, and many measured environmental variables that potentially influence distributions, multivariate ordination and regression techniques have to be applied to help answer these questions. The most common approach has been the use of indirect (e.g., correspondence analysis, CA) and direct gradient analysis techniques (e.g., canonical correspondence analysis, CCA; developed by ter Braak 1986), if taxa are responding to the measured environmental gradients in a unimodal fashion, or principal components analysis (PCA) or redundancy analysis (RDA) if linear techniques are required. The choice of unimodal versus linear techniques can often easily be decided upon by looking at a scatter plot of the data, but to do this objectively it is often recommended to run a detrended correspondence analysis (DCA) to determine the length of the gradient, estimated in standard deviation (SD) units. This is an estimate of the turnover rate of species, since the distribution of most species will rise and fall over four SD units. If the gradient is shorter than about two SD units, linear techniques are more appropriate (Birks 1998).

Detailed statistical discussions of what ordinations such as CCA and RDA do is beyond the scope of this book, but essentially indirect ordination techniques (e.g., PCA, CA) can reveal the underlying patterns within multivariate species data (i.e., many species in many lakes of the training set), whereas direct ordination techniques (e.g., RDA, CCA) can detect those patterns in the species data that can be directly related to measured environmental variables. Direct ordination techniques are analogous to simple multiple regression, with the exception that all species are modeled simultaneously (as opposed to only one at a time) to one or more environmental predictors. They are a form of multivariate regression. The goal is to discover patterns in the community data that are directly related to the measured environmental variables.

Figure 6.4 shows simplified CCA ordination plots for 149 diatom taxa (Fig. 6.4A) in 37 of the Adirondack Park calibration lakes (Fig. 6.4B) used in the PIRLA acidification project (Chapter 7). CCA identified four significant environmental variables (shown by arrows); the length

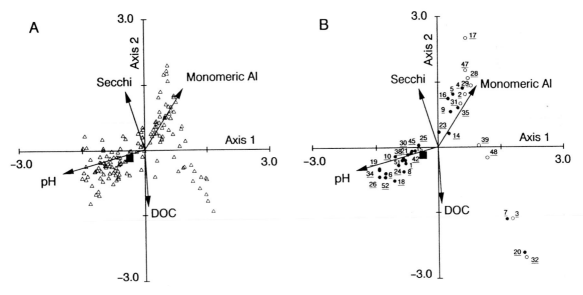

Fig. 6.4 Canonical correspondence analysis (CCA) ordination plots of (A) 149 diatom taxa identified from the surface sediments of 37 Adirondack Park calibration lakes. Individual diatom taxa are shown as triangles. The four significant environmental variables are shown as arrows. (B) The same CCA biplot, but this time showing the position of the 37 calibration lakes (marked by numbers) as they plot on the ordination. Modified from Kingston et al. (1992) used with permission of NRC Research Press.

of an arrow approximates its relative importance in influencing species assemblages, whereas the orientation of the arrows shows their approximate correlations to the ordination axes. In this example, pH, the longest arrow, is closely correlated with axis 1 of the biplot, whilst dissolved organic carbon (DOC) and monomeric aluminum are also important variables, more closely related to axis 2. Diatom species (shown simply as triangles in Fig. 6.4A), which plot on the left-hand side of the ordination, tend to be found in higher-pH waters, taxa that plot towards the bottom right of the ordination tend to be found in low-pH but high-DOC lakes, and taxa plotting in the upper right quadrant tend to be associated with higher concentrations of monomeric aluminum and low pH. Figure 6.4B shows the same CCA ordination, but this time the site scores (i.e., the calibration lakes) are plotted as numbers.

After determining which of the measured environmental variables influence species distributions (and hopefully the variables that are identified in the CCA or RDA are the ones that are of interest for the paleoenvironmental reconstruction, such as pH in the example used above), the next step is to model taxa to these gradients and construct a transfer function. The overall steps are summarized in Fig. 6.5. The original training set of lakes (or other sites, such as samples along a stretch of a river), as well as the associated biological data (i.e., the species data from the surface sediments or other collections from each site) make up the two data matrices used in the transfer function. Once the main environmental gradients (e.g., pH) are established (using, for example, CCA), one should first examine if the measured gradient is relevant to the variable that one hopes to reconstruct, and that a more-or-less even and uniform distribution of sites is available in the training set of lakes (Fig. 6.5, step A). For example, if one wished to study the process of lake acidification down to a pH of about 4.5, it would not be appropriate to use a training set of lakes that simply spanned a pH gradient of 5.5 to 8.5. Moreover, it would not

be appropriate if, say, 50 of the 62 lakes in the training set had a pH between 8.0 and 8.5; an even distribution across the gradient is important to model the species responses that should be tracking this variable.

In the next step (Fig. 6.5, step B) species responses to the environmental variable are explored using regression techniques. In the example shown, five species are exhibiting Gaussian (unimodal) distributions to a lakewater pH gradient. This, of course, shows idealized curves. In nature, the scatter of points is not so clearly defined into neat, unimodal curves. Nonetheless, because many taxa are used simultaneously (although only five are shown in the figure for simplicity, a paleolimnologist may be using several hundred taxa), there is considerable internal redundancy in the data set, and so resulting inferences are much stronger than one might first suspect.

The final calibration step in the development of the transfer function usually involves statistical techniques such as weighted averaging regression and calibration (e.g., using the program WACALIB; Line et al. 1994), or partial least squares (PLS) methods, the details of which are summarized for paleolimnologists in Birks (1995, 1998). These techniques use the information provided in the response curves of the indicators (Fig. 6.5B) to construct a transfer function that relates the indicators and their distributions and abundances to the environmental variable of interest. Using the acidification example from Adirondack Park, Fig. 6.6 shows the performance of the diatom-inferred lakewater pH and acid neutralizing capacity (ANC) transfer functions developed from this lake calibration set.

The predictive ability of the transfer function can be assessed in three different ways. Typically, one first explores how well the surface-sediment assemblages can infer the environmental variable in the training set. This is shown as step C in Fig. 6.5, with the pH inferred from the indicators on the y-axis and the measured lakewater pH in the training set on the x-axis. The resulting coefficient of determination (and related

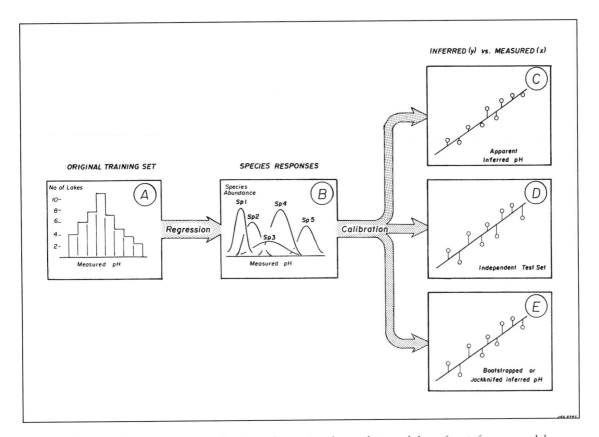

Fig. 6.5 The typical steps used in developing and assessing the predictive ability of an inference model. (A) The selection of a modern set of lakes that span the limnological gradient of interest from which both environmental variables and species assemblages (often from the surface sediments) are collected. (B) The regression step, in which species response curves are estimated based on their distributions in the calibration set of lakes. The calibration steps: (C), where the same set of lakes from which the species responses were estimated is also used to generate the inference model (thus producing an overly optimistic $r^2_{apparent}$); (D), where an independent test set of lakes is used to generate the inference model; or (E), where computer resampling techniques, such as bootstrapping or jackknifing, are used to assess the inference model. Modified from Fritz et al. (1999).

statistics) is often referred to as the "apparent" r^2, or $r^2_{apparent}$, and is an overly optimistic estimate of the robustness of the inference model. The reason for this is that there is some circularity in this type of analysis – one is using the same samples to estimate the species parameters and then the same samples to test the strength of the transfer function.

A more realistic estimate of the robustness of the transfer function can be achieved by running a test set of samples in a form of cross-validation (step D in Fig. 6.5). Using the pH example described above, one would develop the transfer function using 71 surface-sediment samples. In order to independently determine how well the resulting transfer function can actually infer pH, another set of 71 different surface lake sediment samples, across a similar pH gradient, would have to be collected and analyzed. One would then run the original transfer function on this

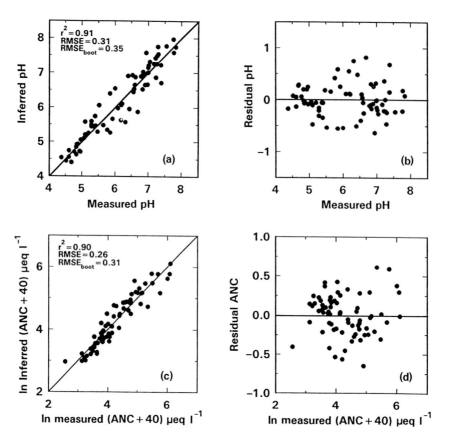

Fig. 6.6 The relationship between measured and diatom-inferred lakewater pH (A) and a plot for pH residuals (B) for a 71-lake surface-sediment calibration set from Adirondack Park, New York. (C, D) The relationships for lakewater acid neutralizing capacity (ANC). Transfer functions were also developed as part of this study for concentrations of lakewater monomeric aluminum (e.g., Al^{3+}) and dissolved organic carbon (DOC). Additional details regarding some of the statistical analyses are discussed in this chapter. From Dixit et al. (1993); used with permission.

new test set of samples, and determine how well pH could be inferred on this new, 71 lake test set. Although statistically robust, this form of model evaluation is very labor intensive. Sampling lakes and identifying and enumerating microfossils in a second series of 71 lakes is not a trivial matter! Furthermore, if one then had a total of 142 calibration lakes, ideally better estimates of species parameters could be made with this expanded lake set, and not simply used for model evaluation.

Fortunately, computer-intensive approaches such as bootstrapping and jackknifing (also known as "leave-one-out" substitution) can pro-

vide more realistic error estimates (Fig. 6.5, step E) without "wasting" data (Birks 1995, 1998). Both these approaches involve creating new training sets as subsets of the original data sets. With jackknifing, a new training set is formed from the original training set, with the exclusion of one of the samples. So in the Adirondack example, the training set now contains 70 lakes, and a new transfer function would be constructed based on this $n - 1$ number of sites. The resulting function will be very similar to the 71-lake set, as only one sample was removed (this is why this approach is sometimes called "leave-one-out substitution").

However, this one sample now forms an independent test sample that can be run, independently, with the 70-lake transfer function, to determine how well one can infer pH for this test sample. The test sample is then returned to the training set, and another sample is removed, and the inferences are run again. One can do this repeatedly until each of the 71 lakes is removed, one by one, from the training set, and errors are independently determined for the transfer function. The resulting coefficient of determination is often referred to as $r^2_{jackknifed}$, and is always lower than the more optimistic $r^2_{apparent}$.

Bootstrapping is a more complex error estimation technique, but it follows similar principles in that it too creates a "test set" as a subset of the original training set. Whilst jackknifing removed one sample at a time from the training set with no replacement (i.e., the new training set would have one less sample in it than the original training set), bootstrapping removes samples with replacement. From the original training set, a "new" subset of test lakes is randomly selected with replacement, so that the "new" set is the same size as the original training set. For this to happen, some samples have to be chosen twice. Because sampling is done with replacement, the samples that are not selected form the independent "test set". This procedure is repeated many times (e.g., 1000), so that many combinations of the "training" and "test" sets are selected.

The strength of a transfer function can be estimated by the coefficient of determination (r^2) of the bootstrapped or jackknifed regression, as well as the root mean squared error (RMSE). These statistics give one an indication of how well an environmental variable can be inferred from a species assemblage. The next step would be to use this transfer function to infer past environmental variables from fossil assemblages. This, of course, presupposes that the assemblages you identified in the training set are similar to those recorded in the downcore samples. It is possible to find different taxa and assemblages in fossil material, as environmental conditions in the past may have been different to some extent

and may not have been captured in the training set (i.e., a "poor" or "no analogue" situation). This problem can be quantified using various analogue matching programs (Birks 1998). As one encounters poorer analogues, inferences become weaker, or simply impossible to make if the assemblages are markedly different. If this occurs, the paleolimnologist must shift to a more qualitative assessment, and infer general trends in environmental variables using any autecological data available for the taxa either from that particular region, or anywhere else such data exist.

6.4 Some new statistical approaches

Like all active areas of science, new ideas and approaches are constantly being proposed for improving paleolimnological transfer functions. For example, Racca et al. (2003, 2004) have suggested that diatom transfer functions may perform better if some of the taxa are removed from the analyses. Attempts at using modern analogue techniques (e.g., Cunningham et al. 2005), Bayesian statistics (e.g., Vasko et al. 2000; Erästö & Holmström 2006), and artificial neural networks (e.g., Racca et al. 2001, 2003; Köster et al. 2004) have also been attempted. As cautioned by Telford and Birks (2005), many of these interesting newer approaches may still require further evaluation before they become routinely used in paleolimnological assessments.

6.5 Assumptions of quantitative inferences

Quantitative reconstructions based on training sets have at least five major assumptions, as outlined by Birks (1995):

1   The taxa used in the training set are systematically related to the environment in which they live.

2   The environmental variable is related to, or is linearly related to, an ecologically important determinant of the ecosystem under study.

3   The mathematical methods used to model the responses of taxa to environmental variables are adequate.

4   Other environmental variables, besides those of interest, have negligible influences on the inferences.

5   The taxa used in the training set are well represented in the downcore paleolimnological assessment, and their ecological responses to the environmental variables of interest have not changed over the time span covered by the paleo-limnological assessment.

This last assumption states that significant evolutionary processes have not occurred in the taxa being used as response variables, and thereby the ecological optima and tolerances of a species estimated from a surface-sediment calibration set will have the same (or very similar) optima and tolerances of the same taxon recovered from, for example, 19th-century sediments. For many biologists who are attuned to evolutionary change, this may seem like a tremendous "leap of faith," but there is ample evidence to show that this appears to be a safe assumption, at least for the dominant indicators paleolimnologists use and for the time frames of most environmental studies (e.g., a few centuries). For example, pale-olimnologists typically use many different taxa in their interpretations, and so there is considerable internal redundancy in the data, which provides "checks" on interpretations. Using the lake acidification example, a paleolimnologist will typically record a shift in a suite of diatom taxa indicating lower pH alongside similar shifts in chrysophyte and invertebrate species, as well as independent changes in sedimentary geo-chemical data. Such multi-proxy studies have repeatedly confirmed that these indicators have been ecologically conservative, and taxa are indicating similar environmental optima and tolerance to pH now (from the surface-sediment calibration sets) as they did in the past (from the downcore sample). Similar examples exist for

other variables, such as nutrients, salinity, and so on.

Birks (1995, 1998) discusses these assumptions in more detail. They may seem daunting, but many studies have shown that these assumptions are reasonable and the resulting transfer functions are robust.

6.6 Exploring the values of interpreting "annually integrated ecology": some questions and answers regarding surface-sediment calibration sets

Question: Do surface sediments collect and preserve everything equally well that lived in the water column?

Answer: No. However, there is a large body of data that shows that this is a reliable approach to estimate annually integrated accumulations of certain biota in most lakes. Moreover, it is important to keep in mind that this is an empirical approach, and the subset of indicators that reach the surface sediments in a recognizable form is a better reflection of what will be eventually preserved in the sediment core, rather than a transfer function developed from living assemblages. Lake sediments accumulate 24 hours a day, every day of the year. They also integrate the biases that are inherent in our approaches (e.g., possible differential deposition, dissolution, etc.). Paleolimnologists are not necessarily saying that whatever makes it to the surface sediments is a total reflection of what is in the lake (although we often know it is a good approximation based on comparisons between living and surface-sediment assemblages).

Question: Are training sets not simply examples of circular reasoning? Paleolimnologists appear to produce strong statistics to back up their inference models, but is this simply because they are using the same training set samples to construct the model as well as to test its robustness?

Answer: No. This argument could have been

made (as discussed above) if the same samples used to develop the model were also used to test its significance (i.e., if only the "apparent" errors were reported, as shown in step C of Fig. 6.5). But if independent error estimates are made by statistical cross-validation, for example using jackknifing or bootstrapping, the process is not circular.

Question: Are paleolimnologists saying that these are the absolute tolerances and optima of taxa? If so, then why is it that some lab experiments suggest that the tolerances may be much broader than those estimated from surface-sediment analyses?

Answer: No, this is not what is being estimated. This question results from a fundamental misunderstanding of what paleolimnologists are estimating when they use surface-sediment calibration sets from different regions. Perhaps the clearest way to describe these concepts is to use G.E. Hutchinson's (1958) concepts of the fundamental and realized niches of taxa (Fig. 6.7), with the realization that biological distributions are influenced by multiple environmental variables, at different temporal and spatial scales, as discussed more fully by Jackson and Overpeck (2000). The environmental optima and tolerances that paleolimnologists are estimating from training sets represent empirical relationships, documenting the distributions and abundances of indicators with respect to the environmental and ecological conditions (e.g., competitive pressures) that are present in the study region. In short, paleolimnologists are attempting to describe the realized niches of taxa, not the fundamental niches (Fig. 6.7). Different systems and regions will have different selection pressures, and so, for example, the environmental optimum and tolerance determined for a diatom species in a laboratory culture flask (e.g., lacking predators, competition from other taxa, etc.) will not be identical to the optimum and tolerance determined for the same taxon in a lake set. Most likely, the environmental tolerance will be much broader in the laboratory experiment than in the field samples, as the

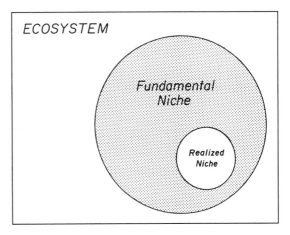

Fig. 6.7 A simplified representation of an ecosystem, and the fundamental and realized niches of a hypothetical taxon. Ecosystems offer a wide spectrum of environmental conditions. Certain taxa will only survive and reproduce under certain environmental conditions, their so-called fundamental niche. However, in nature, there are other restraints on the distributions of taxa, such as competition, predation, and so forth. Therefore, the actual niche that a taxon can exploit successfully is only a subset of its fundamental niche, and may be referred to as its realized niche. With surface-sediment calibration sets, paleolimnologists are primarily trying to define the environmental characteristics of the realized niches of indicator taxa for the region they are interested in, and not the fundamental niches.

former more closely approximates the taxon's fundamental niche, whereas the paleolimnologist is more concerned with the realized niche (Fig. 6.7). As competitive coefficients change from region to region, the optima and tolerances for taxa will also change, but in general will be similar.

Question: Can we be sure that, for example, the surface centimeter of sediment actually corresponds to the past 3 years of water chemistry data?

Answer: No, it is notoriously difficult to estimate age accumulations in the very surface sediments, and so it is simply an approximation. But the statistical treatments paleolimnologists use to estimate the strength of the transfer functions (as noted above) show that we can often estimate, with high precision and with known errors, the environmental optima and tolerances of taxa.

Question: Does this process tell you which season the organism was most common?

Answer: No. You would need seasonal collections for that; lake sediments integrate the seasons (there are exceptions, as sub-annual resolution is possible in some annually layered (varved) sediments). However, some researchers (e.g., Bradbury 1988; Siver & Hamer 1992) have more fully exploited the known seasonal patterns of indicators to further refine paleolimnological inferences, and such data can be very useful in understanding changes in taxa in sediment cores.

Question: Is this the best and most feasible way to provide strong and highly defendable inference equations for species distributions?

Answer: Absolutely. No other approaches that I am aware of have even begun to approach the robustness of these techniques.

Question: What happens if one is undertaking a paleolimnological study in a region or for a variable where no transfer function exists? Can inferences still be made, on a more qualitative basis?

Answer: As will be shown by various examples in this book, many elegant studies have been completed without the benefit of transfer functions or other detailed quantitative assessments. The "training set approach" is, indeed, a powerful one, but it is not the only approach. Quantitative transfer functions may allow for a more objective analysis of complex data but, depending on the question being addressed, sometimes much simpler, qualitative approaches can be very powerful as well.

Similar to the other topics dealt with in this book, it is important to keep perspective and realize what it is that we are measuring and how

the data should be used. In my opinion, though, the statistical robustness and success of these applications speak for themselves.

6.7 Some personal thoughts on more fully exploiting paleolimnological calibration data

Paleolimnologists are justifiably proud of the accuracy and precision of many of their transfer functions. However, I worry that we have become satisfied with our calibration work once we have produced a statistically defendable equation, but meanwhile we are neglecting enormous amounts of ecological data that still remain untapped and un-interpreted in our spreadsheets. In most cases, paleolimnologists have exerted considerable effort, not least of which has included the careful and painstaking identification and enumeration of indicators, to develop elegant mathematical summaries of their ecological distributions. Yet so much of our hard-gained ecological data remains unexplored and under-exploited. An effective use of our time might be to further mine these data sets to garner as much ecological information as possible, instead of simply developing new transfer functions.

6.8 Summary

A large number of biological taxa are preserved in lake sediments. However, before environmental conditions can be reconstructed, the ecological optima and tolerances of these indicators have to be estimated. In many cases, this information is qualitative. For example, a paleolimnologist can ascertain from past ecological or experimental studies that a particular taxon is more common in acidic or high nutrient waters. Such qualitative assessments are still useful in many paleolimnological contexts. However, considerable progress has been made in the development and application of surface-sediment calibration sets (i.e., training sets), which can be used to provide more quantitative estimates of the environmental optima and tolerances of indicators that can then be used to construct transfer functions to infer environmental variables of interest. In some cases, such as diatom-based functions to infer lakewater pH, very robust and ecologically sound transfer functions have been developed. In other cases, inferences are less robust, and so this approach is more readily used to infer overall trends in some environmental variables.

7

Acidification:
finding the "smoking gun"

Eighty percent of pollution is caused by plants and trees.
Ronald Reagan (1987)

7.1 Acidic precipitation: definition and scope of the problem

In 1852, R. Angus Smith coined the term "acid rain" to refer to the effect that 19th-century industrial emissions had on the precipitation of the English Midlands (for an historical review, see Gorham 1989). As summarized by Schindler (1988), Smith could hardly have dreamed that about 130 years later the topic would be studied by thousands of scientists, would utilize many hundreds of millions of dollars of research funds, and would become a topic of extreme economic and international political importance (e.g., NAPAP 1991). The data generated by this interest frequently documented the adverse effects of acidification on lake ecosystems. Significant environmental legislation was enacted in the early 1990s in many regions to limit acid emissions, although recovery has been substantially less than expected in most instances. Since the 1980s, research on acid rain has been decreasing. Nonetheless, the problem is still one of considerable importance and many questions remain (American Fisheries Society 2003).

Acidic precipitation had far-reaching ecological and environmental repercussions. Ever since humans learned to burn fossil fuels, a large, unanticipated "experiment" has been under way on this planet. As sulfur and nitrogen oxides were released into the atmosphere, transported through the atmosphere and converted to sulfuric and nitric acids, and then precipitated back to Earth, a large-scale and unintentional "acid titration" was under way. Suspicions of a possible problem were aroused by the 1950s, and certainly by the 1960s and 1970s (Odén 1968; Beamish & Harvey 1972) it was clear that many lakes and rivers were being adversely affected. Water chemistry was changing as pH dropped in the poorly buffered systems, scientists and anglers began to notice that fish populations showed signs of stress and then started disappearing, the size and composition of algal assemblages and other primary producers were changing rapidly, and virtually every aspect of the ecosystem was being altered. Less obvious changes were also occurring. For example, inorganic monomeric aluminum (e.g., Al^{3+}) levels, which at certain concentrations are toxic to fish and other organisms, were increasing in many acidic lakes, presumably due to increased mobilization from soils and sediments at lower pH.

Acidic precipitation had been occurring, at various levels, for over a century, and so many of these changes were gradual, largely imperceptible at the time scales familiar to most managers and scientists (e.g., a few seasons or years). Long time-series data were necessary to track the timing and progress of this problem, as well as its effects on biota. Such monitoring data were not available. One option was to use experimental interventions to mimic the acidification process. Perhaps the most dramatic example of this approach was the whole lake manipulation of Lake 233 in the Experimental Lakes Area of northwestern Ontario, where Schindler et al. (1985) showed large-scale ecosystem changes related to the artificial acidification of this lake (Box 7.1). A

Box 7.1 Ecosystem changes related to lake acidification: the experimental manipulation of Lake 223

Ecosystem-level responses to stressors such as acidification are difficult to document, because most systems are too large or complex, individual stressors cannot be readily identified or isolated from other perturbations, and documentation of ecosystem structure, function, and natural variability is woefully inadequate due to the lack of long-term monitoring data. Schindler et al. (1985) attempted to overcome some of these problems by conducting an ecosystem-scale experiment in Lake 223, one of the lakes set aside by the Canadian government as part of the Experimental Lakes Area (ELA) in northwestern Ontario. This small Precambrian Shield lake, surrounded by virgin boreal forest, was chosen as a representative site typical of the thousands of poorly buffered softwater lakes that were being threatened by acidification. The ELA region was not subjected to high acidic deposition rates, and so Lake 223's limnological characteristics in 1974 and 1975 (i.e., the two years of study before the manipulation, when the lake had a pH of about 6.8) were used as reference conditions.

In 1976 sulfuric acid began to be added in large quantities to mimic the acidification process. Nearby lakes were also monitored to study natural variability. Schindler et al. (1985) recorded many striking physical, chemical, and biological changes in the lake from the start of the manipulation in 1976 until it stopped in 1983, when the lake was at a pH of about 5.1, about 1.7 pH units lower than its original pH. Amongst the spectrum of biological repercussions of the acidification were changes in phytoplankton populations, particularly an increase in dinoflagellates; massive increases in filamentous green algae (*Mougeotia*) in the littoral zone, seriously affecting some fish spawning grounds; disruption and then cessation (at pH 5.4) of fish reproduction; and marked changes in invertebrate communities (especially some cladoceran species and benthic Crustacea; for example, crayfish were infested with the microsporozoan parasite, *Thelohania*). The food web of lake trout (*Salvelinus namaycush*), a top predator in the system, was severely altered by the extirpation of key prey species. Below pH 6.0, the opossum shrimp (*Mysis relicta*) and the fathead minnow (*Pimephales promelas*) stopped reproducing. The decline in these key species caused lake trout to starve, as seen in a comparison of photographs taken of lake trout in 1979 (when the pH was 5.4) and in 1982, when the pH was 5.1 (Fig. 7.1). Overall, the number of species in the lake at pH 5.1 was about 30% lower than in the pre-acidification years of 1974 and 1975.

This study showed that large-scale ecosystem changes occur during the acidification process. After 1983, researchers stopped adding acid to Lake 223, and the ELA staff monitored the recovery process. The lake began to return to pre-acidification conditions, showing that lakes can recover if acid inputs are stopped.

Fig. 7.1 Lake trout taken from Lake 223 in 1979 when the pH was 5.6 (A), and one taken in 1982 when the pH was 5.1 (B). Based on work described by Schindler et al. (1985). Photographs supplied by K. Mills and D.W. Schindler.

second option was to use real-time data, which showed acidification and allowed measures of rates of change (Dillon et al. 1987), although often on quite short time frames (e.g., a decade or less). Another option was to use dynamic bio-geochemical models (e.g., MAGIC, discussed in Section 7.18). The fourth option was to reconstruct the missing monitoring data using lake sediments.

Acid precipitation, and its associated limnological and ecological impacts, were environmental issues that closely involved paleolimnologists. I therefore use acidification as the opening chapter of examples in this book. The paleolimnological research on acid rain left an important legacy of protocols and techniques that were relatively easily transferred to the study of other environmental problems, which will be discussed in the following chapters.

7.2 The four steps in the trans-boundary problem of acidic deposition

Although the term "acid rain" is widely used, a more appropriate term is "acidic deposition," as deposition may take many forms, aside from rain (e.g., snow, fog, sleet, haze, dry deposition, etc.). Figure 7.2 summarizes the major steps that occur in the formation and deposition of acidic precipitation, although the problem is of course far more complex.

The formation and deposition of acidic precipitation can be divided into four major stages: emissions, transport, transformation, and fallout or deposition.

1   *Emissions.* There are a myriad of sources of acidic precipitation, both natural and anthropogenic,

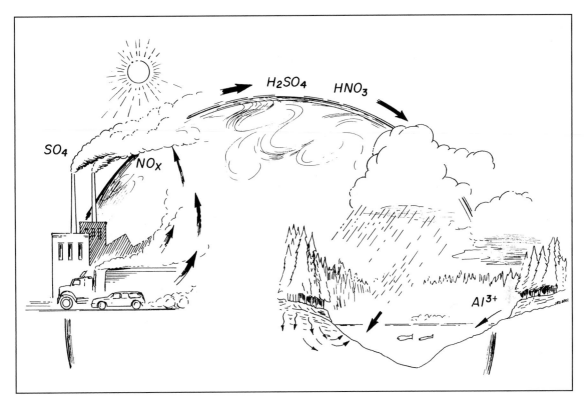

Fig. 7.2 A schematic diagram showing the major steps involved in the acid precipitation problem.

with the key components of anthropogenic emissions being sulfur and nitrogen oxides. Sulfur emissions may originate from natural sources (e.g., volcanoes, sea spray, biological processes, etc.), but it was the additional anthropogenic sources that resulted in enhanced acidic precipitation (e.g., coal and oil combustion, smelting of ores). Most of the research and legislation focused on sulfur emissions, as often these came from "point" sources. Natural sources of nitrogen emissions include biological decay, but fossil fuel combustion and agricultural processes represent significant anthropogenic sources of additional emissions.

2   *Transport*. Because most emissions are in a gaseous form, it is not surprising that long-range transport (sometimes thousands of kilometers) was a key factor in the acid precipitation problem. Many millions of dollars were used to study the transport issue in North America and Europe.

3   *Transformation*. Most anthropogenic emissions are in the form of sulfur oxides (SO_2) and nitric oxides (NO_x). Elementary chemistry shows that, in

combination with water (H_2O), these oxides will produce sulfuric (H_2SO_4) and nitric (HNO_3) acids.

4   *Deposition or fallout*. Eventually, what goes up must come down. Following the transport and transformations of the emissions, acidic precipitation is deposited. A large number of factors will determine what the environmental effects may be of deposition; for example, local geology (e.g., acid-sensitive granite versus well-buffered limestone), depth and type of soils, slope, and vegetation.

The above steps often resulted in what I like to refer to as the "irony of acid rain." Quite often, regions producing acidic emissions were not severely affected, but instead regions distant from the sources of pollution had to suffer the environmental repercussions. For example, several of the major source areas of acidic emissions were on well-buffered bedrock. More often, because of the long-range transport step (Fig. 7.2), regions or countries producing acidic emissions were not directly affected by these emissions, but instead deposition occurred on acid-sensitive areas, sometimes thousands of kilometers downwind. Air masses do not carry passports. Not surprisingly, this resulted in significant international tensions. Many critical questions were raised; paleolimnology provided many of the answers.

Acidification is a dynamic process, and so effective management will require analyses of trends over different time scales, including estimates of pre-acidification conditions. However, S.P.L. Sörensen only invented the pH function in 1909, and measurements of pH were rarely made until the middle part of the 20th century, and even then for only a few sites, and rarely on any systematic and continuous basis. Yet, many if not all environmental issues that deal with lake acidification have some temporal component, and so it is critical to be able to reconstruct past trends in lakwater pH.

In the following sections, I will primarily use three large-scale paleolimnological acidification projects as examples of the approaches used: (i) the Paleoecological Investigation of Recent Lake Acidification (PIRLA) I and II projects, which tracked acidification in the United States (Charles & Whitehead 1986; Charles & Smol 1990; Whitehead et al. 1990); (ii) the paleolimnological component of the Surface Waters Acidification Programme (SWAP), which used similar techniques in the United Kingdom and parts of northwestern Europe (Battarbee et al. 1990); and (iii) paleolimnological studies in the Sudbury region of Ontario (Canada), which tracked the effects of dramatic point source acidification and subsequent recovery from major smelting operations (Dixit et al. 1995).

7.3 Can we use paleolimnological indicators to infer lakewater acidity and related variables?

Fortunately, the distributions of many bioindicators are closely related to lakewater pH, and statistically robust and ecologically sound transfer functions have been developed that can infer acidification-related variables. Diatoms (Battarbee et al. 1999) and, to a lesser extent, scaled chrysophytes (Smol 1995b) have taken the lead in these quantitative analyses. As described in Chapter 6, surface-sediment training sets were used to develop these functions.

It has long been recognized that the distributions of certain diatom taxa were closely linked to pH (for an historical review, see Battarbee et al. 1986, 1999). Much of the early work was stimulated by Hustedt's (1937–9) pioneering treatise on the diatoms of Java, Bali, and Sumatra, where he enumerated diatoms from over 650 localities, and concluded that hydrogen ion concentrations had the greatest influence on species distributions. He classified the diatoms he recorded into five categories:

- Alkalibiontic: occurring at pH values > 7.
- Alkaliphilous: occurring at pH values of about 7, with widest distributions at pH > 7.
- Indifferent: equal occurrences on both sides of pH 7.
- Acidophilous: occurring at pH of about 7, with widest distributions at pH < 7.
- Acidobiontic: occurring at pH values < 7, optimum distribution at pH = 5.5 and under.

In many studies, these categories are still used to describe the distributions of taxa, with often the addition of "circumneutral" taxa, which are those most commonly found near pH 7.

A variety of empirical indexes were developed based on the above pH categories, which were used extensively in the initial phases of paleo-acidification work (reviewed in Battarbee et al. 1986, 1999; and chapters in Smol et al. 1986). For example, Nygaard's (1956) various indexes, especially Index α, were used in much of the early work. He believed, quite logically, that acidobiontic taxa were better indicators of acid conditions than, for example, acidophilous taxa, and so the former were weighted more strongly in his equations. Index α was defined as the percent acidophilous taxa plus five times the percentage of acidobiontic taxa, divided by the sum of the percentage of alkaliphilous taxa plus five times the percentage of alkalibiontic taxa.

The next major advance in quantification was from Meriläinen (1967), who evaluated the usefulness of Nygaard's indexes to Finnish lakes. Meriläinen used linear regression analysis to establish quantitative relationships between log Index α and measured pH values. A variety of multi-linear regression approaches followed (e.g., reviewed in Charles et al. 1989), which eventually were replaced by direct gradient analysis techniques (estimates of individual taxa to pH), such as weighted averaging regression and calibration (see Chapter 6).

Many examples are available showing how diatom species composition and other indicators (e.g., chrysophyte scales) can be used to quantitatively infer lakewater pH. These surface-sediment calibration studies have been used in many lake regions of Europe and North America. For example, Birks et al. (1990b) combined the diatom data from the SWAP calibration sets from England, Norway, Scotland, Sweden, and Wales, and constructed robust transfer functions to infer lakewater pH. As part of the PIRLA project, a surface-sediment calibration set of Adirondack Park drainage lakes was used to explore the environmental optima and to develop transfer functions for diatoms (Fig. 6.6; Dixit et al. 1993), chrysophyte scales (Cumming et al. 1992b) and cysts (Duff & Smol 1995). These analyses provided quantitative transfer functions that could be used to infer limnological variables relevant to acidification studies in the region, such as lakewater pH, acid neutralizing capacity (ANC), monomeric aluminum (which is often associated with fish loss), and dissolved organic carbon (DOC). Working near the metal smelters of Sudbury, Dixit et al. (1989, 1991) used chrysophyte scales and diatoms, respectively, in the surface sediments of 72 lakes, to construct pH transfer functions, as well as relationships to estimate past trends in metals such as [Ni] and [Al]. Examples of other calibration sets are

summarized in Charles et al. (1989) and Battarbee et al. (1999).

In some ways, from my personal perspective, and given the benefit of hindsight, paleolimnologists were "lucky" to begin developing many of these transfer functions at the time of the acid rain debates. I say this for two major reasons. First, high-quality data on historical trends were required, and required fast (given some of the legislative deadlines), and because acid rain was such an important environmental problem, significant resources were made available to the paleolimnological community. This allowed researchers to collect high-quality calibration data, with stringent quality assurance/quality control guidelines. Such data attracted some outstanding scientists with strong statistical backgrounds to work with paleolimnologists and their data sets (Cajo ter Braak and H. John B. Birks come immediately to mind). These partnerships proved to be immensely productive.

Second, I feel we were "lucky" to start quantitative paleolimnology with acidification issues, because most of our indicators are very strongly related to lakewater pH and related variables. This is something we have come to appreciate as we tackle tougher environmental inference problems, such as nutrients. I am not saying that the acidification work was easy, straightforward, or uncomplicated. However, given the benefit of hindsight, I think pH and related variables were "easier" to infer than many problems we are now dealing with.

From the surface-sediment training sets, it was clear that ecologically sound and statistically robust inference models were available to reconstruct past lakewater pH, and weaker models were available to study trends in variables such as dissolved organic carbon (DOC), inorganic monomeric Al, and some metals related to smelting operations. Other approaches were also developed, such as the use of sulfur isotopes to track the acidification process (Nriagu & Coker 1983; Fry 1986). The next step was to apply these transfer functions and other approaches to detailed analyses of lake sediments.

7.4 Have European lakes acidified? And if so, what are the causes?

Once transfer functions were developed, detailed paleolimnological analyses could be undertaken. Many examples are available (for a review, see Battarbee et al. 1999), and only a few examples will be given here.

In Europe, the SWAP Palaeolimnology Programme assessed several hypotheses as to the causes and timing of acidification (Battarbee et al. 1990; Renberg & Battarbee 1990). Using the diatom-transfer functions described above, as well as other approaches, SWAP paleolimnologists reconstructed trends in lakewater pH and related variables in a number of specially selected lakes in Norway (two lakes), Sweden (one lake), and Scotland (four lakes). Diatoms, chrysophytes, cladoceran, and chironomid analyses were used to track past lakewater changes, whilst documentary sources and palynology were used to study catchment histories. Trace metals (e.g., Pb), sulfur, polycyclic aromatic hydrocarbons (PAH), carbonaceous particles, and sediment magnetism were used to track atmospheric contamination.

The 1980s was a period of considerable debate as to the reasons why certain lakes were currently acidic. Because of the lack of long-term monitoring data, competing hypotheses were presented, but were difficult to assess using the sparse limnological data available for most lakes. Four major hypotheses were actively debated, the first three of which are as follows: (i) the lakes are acidic as a result of natural, long-term soil acidification over the postglacial period (approximately the past 10,000 years) as suggested by, for example, Pennington (1984); (ii) organic acids were produced naturally by vegetation, which acidified waters in the catchment (Rosenqvist 1977, 1978); and (iii) acidic precipitation increased as the combustion of fossil fuels increased, and this resulted in acidic precipitation. The fourth hypothesis, debated especially in the United Kingdom and Scandinavia, but also to some

extent in North America (Dobson et al. 1990), was that acidification was occurring as a result of afforestation of upland moorlands or with the decline in the use of pastures over the past two centuries (Rosenqvist 1977, 1978; Harriman & Morrison 1982). As noted by Renberg and Battarbee (1990), a complicating factor was that some of these processes would have occurred together over similar time frames. As a result, SWAP scientists had to very carefully choose sites and research designs to disentangle these potential multiple stressors.

To answer the above questions, SWAP paleolimnologists assessed the influence of catchment vegetation and land-use management on lake acidification by studying afforestation in Scotland, spruce expansion in Sweden, grazing and burning in Norway and Scotland, and then resettlement in the Swedish Iron Age as an analogue for postwar agricultural change. The acidification of some hilltop lakes with very small catchments in Norway further addressed the influence of drainage basin changes on lakewater pH and related variables. The SWAP paleolimnology program was a major contribution to the acid rain debates, and some key results of this coordinated research effort are summarized below. More detailed summaries of this work can be found in Battarbee et al. (1990) and Renberg and Battarbee (1990).

7.5 How have different levels of acidic deposition affected European lake ecosystems?

The paleolimnological results from the seven integrated SWAP studies noted above showed that lakes in areas of low acid deposition (i.e., northwestern Scotland, central Norway) had very low sediment concentrations of airborne pollutants and had not undergone any significant declines in lakewater pH over the past two centuries. The paleolimnological data indicated quite different trends in areas (i.e., southwestern Scotland,

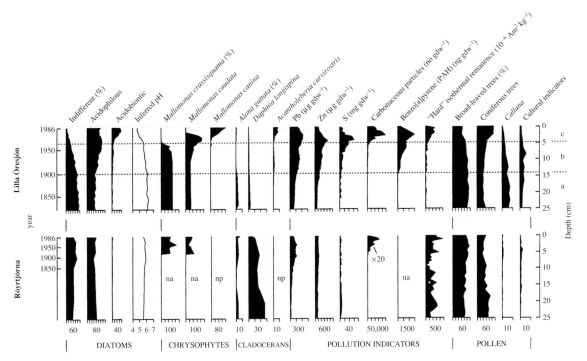

Fig. 7.3 A comparison of paleolimnological data from Lilla Öresjön (southwest Sweden, high sulfate deposition area) and Röyrtjörna (central Norway, low deposition area). (From Renberg and Battarbee 1990, used with permission.)

southern Norway, southwestern Sweden) receiving high acidic precipitation – these sites recorded high concentrations of airborne pollutants and marked declines in lakewater pH, concomitant with the time period of acidic deposition. In all cores, there was a strong correlation between increasing trends of airborne pollutants and decreasing diatom-inferred pH.

Renberg and Battarbee (1990) summarize some of the above data in Fig. 7.3, where the recent histories (approximately the past two centuries) of two Scandinavian lakes are compared. Lilla Öresjön, which is situated near the west coast of Sweden, is in an area of high acidic deposition, where lakes are typically acidic (Renberg et al. 1990). Meanwhile, Röyrtjörna is situated near the coast in central Norway, where there is little acidification.

There are striking differences between the two profiles (Fig. 7.3). In Lilla Öresjön there is

a period of little change in the 1800s (Zone a in Fig. 7.3), but beginning in the early 1900s (Zone b), acidophilous diatoms begin to increase in relative frequency, and inferred pH declines slightly. The acidification trend increases strikingly after about 1960, with increases in acidobiontic diatoms and a drop in inferred pH (Zone c) to about 4.6 in the recent sediments (closely matching the present-day measured pH), as opposed to a background pH of about 6.2 in the 19th century.

Because this was a multidisciplinary study, changes in other biota and indicators could also be assessed. For example, concurrent with the diatom changes, there were striking shifts in scaled chrysophyte assemblages, with circumneutral and generalist taxa, such as *Mallonomas crassisquama*, being replaced by acidobiontic species such as *M. canina*. Changes in invertebrate assemblages were also recorded (Fig. 7.3), and overall diversity decreased.

The biological shifts noted above coincided with changes in proxies of atmospheric contaminants in the sediments. For example, metal concentrations (e.g., Pb and Zn) increased after the turn of the 20th century, and reached the highest concentrations during the period of greatest inferred acidification. Spheroidal carbonaceous particles and polycyclic aromatic hydrocarbons (PAH), which are indicators of coal and oil combustion, followed similar patterns and peaked between 1970 and 1980. SWAP scientists also measured the magnetic mineralogy of the sediments, which should reflect the "hard" isothermal remanence component of the sediment, to track past fly ash and other ferromagnetic pollutants from the combustion of fossil fuels and other industrial sources. These data also revealed increased contamination of the lake, especially in Zone c (Fig. 7.3).

The recent environmental history of Röyrtjörna presents a quite different trajectory, where there was relatively little change in the various paleolimnological indicators, and diatom-inferred pH remained relatively constant between 5.6 and 5.9 (Fig. 7.3).

The above data indicated that Lilla Öresjön, situated in an area of high atmospheric sulfate deposition, has acidified because of this precipitation, whereas Röyrtjörna, which receives little sulfur deposition, has not acidified. However, because of some land-use changes, these results, although convincing, were not totally unambiguous. As noted earlier, a variety of land-use hypotheses were also proposed to account for the large numbers of presently acidic lakes. Although Röyrtjörna's catchment had been relatively undisturbed over the last two centuries (as indicated by the pollen data and historical records), Lilla Öresjön had considerable changes in its drainage basin concomitant with the period of increased sulfate deposition. For example, as summarized in Renberg and Battarbee (1990), Lilla Öresjön's catchment was primarily a heathland until a pine and spruce forest was established in the 20th century. Lumbering began in the 1980s and, although agriculture is still ongoing, grazing

in the surrounding forests declined in the 1930s and stopped totally in the 1940s. The "weight of evidence" pointed to sulfate deposition as the root cause of lake acidification in Lilla Öresjön, but it was still not definitive, as skeptics such as Rosenqvist (1977, 1978) might argue. The subsequent SWAP projects addressed these issues by a series of strategic paleolimnological studies.

7.6 Have European lakes acidified simply as a result of natural, long-term acidification?

The above examples documented some marked patterns of limnological change over the past few centuries in lakes receiving acidic deposition. However, arguments were being made that currently acidic lakes may simply be the "end products" of natural, long-term acidification processes. Pennington (1984) used paleoecological and geochemical data from the postglacial histories of Lake District (Cumbria, UK) upland lakes to argue that progressive acidification had occurred since before 5000 ^{14}C years ago. The mechanisms she suggested for this acidification included the removal of base cations from the soils by rainfall (i.e., base cation leaching), a process that would have been intensified after the removal of the deciduous forests by humans. These forests were replaced by heather (Calluna) and other plants, whose decomposition products could have acidified the soils and the waters. Additional "natural acidification" could potentially result from the Sphagnum mosses that colonized catchments where drainage was eventually impeded by paludification processes. Subsequent acidification of the soils would likely have occurred by decomposition products of vegetation matter. Indeed, paleolimnological data did exist that confirmed natural, slow, long-term (on the scales of thousands of years) patterns of acidification (studies reviewed in Charles et al. 1989). However, the questions of magnitude, scale, and timing were critical to the overall debate.

SWAP scientists assessed these competing hypotheses by conducting their integrated analyses on a series of cores that spanned the full postglacial sequences of sediment accumulation in the SWAP study regions. As an example, I will return to Renberg's (1990) study of Lilla Öresjön, where paleolimnological data indicated recent acidification, concomitant with acidic deposition in southwest Sweden. Most paleolimnological studies that study environmental trends over several millennia analyze their sediment profiles at relatively coarse sampling intervals (e.g., every 10 cm). In a somewhat "legendary" study for paleolimnological effort, Renberg (1990) analyzed diatoms in 700 contiguous samples, at 0.5 cm intervals, over the entire length of the 3.5 m sediment core (Fig. 7.4)! The results

were worth the effort. Renberg classified the diatoms into Hustedt's four pH categories, and also calculated the diatom-inferred pH based on the SWAP transfer functions. He identified four major "pH periods" in the lake's c. 12,600-year history: Period I was an alkaline period (12,600–7800 yr BP) following deglaciation, beginning with a diatom-inferred pH of about 7.2, but dropping to around 6.0 by 7800 yr BP; Period II was a naturally acidic period (7800–2300 yr BP), when the diatom-inferred pH decreased from 6.0 to 5.2; Period III was a zone of elevated pH (> 6) that coincided with the period of expanded agriculture (2300 yr BP to the start of the 20th century); and then Period IV was a time of marked acidification beginning around AD 1900, and becoming especially severe during

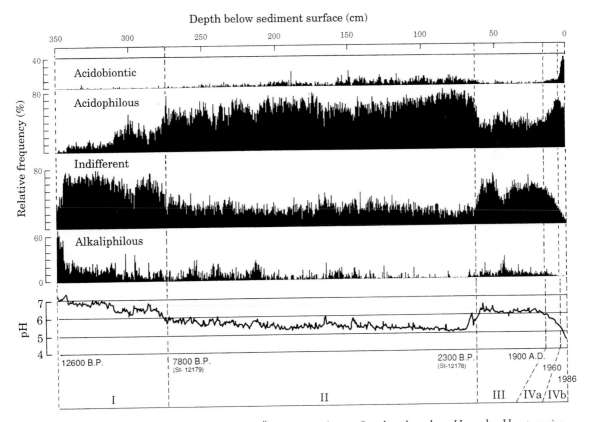

Fig. 7.4 The diatom-based pH history of Lilla Öresjön, southwest Sweden, based on Hustedt pH categories and diatom-inferred pH using weighted averaging (bottom panel). From Renberg (1990); used with permission of the Royal Society of London.

the 1960s, dropping to a pH of about 4.5, as described above in the detailed short core analyses (Fig. 7.4).

The changes occurring in the post-1960 sediments were unprecedented in the lake's 12,600-year history, and once again indicated that acidic deposition was a key factor in the lake's highly acidic state. However, Renberg's (1990) study also showed how pH can change on long-term time scales, without any deposition (Fig. 7.4). For example, the very slow and natural long-term acidification trend occurred over about 5000 years, but clearly the inferred pH never dropped below 5 until the lake's most recent sediments. This study also showed that pH can fluctuate widely, even without human influences. He showed that pH could change several times with an amplitude of up to 0.5 pH units. These natural fluctuations provided a more realistic assessment of "natural variability."

Finally, one of the most interesting aspects of this study was Period III (Fig. 7.4), when the diatom-inferred pH increased from about 5.2 to 6.3. The start of this zone coincided with a number of key changes in the fossil pollen. For example, cereal and other pollen grains indicative of agriculture increased, whilst oak pollen declined. These data would support the hypothesis that land-use changes were indeed affecting the pH of the lake. These questions would be addressed more fully in a subsequent investigation, and are discussed later in this chapter.

In addition to the above study, SWAP scientists analyzed postglacial cores from England (Devoke Water; Atkinson & Haworth 1990) and Scotland (Loch Sionascaig – Atkinson & Haworth 1990; Round Loch of Glenhead – Jones et al. 1989; Birks et al. 1990b). Similar to Lilla Öresjön, the pH trajectories of Devoke Water and Loch Sionascaig show early phases of alkaline conditions, followed by some natural acidification to between pH 6–7 for most of the remainder of the sequence. Round Loch at Glenhead shows more acid conditions throughout its sequence. As with Lilla Öresjön, none of these sites indicate very acidic pH values (i.e.,

< 5) prior to the time of acidic deposition, but clearly that slow, long-term, natural acidification can occur.

In the Round Loch example noted above, Jones et al. (1989) were able to specifically assess the influence of peat formation on lakewater acidity. By using both diatoms (to indicate lakewater pH) and pollen (to indicate terrestrial changes, including the development of an acidic, blanket mire peatland dominated by *Calluna*), Jones et al. showed that the lake was slightly acid (pH = 5.5–6.0) *before* peat formation, and that no acidification occurred while peat was developing in the catchment. Moreover, acidification to pH values below 5.0 only occurred in the post-1850 sediments, concurrently with periods of elevated acidic deposition.

The main conclusion of the above postglacial studies was that, although some very slow, natural, long-term acidification can occur, and that natural variability can be significant, these natural changes could not account for the presently highly acidic state of the lakes. These long-term acidification processes could, however, predispose lakes to become more acidic with the added onslaught of atmospheric deposition of strong acids.

7.7 How have land-use changes affected lakewater pH?

As introduced in the Lilla Öresjön example above (Renberg 1990), and followed up by studies of other lakes (Renberg et al. 1993a,b), land-use changes may potentially affect lakewater pH. The relative roles that land-use changes may have on lakewater acidity have been hotly debated and characterized by polarized discussions. Rosenqvist (1977, 1978) was often cited as the main proponent of the importance of terrestrial changes, as he had argued that acidification in Norway had occurred as a result of a decline in the traditional forms of agriculture and burning. Basically, the argument put forward was that the acidification of Norwegian lakes was more closely related to terrestrial changes than to atmospheric deposition. Rosenqvist argued that the replacement of traditional agricultural practices led to the regeneration of natural vegetation, and hence an increase in the raw humus content of soil profiles. The resulting changes in ion-exchange processes would lead to acidification of surface waters. Arguments for natural causes of lake acidification were also being made in North America (e.g., Krug & Frink 1983; Dobson et al. 1990), but these will be discussed later.

Because many of these catchment changes occurred synchronously with the time of acidic deposition, disentangling these possible multiple stressors can be tricky. However, by carefully choosing lakes, comparative paleolimnological studies could be performed to evaluate these hypotheses.

The general arguments posed by the land-use hypotheses was that the decline of traditional agriculture has resulted in acidification of soils and hence waters draining these catchments. SWAP addressed these problems in several ways. Anderson and Korsman (1990) applied an historical analogue approach. Suitable modern analogues to evaluate Rosenqvist's hypotheses were difficult to find, because these sites typically also receive high amounts of acidic deposition. They argued that the depopulation of farms and villages in Hälsingland (northern Sweden) during the Iron Age, an event that was well documented by archeologists, was a good analogue for the land-use changes that were being cited as potential causes for lake acidification, without the complicating factor of contemporaneous acidic deposition. The area was abandoned c. AD 500 during the so-called Migration Period, and was not resettled before the Middle Ages (c. AD 1100). They found that, although there was strong evidence for re-vegetation of the forests from the pollen data, the diatom data did not indicate any concurrent acidification. Importantly, though, one of the study lakes did show evidence of recent acidification, as would be expected with the hypothesis of acidic deposition.

The acidifying potential of conifers such as Norway spruce (*Picea abies*) in a catchment was specifically addressed by Korsman et al. (1994). Norway spruce forests have expanded significantly in Sweden over the 20th century, both due to natural spreading and to afforestation practices. Once again, though, areas of afforestation in Sweden typically overlap with areas receiving high acidic deposition, and so it is difficult to assess the relative roles of land-use changes and atmospheric deposition on lakewater acidity. Korsman et al. (1994) again employed a historical approach, as the migration of Norway spruce into Sweden about 3000 years ago provides a suitable analogue for assessing the acidification potential of this vegetation change. Eight acid-sensitive lakes were chosen for study. Using pollen to track past vegetation changes, and diatoms as quantitative indicators of lakewater acidity and DOC changes, they showed that at four sites there were significant changes in the diatom assemblages with the arrival of spruce, as one might expect, but none of the study lakes acidified. Diatoms did indicate that DOC increased in three of the sites, likely as a result of the accumulation of raw humus.

Another approach employed to disentangle the effects of atmospheric deposition and land-use changes was to find lakes perched on hilltops with very small catchments, which, by their very nature, could be little influenced by land-use changes. Birks et al. (1990a) used this approach on two lakes in southwest Norway that were subjected to high sulfate loading. Paleolimnological analyses of diatoms and chrysophytes indicated that, prior to c. 1914, both hilltop lakes were somewhat naturally acidic, but then both sites acidified further as levels of acidic deposition increased.

There is evidence, however, that changes in forests can also affect lakewater pH. For example, Kreiser et al. (1990) used a comparative paleolimnological approach to assess the effects of modern forestry practices in the United Kingdom on lake acidification trajectories. More specifically, they asked whether afforestation could lead to lake acidification. They compared the acidification histories of an afforested and non-afforested (moorland, control) site in an area of Scotland receiving high acidic deposition with an afforested and a similar control, non-afforested site in a region of relatively low deposition. Diatom data indicated all four sites acidified starting in the mid-1800s, regardless of land-use practices. However, the patterns of acidification during the 20th century varied considerably amongst the four study lakes. In the area of high deposition, the control site did not change markedly in diatom-inferred pH after c. 1930, but the afforested site had a significant acidification trend beginning around 1960. Because no changes were recorded in the control site, and because atmospheric input has been broadly similar for both lakes, the authors concluded that the accelerated acidification was linked to the planting of conifers in the catchment in the 1950s. However, any acidification caused by forest growth alone has been minimal when compared to the combined effects of forestry and acidic deposition. As summarized by Battarbee et al. (1999), afforestation of sensitive catchments can seriously exacerbate acidification, but only in areas receiving high levels of sulfur deposition.

7.8 Acidification in North America: different lakes and histories, but similar political and scientific debates

Concurrent with the work in Europe, the PIRLA-I and II projects (Charles & Whitehead 1986; Charles & Smol 1990), as well as related research programs, were asking similar questions in North America. For example: Were lakes acidic because of acidic deposition, or were some lakes naturally acidic? Is acidification part of a natural, long-term cycle or trend? What are the critical loads for sulfate deposition? Were some lakes naturally fishless? Are lakes recovering from decreased deposition loads? As in Europe,

answers to these questions required an historical perspective.

7.9 Trajectories of acidification in selected North America lakes: The PIRLA-I project

In the PIRLA-I project, paleolimnologists used detailed analyses of short sediment cores in several acid-sensitive regions of the United States (i.e., Adirondack Park – Charles et al. 1990; northern New England – Davis et al. 1994; the Upper Midwest – Kingston et al. 1990; Florida – Sweets 1992; Sweets et al. 1990; and complementary work in the Sierra Nevada region – Whiting et al. 1989) to track changes in specific lakes. Similar to the SWAP lakes, this typically showed that many currently acidic lakes had acidified, and the timing of acidification clearly fingered acidic deposition as the primary cause. Diatom and chrysophyte taxa were the main indicators used to infer past pH levels, whilst a suite of other chemical and biological proxy data were used to track related limnological and environmental changes. A key study area was Adirondack Park in northern New York State, which is a large, forested area, containing about 3000 lakes. Many of the lakes are presently acidic, and it was also known that this region received a large amount of acidic precipitation. However, the link between acid rain and acidic lakes was difficult to prove, as long-term pH data were not available. Many examples are available from this region (e.g., Charles et al. 1990), but due to space limitations, only the acidification history of Deep Lake (New York) is described here.

Deep Lake is a headwater lake in the Adirondack Mountains that was known to be acidic (pH ≈ 4.8) and fishless in the 1980s. But was this lake naturally acidic, or did it become more acidic as a result of acidic precipitation? A detailed paleolimnological analysis of this site provided these answers (Charles et al. 1990).

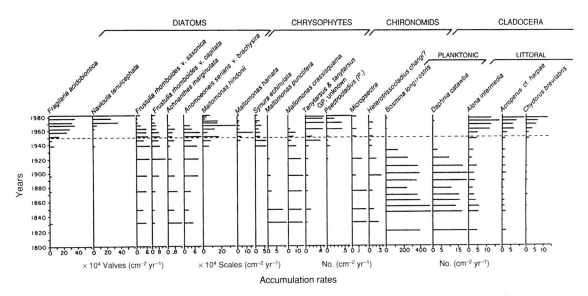

Fig. 7.5 Accumulation rates of selected diatom, chrysophyte, chironomid, and Cladocera taxa from a ^{210}Pb-dated sediment core from Deep Lake (Adirondack Park, New York). From Charles et al. (1990); used with permission.

The major stratigraphic changes recorded from the multiple indicators recovered from a ^{210}Pb-dated sediment core from this site are summarized in Figs 7.5–7.7.

The siliceous algal remains of diatoms and chrysophytes were the primary indicators used to track lakewater acidity changes (Fig. 7.5). Even a cursory glance at the diatom and chrysophyte species profiles indicates that a major change in the biological community began to occur in Deep Lake around 1920, followed by unprecedented algal changes occurring in the post-1950 sediments (Fig. 7.5). For example, diatoms characteristic of low-pH waters, such as *Fragilaria acidobiontica* and *Navicula tenuicephala*, and chrysophytes such as *Synura echinulata* and then the more acidic taxa *Mallomonas hindonii* and *M. hamata* increased, replacing circumneutral taxa such as *M. crassisquama*. Changes in fossil midge (chironomid) larvae and cladoceran invertebrates also occurred, with the near extirpation of the planktonic Cladocera taxa.

Using these species changes in conjunction with the transfer functions constructed earlier from the surface-sediment calibration set (see pp. 79–83),

Charles et al. (1990) were able to provide quantitative estimates of past pH changes (Fig. 7.6). These inferences showed that Deep Lake had a slightly acidic background pH, but then acidified further during the 20th century, coincident with increases in the deposition of strong acids (Fig. 7.6). The source of this pollution was confirmed with the other paleo-indicators studied in the Deep Lake core. For example, coal and oil soot particles, polycyclic aromatic hydrocarbons (PAHs), lead (Pb), and vanadium (V) provided a clear record of increased atmospheric input of materials associated with the combustion of fossil fuels beginning in the late 1800s and early 1900s (Fig. 7.7). As described later in this chapter, Uutala (1990) was also able to track the concurrent demise of fish populations from this lake using *Chaoborus* mandibles. In summary, data such as these convincingly showed that Deep Lake and many other PIRLA-I lakes were adversely affected by acidic precipitation. As serious and damning as these data were, these studies focused on a relatively small number of individual lakes. Can one use paleolimnology to make regional assessments of acidification?

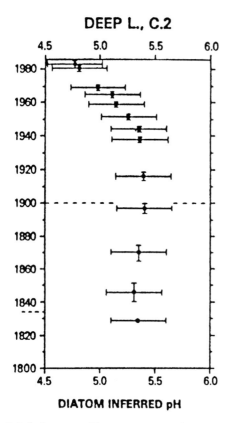

Fig. 7.6 Lakewater pH reconstruction for Deep Lake using the diatom taxa percentages and the Adirondack Park diatom-pH inference model. From Charles et al. (1990); used with permission.

7.10 Regional assessments of lake acidification: the PIRLA-II project

If lakes have acidified, then by how much? What percentage of lakes in a particular region have acidified? These questions are, in many ways, related to the one we just explored. However, the important examples provided above related to individual lakes. For many management issues, it is important to determine regional trends, and not just site-by-site studies.

The PIRLA-II project (Charles & Smol 1990) began to address some of these regional issues. Detailed, centimeter-by-centimeter paleolimnological analyses of sediment cores are time-consuming, but the so-called "top/bottom" approach (Box 4.1) allows one to analyze sediments from a large number of lakes to compare pre-impact to present-day conditions. Cumming et al. (1992b) adopted this approach to answer the questions: What proportion of the Adirondack Park lakes has acidified since the 1800s? If they have acidified, then by how much? What proportion of these lakes was naturally acidic? These were critically important scientific *and* political questions at the time.

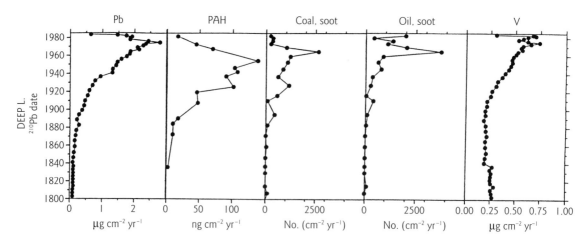

Fig. 7.7 Measured concentrations of lead (Pb), polyclyclic aromatic hydrocarbons (PAH), coal soot, oil soot, and vanadium (V) from the Deep Lake sediment core. From Charles et al. (1990); used with permission.

In order to make regional assessments of pH change, lake selection had to be done carefully, following a stratified, random, statistical design. A total of 37 lakes were chosen for analysis from EPA's Direct Delayed Response Program covering the spectrum of lake conditions (although restricted to lakes with an acid neutralizing capacity of less than 400 µeq l^{-1}) present in the Park. Diatom valves and chrysophyte scales were chosen as the primary indicators, as robust transfer functions were available for these species assemblages. Various lines of evidence confirmed that the "bottom" sediments (taken from the sediment profiles at a depth below 20 cm and usually below 25 cm depth) pre-dated the time of marked anthropogenic acidification, and approximately reflected limnological conditions from the early 1800s. This population-based, top/bottom study clearly showed that recent acidification was a major problem in the Park. A summary of the diatom data are presented in Fig. 7.8, with the lakes listed in order of increasing present-day pH, with the lakes currently most acidic appearing at the top. The histograms show

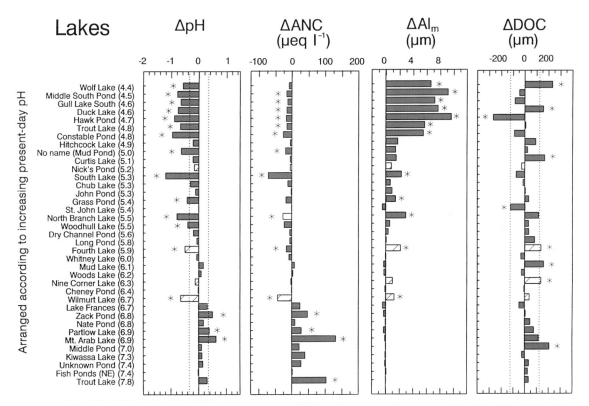

Fig. 7.8 Diatom-based estimates of historical change (pre-industrial to the present) in pH, acid neutralizing capacity (ANC), total monomeric aluminum (Al$_m$), and dissolved organic carbon (DOC) from the 37 randomly selected Adirondack Park (New York) lakes. The estimates are presented as the differences in inferred water chemistry between the top (0–1 cm) and bottom (> 20 cm, usually > 25 cm) sediment core intervals. The lakes are arranged according to increasing present-day measured pH. Lakes indicated by hatched bars were limed within 5 years before coring and/or assessment of water chemistry. The dotted lines represent plus or minus the bootstrap root mean squared error (RMSE$_{boot}$) of the prediction for pH and DOC. Estimates of change in ANC and Al$_m$ are presented as differences between backtransformed estimates, since these models were developed on log$_e$ data. Asterisks denote changes that are > RMSE$_{boot}$ for the various inference models. From Cumming et al. (1992b); used with permission.

the diatom-inferred changes since pre-industrial times in lakewater pH, acid neutralizing capacity (ANC), monomeric aluminum (Al_m, which may be toxic to fish), and dissolved organic carbon (DOC), a potential source of natural acidity. It is evident that all presently acidic lakes (present pH < 6.0) have undergone large declines in pH and ANC, and concurrent increases in monomeric aluminum since pre-industrial times. Conversely, there was no clear, unidirectional change in DOC.

The above "before and after" inference data can be shown in various ways. For example, Cumming et al. (1992b) also determined if any spatial patterns were evident by projecting these data onto maps of Adirondack Park. One figure mapped pre-industrial lakewater pH inferences for the study lakes, and showed that the lakes were typically circumneutral or only slightly acidic in the early 1800s. A second map showed the diatom-inferred changes in lakewater acidity since the early 1800s. This second spatial analysis graphically showed that most of the acidification occurred in the southwestern portion of the Park, a region that is relatively high in elevation, has poorly buffered catchments, and where total loading of acid is high, mainly because the amount of precipitation is higher. Taken together, this was compelling evidence that acidic precipitation was the root cause of the large number of acidic lakes in the Park.

Because of the statistical design used for the choice of study lakes, the results of the Cumming et al. (1992b) investigation could be extrapolated to a predefined population of 675 low-alkalinity Adirondack region lakes. This probability-based projection showed that approximately 25–35% of the target population had acidified. Cumming et al. further estimated that approximately 80% of the target population of lakes with a current pH of ≤ 5.2 and 30–45% of lakes with pH between 5.2 and 6.0 have undergone large declines in pH and ANC, and concomitant increases in Al_m.

So, based on the studies described above, apparently many Adirondack Park lakes had

acidified. But when did they acidify? This was the second major question posed by PIRLA-II scientists (Cumming et al. 1994). A detailed, close-interval analysis of fossil chrysophyte scales from 20 low alkalinity Park lakes produced evidence consistent with the hypothesis that lakes had acidified due to anthropogenic deposition since pre-industrial times (Cumming et al. 1994). Four categories of lakes were identified, based on their pH trajectories: (i) lakes that showed little evidence of acidification (< 20% of lakes); (ii) lakes with background pH values between 5 and 6 that began to acidify c. 1900; (iii) lakes that began to acidify c. 1930–50; and (iv) naturally acidic lakes that acidified further starting in around the 1900s. The timing of the acidification of these lakes was consistent with their location, watershed characteristics, and levels of sulfate deposition. Furthermore, the timing of changes in pH was not consistent with alternative mechanisms, such as natural (changes in vegetation and soils resulting from "blow-down" episodes, where a large number of trees were uprooted as a result of strong winds) or anthropogenic (e.g., logging) processes. Acid precipitation was causing lakes to become acidic.

7.11 Have seepage lakes also acidified?

The Cumming et al. (1992b) study was (purposely) restricted to drainage lakes (i.e., lakes that typically have inlets and outlets and receive most of their hydrological input from streams). However, many acid-sensitive regions also have large numbers of seepage lakes, which are defined as lakes that do not have surface water inlets and outlets, and receive most of their water from direct precipitation and/or groundwater seepage. Dixit and Smol (1995) reconstructed lakewater pH from the top and bottom sediments of 13 Adirondack Park seepage lakes and showed that, indeed, these lakes had also acidified since the early 1800s. However, the magnitude of acidification was less than the levels recorded in the drainage lakes, partially because pre-industrial pH values were already more acidic in the seepage lakes.

7.12 Is atmospheric deposition the only possible cause of lake acidification?

Although widespread acidification was a common conclusion for most paleolimnological studies in acid-stressed regions, some naturally acidic sites were also identified in both the European and North American data sets. For example, Korsman (1999) did not find evidence of widespread acidification from his top/bottom sediment study of 118 lakes in northern Sweden. Unlike many acidic lakes in southern Sweden, where strong inferences of recent anthropogenic acidification were recorded (e.g., Renberg et al. 1993a), Korsman showed that many of these northern lakes only showed minor changes since pre-industrial times in pH, alkalinity, and color, with a mean pH decline of only 0.04 of a pH unit. Detailed analyses of long sediment cores from five of the presently acidic sites showed that these lakes had experienced long-term natural acidification beginning thousands of years ago, due to soil-forming processes and natural changes in vegetation. Other examples of naturally acidic lakes include a small hilltop lake in southern Finland, where Korhola and Tikkanen (1991) used cladoceran fossils to infer natural acidification from about 8500 years ago, some lakes in North America (e.g., Winkler 1988; Ford 1990; Ginn et al. 2007), as well as other regions.

Cameron et al. (1998) used paleolimnological techniques, including diatom analyses, to show that the acidic status of a northern Quebec (Canada) lake, which lost its fish population in the 1930s, was a result of exposure to sulfide-rich graphite shales, and not, as was initially proposed, due to the use of explosives by prospectors. They showed that the lake was undergoing a long-term, natural acidification due to the local geology, and that pH levels had dropped past the tolerable limit for fish by about 1930.

Other natural processes can also produce acidic lakes, although at times human activities accelerate these processes. For example, Lake Blåmissusjön in northern Sweden currently has a pH of about 3.0. Renberg (1986), using diatom inferences from a 26-cm long sediment core, showed that the acidification of this lake was not due to atmospheric acidic deposition, but from natural processes occurring in the catchment soils. The watershed is rich in sulfides derived from marine sediments that oxidize (and so form strong acids) once exposed to air. This natural process, however, was accelerated with agricultural and ditching activities, with marked acid pulses occurring at the turn of the 20th century.

The identification of naturally acidic lakes has important management implications. Once the acidification problem was recognized, several regions, and some countries such as Sweden (Appelberg & Svenson 2001), embarked on comprehensive mitigation programs that involved lake and/or watershed liming. Heated arguments have ensued as to the efficacy and ecological soundness of some of these programs. However, regardless of the procedures, it makes no ecological or environmental sense to lime naturally acidic systems. Without the historical perspective that paleolimnology can provide, many naturally acidic lakes may be unjustifiably limed, resulting in massive alterations to specialized ecosystems and food webs that have persisted for thousands of years in a naturally low pH state. To use the medical analogies used at the start of this book, one should be certain a system is sick before one prescribes a treatment.

7.13 Did acidification result in losses of fisheries?

One of the most distressing symptoms of lake acidification was the apparent decline in fish populations. Nonetheless, it was often difficult to link unequivocally the decline in fish populations with acidification. Fish are notoriously difficult to track in the paleolimnological record, as they rarely leave fossils in sufficient numbers to reconstruct populations. Other, more indirect, approaches were developed.

In eastern North America, the chitinized mandibles of Chaoboridae (Diptera) larvae, more commonly referred to as the phantom midges, were used as a relatively blunt but still useful tool to track past fisheries. Continuing with the North American examples described above, five species of *Chaoborus* larvae are typically found in Adirondack Park or Sudbury lakes. Larvae from three of these species (*C. punctipennis*, *C. albatus*, and *C. flavicans*) occupy sediments or profundal waters during the day to avoid planktivorous fish, but migrate into the surface waters at night to feed. This behavioral pattern allows them to coexist with fish. The key species, however, is *C. americanus*, whose larvae, which are large and do not undertake diurnal vertical migrations, can rarely, if ever, coexist with fish (Uutala et al. 1994; Sweetman & Smol 2006). *Chaoborus trivittatus* larvae, on the other hand, only undergo weak vertical migrations and so can only withstand slight planktivory. Although less reliable, increases in this fish-sensitive taxon may potentially be used as a "warning indicator" that fish populations are declining (i.e., fish predation is decreasing).

Chaoborus mandibles are typically well preserved in lake sediments, and since they can be identified to the species level, they are potentially powerful indicators of the past status of fisheries (Uutala 1990; Lamontagne & Schindler 1994), and have been used in a number of North American acidification studies. For example, Uutala (1990) and Kingston et al. (1992) identified *Chaoborus* mandibles in the sediments of many of the PIRLA Adirondack Park cores. They showed that many of the PIRLA lakes became fishless at the same time that the lakes acidified and increased in monomeric aluminum, as inferred from diatoms and chrysophytes. Lakes that were naturally well buffered from acidification continued to maintain healthy fish populations, and *C. americanus* could never colonize the lake.

Although rare in Adirondack Park lakes, diatoms and chrysophytes also identified a few sites that appeared to be naturally acidic. These lakes are currently fishless. Uutala (1990) confirmed that lakes such as Upper Wallface Pond (New York) were not only naturally acidic (based on diatom data), but also naturally fishless (based on the continued presence of *C. americanus* mandibles in the sediments) even before the period of potential human impact. Such naturally acidic and naturally fishless sites should not be considered candidates for lake liming or fish-stocking programs, as these ecosystems and the food webs they support have developed, over thousands of years, in naturally acidic waters and without fish predators. Nonetheless, several of these sites have in fact been limed and stocked in unsuccessful bids at "remediation." A historical perspective, such as one attainable from the paleolimnological record, would have saved both money and unwanted ecological repercussions from inappropriate interventions.

These types of studies can also pinpoint the time of fish recolonizations. For example, as part of a multidisciplinary paleolimnological investigation, Uutala et al. (1994) identified the timing of fish extirpation based on *Chaoborus* fossils, which coincided with lake acidification, as inferred from diatom and chrysophyte indicators, from a series of cores in the Dorset region of Ontario. However, they also tracked the recolonization of fish (via stocking programs) once the lakes began to recover (Uutala et al. 1994). For example, the dominance of migratory *Chaoborus* taxa in the bottom part of the Shoelace Lake core confirmed the long-term historical presence of fish in this lake. The appearance of *C. americanus* mandibles in the post-1960 sediments documents the extirpation of fish, which coincided with the later stages of acidification, as reflected by the diatom and chrysophyte species changes (Fig. 7.9). The lake was restocked with brook trout in 1977; this food web manipulation is clearly marked in the *Chaoborus* profile by the post-1977 decline in *C. americanus* mandibles. This is further evidence of the sensitivity

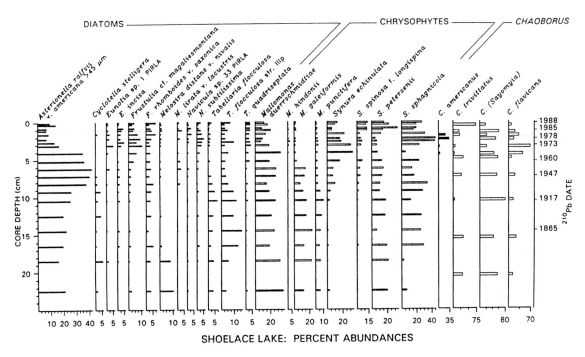

Fig. 7.9 Profiles of diatom valves, chrysophyte scales, and *Chaoborus* mandibles in a ^{210}Pb-dated sediment core from Shoelace Lake (Ontario). The diatoms and chrysophytes indicate a slight acidification trend beginning in the 1930s, but increased acidification and elevated monomeric aluminum concentrations in the 1960s. This coincided with the decline in fish populations, as marked by the increase in *Chaoborus americanus*. The decline in this taxon in the post-1977 sediments coincides with a fish-stocking program. From Uutala et al. (1994); used with permission of Elsevier.

of *C. americanus* to the absence and presence of fish.

Similar studies have been completed in other acid-sensitive and fishless regions. For example, paleolimnological studies in Sudbury showed that lakes became fishless concurrent with acidification (Uutala & Smol 1996).

7.14 Are lakes beginning to recover as a result of decreases in anthropogenic emissions?

Up until a few years ago, mining and smelting facilities in the Sudbury region (Ontario, Canada) were believed to be the world's largest point sources of sulfur emissions. Copper- and nickel-containing ore (unfortunately, most of it also containing high concentrations of sulfides) was discovered in the 1880s while the railroad was being built. Soon after, open roasting beds, followed by smelters, were operating in the region, with devastating effects on the landscape. At the peak of its operation in the 1960s, the Sudbury stacks discharged enough SO_2 to fill about 80 train tanker cars with concentrated sulfuric acid each day! Terrestrial vegetation was so negatively impacted by the fumigations that the area was often referred to as a "moonscape" (Fig. 7.10), and the Sudbury region could easily be identified on LANDSAT satellite images. Not surprisingly, Sudbury was one of the first places where the problem of lake acidification was recognized (Gorham & Gordon 1960).

Fig. 7.10 The Sudbury landscape was so devastated by the acidic and metal deposition that it was often referred to as a "moonscape." In fact, in May 1971, NASA Apollo astronauts trained in the area for space missions. From NASA 71-40732; used with permission.

A large volume of paleolimnological data has been amassed (reviewed in Dixit et al. 1995) showing how dramatically this region had been affected by massive acidification. For example, using a top/bottom approach in 72 lakes within a 100 km radius of Sudbury, S.S. Dixit et al. (1992) showed that extensive acidification had occurred (sometimes by over two pH units) in presently acidic (pH < 6.0) lakes. As in other studies, they identified a few naturally acid lakes, but these sites acidified further since the 1800s.

As infamous as Sudbury became as the "acid rain capital of the world," it is now also well known for its remarkable record of declining emissions. Following provincial orders issued in 1969 and 1970, attempts were first made at improving local conditions. This was initially accomplished by the construction in 1972 of the 381 m high "Superstack," the world's tallest smokestack at the time, which alleviated local problems, but did this by dispersing the pollutants over a larger geographical range. In the same year, Inco's Copper Cliff smelter and Falconbridge's iron ore sintering plant were permanently closed. Combined with other reductions in emissions, this resulted in a > 50% decrease in sulfate deposition in Sudbury (Potvin & Negusanti 1995). Further reductions were implemented in 1978, followed

by other mitigation programs that continued to lower emissions to the present day (emissions are currently estimated to have been reduced by about 90%). Can lakes that have been subjected to such massive amounts of acidification still recover following equally striking declines in emissions?

As depressing as the effects of acidification were in Sudbury, the precipitous chemical recovery that many lake systems experienced once deposition decreased was encouraging (Gunn 1995). Also, because local government and other scientists had recognized the acidification problem early on in Sudbury, several decades of water monitoring data were available for a number of lakes. This allowed paleolimnologists to evaluate the veracity of their reconstructions by comparing, for example, diatom-based pH inferences to known changes in lakewater pH. One such example is Baby Lake, which is located about 1 km southwest of the Coniston smelter, which operated from 1913 to 1972. Lakewater pH was measured in the lake from 1968 to 1984 by Hutchinson and Havas (1986), who showed that pH had exhibited a striking recovery following closure of the Coniston smelter (Fig. 7.11). Diatom- and chrysophyte-inferred paleolimnological data showed that the background pH of

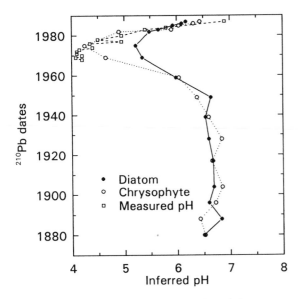

Fig. 7.11 Chrysophyte (open circles) and diatom (closed circles) inferred pH changes from a ^{210}Pb-dated sediment core from Baby Lake, Sudbury. The measured pH data (squares) for the lake dating back to 1986 are from Hutchinson and Havas (1986). From S.S. Dixit et al. (1992); used with permission.

Baby Lake was circumneutral (~ pH 6.5), but then began to drop precipitously around 1940 to a low of about 4.2 by 1975, the period of massive acidic deposition in this area (A.S. Dixit et al. 1992). Equally striking was the recovery in diatom- and chrysophyte-inferred pH after the smelter closed, and the fact that the trends in the paleolimnological data closely matched the trends in the measured pH data.

Studies similar to the example cited above were completed for many other Sudbury area lakes. Paleolimnological inferences repeatedly showed that marked acidification had occurred, generally followed by recovery coincident with declines in emissions. Because many of these sites were also part of lakewater monitoring programs, the fidelity of the paleolimnological record could repeatedly be confirmed. Dixit et al. (1995) summarized some of these patterns on a spatial scale by mapping diatom-based pH inferences from 22 Sudbury area sediment cores for four time slices (Fig. 7.12): recent times,

represented by surface sediments deposited in around the 1990s (A); in c. 1970, during the peak in deposition (B); in c. 1930, early in the acidification process (C); and pre-1850, representing background conditions, before the ore was discovered (D). Prior to 1930, none of the lakes had a pH less than 5.0, but 13.6% of the lakes were in this category by 1970. In 1970, 22.7% of lakes were in the pH range of 5.6–6.0, whilst only 9.1% of these lakes were at this range by the mid-1990s (i.e., recent times), attesting to the recovery patterns. Interestingly, almost all the naturally high-pH lakes had actually increased in pH over time, a pattern observed in other paleolimnological assessments (e.g., Cumming et al. 1992b), and confirmed by ongoing water-monitoring programs.

Sudbury provides a very optimistic picture of chemical recovery from acidification. However, not all lakes are showing recovery, even with declines in emissions. For example, as part of the PIRLA-II project examining lake acidification in Adirondack Park (New York), Cumming et al. (1994) undertook a detailed, close-interval paleolimnological study of 22 lakes. It was well established that deposition in the northeastern United States had declined since the 1970s (Husar et al. 1991). Were the Adirondack Park lakes responding to these declines in emissions? Are they still acidifying? Or are they recovering? Cumming et al. found that although a few lakes were beginning to show signs of recovery (based on a return of less acidic chrysophyte taxa in the most recent sediments of the sediment cores), pH changes in the Adirondack lakes since the post-1970 declines in sulfate deposition have been small and variable. These patterns are discussed within a more regional context later in this chapter.

Some lakes near other mining sites have also been slow to recover. Ek and Renberg (2001) undertook a paleolimnological study of 14 lakes near the now abandoned site of the Falun copper mine, in central Sweden. They used diatoms from sediment cores to reconstruct past lakewater pH, and analyzed the sediments for metals and

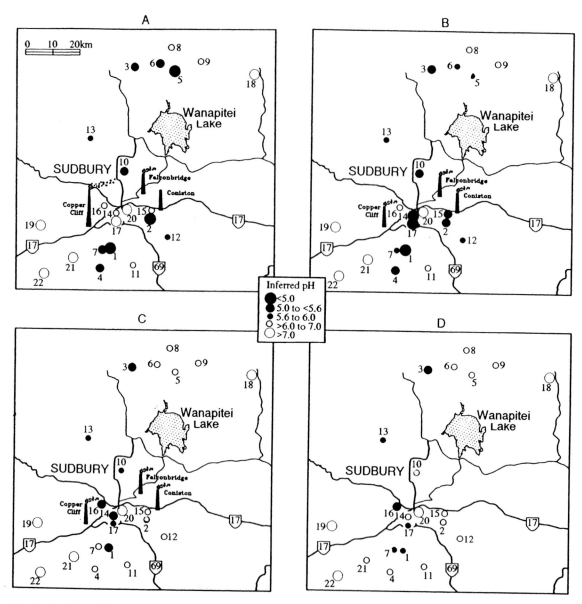

Fig. 7.12 Spatial and temporal distribution of diatom-inferred lakewater pH from 22 Sudbury area lakes. Four time slices are shown, capturing (A) recent (i.e., around the 1990s), (B) c. 1970 (the time of maximum deposition), (C) around the 1930s (the period of early industrialization), and (D) pre-industrial conditions (i.e., pre-1850). Lakes are grouped into five pH categories, as shown in the legend at the center of the figure. From Dixit et al. (1995); used with permission.

$^{206}Pb/^{207}Pb$ isotope ratios (see pp. 131–4) to decipher the historical and geographical distribution of airborne pollution. The Falun copper mine site is somewhat unique, as it is one of Europe's oldest mines, being in continuous use from at least 1000 years ago until it closed in 1993. Its production peaked in the 17th century, when it supplied two-thirds of the world's copper. As such, the sediments in the lakes around the Falun site provide an excellent opportunity to study both long-term, millennial-scale pollution, as well as any possible recovery.

Of the 14 lakes studied, eight had acidified, some beginning as early as the late 17th century, when sulfur oxide emissions were estimated to be about 40,000 tonnes per year. This is probably the oldest record of anthropogenic acidification on this planet. However, despite the very high level of emissions, lakewater pH had only decreased from about 0.4 to 0.8 pH units, with the lowest diatom-inferred pH of 5.8. Moreover, the geographical extent of lake acidification was limited, probably because Falun is situated in a valley and the pollution came and remained at ground level. Hence, most of the emissions were not transported far from the mine. These acidification patterns are in marked contrast to the Sudbury examples discussed earlier, where a "superstack" was built in the early 1970s. Furthermore, land-use changes, such as the forestry and agriculture that followed the expanding Falun mine operation, may have added additional buffering to the lakes (Renberg et al. 1993a,b), and so dampened the impacts of acidic deposition.

Also, in marked contrast to Sudbury, despite the declines in emissions and the eventual closure of the Falun mine in 1993, there is still no evidence of recovery in lakewater pH that can be inferred from the diatom assemblages. This is likely due to the large amounts of sulfur still stored in the soils in the lakes' catchments (three times more than the southern Swedish average), which are continuing to acidify the lakes and delay the recovery process. Forest regeneration and other catchment changes may also be slowing down the pH recovery process.

This study showed how important certain variables, which are perhaps not inherently obvious, can determine the type and magnitude of limnological changes that may occur in lakes affected by acid deposition. Paleolimnological data such as these also illustrate how difficult certain modeling exercises can be, as many of the factors identified as making important contributions to the lakes' conditions are rarely included in computer-based models.

7.15 Can we predict which lakes are more likely to recover?

It is now recognized, due in part to the powerful inferences available from paleolimnological studies, that many acid-sensitive lakes have acidified as a result of acidic precipitation. It is also generally recognized that, with recent cutbacks in deposition, *some* lakes are showing signs of recovery. It is not entirely clear, however, why some lakes are still acidifying or appear to be in "steady-state," and not showing any positive signs of recovery. Can the paleolimnological record assist lake managers by predicting which lakes are more likely to recover at a certain depositional load, whilst other lakes may require even further cutbacks in deposition?

Smol et al. (1998) attempted to answer the above question by comparing their 12 years of paleolimnological studies in the Adirondack Park (i.e., the PIRLA II lakes) and in the Sudbury region. In total, they had detailed chrysophyte-inferred pH paleolimnological analyses, spanning approximately the past two centuries of sediment deposition, from 36 Sudbury area lakes and 20 Adirondack Park lakes. Monitoring data had been suggesting that the Sudbury area lakes were experiencing striking recoveries from acidification, but that the Adirondack Park lakes showed disappointing, if any, signs of recovery. It was now well known that both regions had contained very acidic lakes, and that acidic deposition was the root cause of the problem. Both regions were enjoying reduced deposition as a result of legislation and mitigation efforts. Why were the Sudbury lakes recovering so much faster than the Adirondack Park sites?

Figure 7.13 summarizes the main features and conclusions of the Smol et al. (1998) comparative study. The top panel compares the pre-industrial (i.e., pre-1850) inferred pH values for the

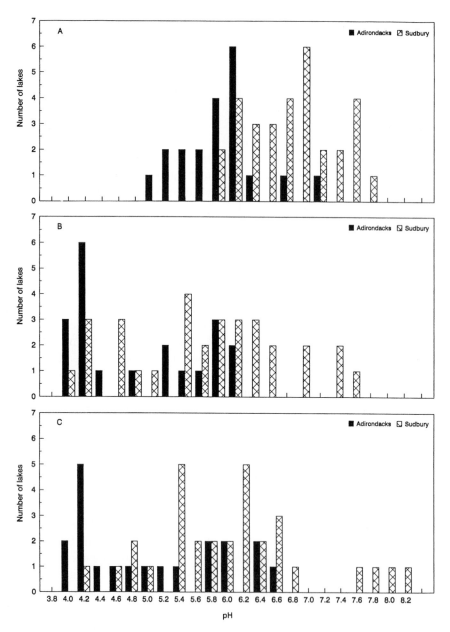

Fig. 7.13 A comparison of frequency distributions between the Adirondack (solid bars) and Sudbury (hatched bars) study lakes for (A) the pre-industrial chrysophyte-inferred pH, (B) the lowest chrysophyte inferred pH for each core, and (C) the chrysophyte-inferred pH for the top (recent sediments) of each core. The frequency distributions were calculated based on 0.2 pH increments from chrysophyte-inferred pH values from 4.0 to 8.2. From Smol et al. (1998); used with permission of Blackwell Publishing.

Adirondack and Sudbury lakes. It is immediately evident that the Sudbury lakes, in general, were naturally more buffered than the Adirondack lakes, as the former had higher pre-industrial pH. The middle panel shows the minimum chrysophyte-inferred pH for each lake, documenting the most acidic period in each lake's history. Clearly, both the Adirondack and the Sudbury area lakes acidified greatly from their pre-industrial pH values; many Adirondack lakes record very low minimum pH values. The bottom panel shows the chrysophyte-inferred pH for the tops (recent sediments) of each core. The Sudbury histograms generally shift to the right, showing recovery from the most acidic period. Although some recovery is seen in the Adirondack lakes, by and large there is little recovery from the most acidic period. The Sudbury lakes are indeed recovering much more than the Adirondack lakes.

With these scenarios in mind, one can then explore reasons for these differences. In general, lakes with higher background pHs and natural buffering capacities showed the greatest recovery. Conversely, lakes such as those in Sudbury appear to have higher critical loads than the more naturally acidic lakes in Adirondack Park. Certainly other factors, such as differences in deposition trajectories and hydrology between the two regions, are also likely important (Smol et al. 1998). By knowing pre-disturbance conditions, and then tracking the effects of stressors on these lake systems, we can better understand lake processes and also better predict the effects of remedial strategies.

7.16 Can paleolimnological data be used to set critical loads?

Using paleolimnological data such as those generated by the Cumming et al. (1994) study, and comparing the inferences to known loading levels of sulfate deposition since the 1800s (Husar et al. 1991), one has data on both the lake

responses (paleolimnological data) as well as the stressors (sulfate loading). A simple comparison of these two times series allows one to estimate critical load values (i.e., the highest load of a substance that will not cause chemical changes leading to long-term harmful effects; Box 2.3).

Using the example of Twin Pond (New York) from the Cumming et al. (1994) Adirondack Park study, the chrysophyte-inferred pH data indicated that the lake began to acidify about 1930 (i.e., it had surpassed its critical load; bottom part of Fig. 7.14). As sulfate loading data are

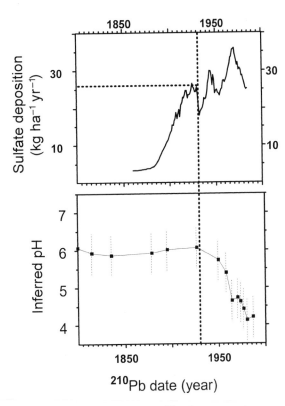

Fig. 7.14 The historic wet sulfur deposition data for the Adirondack region (modified from Husar et al. 1991) is shown in the top figure, with the chrysophyte-inferred pH profile from Twin Pond shown below. Based on these paleolimnological data, Twin Pond began to acidify in about 1930. By extrapolating this time period to the sulfate loading curve (dotted line), this would suggest that the critical load of sulfate deposition for this site is about 26 kg ha^{-1} yr^{-1}. Modified from Cumming et al. (1994).

available for this time period (Husar et al. 1991; top part of Fig. 7.14), one can simply extrapolate the loading data to the timing of the inferred pH change, and estimate that the critical sulfate loading for Twin Pond is about 26 kg ha^{-1} yr^{-1}. By doing this for many lakes, it became clear that critical loads were quite different for the various lakes used in the study, and critical loads were often significantly lower than the suggested reductions proposed in legislation for the region. For example, Cumming et al. (1994) estimated that critical loads for Adirondack Park lakes that started to acidify c. 1900 and c. 1950 would be about 5–10 and 20–25 kg ha^{-1} yr^{-1}, respectively.

Battarbee et al. (1996) used a different paleolimnological approach to estimate critical loads in Europe. Taking advantage of their large data set of diatom-inferred acidification profiles from the United Kingdom, they developed an empirical model based on a dose–response function. This function can be used to set critical load values for a site based on a knowledge of the ratio of Ca^{2+} of the water (sensitivity) to modeled sulfur deposition (i.e., loading) at that site. The diatom-based model is best suited for setting the baseline critical load for a site, as it identifies the point in time that biological change, as a result of acidification, begins to occur.

7.17 Assessing the efficacy of liming programs

Although reducing emissions is generally considered to be the only real, long-term solution to the acid rain problem, in some cases lake restoration efforts have included liming programs where calcium carbonate (CaCO$_3$) is added directly to water bodies or their catchment to raise the acid neutralizing capacities (Fig. 7.15). For example, since the 1970s, over 8000 lakes and 12,000 km of running waters have been limed in Sweden alone (Appelberg & Svenson 2001).

Although some post-liming monitoring programs have been implemented, quite often any subsequent effects of mitigation are poorly documented, and it is not clear what, if any, long-term effects liming has had. In one of the first high-resolution paleolimnological investigations of the effects of liming, Renberg and Hultberg (1992) studied the recent history of Lysevatten in southwestern Sweden, a forest lake that was limed in 1974 and again in 1986. The diatom-inferred pH data agreed well with the available historical pH data, with acute acidification beginning in the 1960s, depressing the natural pH of about 6–7 down to inferred pH values below 5. The liming in 1975 was clearly recorded in the diatom-inferred pH values, with an increase in pH to about 7.5. However, the diatom species composition did not return to pre-acidification assemblages, as the liming created an "artificial assemblage" not previously recorded in the lake. The paleolimnological data show that the effect of the 1975 liming was short-lived, as the lake rapidly re-acidified. The lake was re-limed in 1986.

However, as shown by many examples in this book, different types of lakes may respond differently to similar stressors and mitigation efforts. To continue with the example of the Swedish liming program, Guhrén et al. (2007) compared the paleolimnological histories of 12 limed lakes from pre-industrial times to the present, with the goal of addressing questions concerning the natural or background conditions of these lakes, as well as the effects of early human impacts, acidification, and subsequent liming. Liming elicited different limnological trajectories in the study lakes, including a return to pre-acidification diatom communities in some sites, but also shifts to new biological assemblages that were not previously recorded in the lakes' histories. Furthermore, as some Swedish lakes have been affected by land-use changes prior to the 19th century (e.g., Renberg et al. 1993a,b), the Guhrén et al. (2007) study also challenged the concept of simply using limnological conditions inferred from the 1800s as an appropriate management goal

Fig. 7.15 Liming of Whirligig Lake (Ontario, Canada), using a helicopter. Photograph taken by W. Keller.

for lakes such as these. They concluded that knowledge of the long-term histories of these lakes can provide more realistic mitigation targets, and also set more appropriate restoration goals for the European Water Framework Directive (see p. 19).

As many lakes in acid-sensitive regions have now been limed, some over 50 years ago, paleolimnological studies such as these provide an efficient tool to study the efficacy of these mitigation programs, as well as their limnological and ecological repercussions.

7.18 How much confidence can we put in computer models to understand the acidification problem?

Computer models, which are often used to predict future environmental changes (Box 2.4), are important tools for many environmental issues, including those dealing with lake management issues. The outputs of these models can often be used for important policy decisions. Questions often remain, however, as to how reliable these models are, and how effectively they can be evaluated. One way is to simply wait several decades, and find out if the model predictions were correct! However, this is clearly not a practical method. Another evaluation procedure is to hindcast the models back in time. Although we do not have empirical data on future conditions, paleoenvironmental techniques can be used to reconstruct past conditions. By comparing model hindcasts of, for example, past lakewater pH with paleolimnological assessments of past pH, an evaluation of the results can either provide confidence in the model being used (Jenkins et al. 1990), or may suggest ways to modify and improve it for a particular region (Sullivan et al. 1996). I will use the latter Adirondack studies as an example.

The Model of Acidification of Groundwater in Catchments (MAGIC; see Cosby et al. 1985) has been widely used in North America and Europe, and it was the principal model used in the NAPAP assessment to project the response of surface waters to changing levels of sulfur deposition (NAPAP 1991). The original MAGIC model, however, did not include organic acids, which were a potential source of acidity. In fact, the original MAGIC model hindcasts (Sullivan et al. 1996) yielded pre-industrial pH values that were substantially higher than the PIRLA diatom-based estimates of lakewater pH (Cumming et al. 1992b) in the Adirondack data set (Fig. 7.16A). The discrepancy was greatest in the most biologically sensitive part of the pH range (i.e., pH 5.0–6.0). Moreover, MAGIC hindcasts were consistently greater than a pH of 6, whilst the diatom-based inferences suggested more acidic natural conditions in some lakes.

The MAGIC model was then modified to include a dissolved organic carbon (DOC) component. The agreement between the diatom-inferred pre-industrial pH values and the modified MAGIC model hindcasts were much closer (Sullivan et al. 1996; Fig. 7.16B). This synergy between modeling efforts and paleolimnological assessments was an excellent example of gathering information from quite different techniques

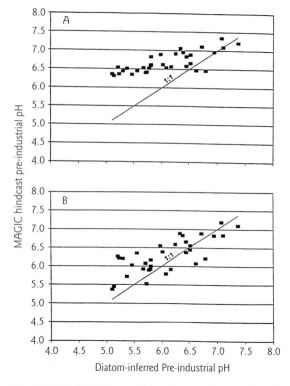

Fig. 7.16 MAGIC model hindcasts of pre-industrial pH versus diatom-inferred pH for 33 statistically selected Adirondack lakes. (A) The hindcasts that did not include organic acids in the MAGIC simulations. (B) The much stronger relationship that includes a triprotic organic acid analogue model in the MAGIC simulations. From Sullivan et al. (1996); used with permission.

to improve and strengthen both disciplines, as well as to get more realistic environmental assessments.

7.19 From rainforest to wasteland in 100 years: Acidification in the Southern Hemisphere

All the acidification examples included thus far have been from the Northern Hemisphere, as most of the world's acidic precipitation is produced and deposited north of the Equator. However, a recent example, the first from the Southern Hemisphere, has shown how acidic emissions and metal pollution from the smelters of European mineral prospectors, beginning in the late 1800s, dramatically altered parts of the western Tasmanian (Australia) rainforest. This ecosystem had been home to Aboriginal tribes for more than 36,000 years (Fig. 7.17). Hodgson et al. (2000) used paleolimnological techniques on Owen Tarn (unofficial name) to track the effects of acid rain, deforestation, and soil erosion on this small mountain lake that is located about 5 km from the former smelters at Queenstown to the west (which smelted sulfidic copper from 1896 to 1969) and about 10 km from the former smelters at Crotty to the east. Pollen analyses chronicled the industrial deforestation related

Fig. 7.17 The rainforests of southwest Tasmania (A), the Queenstown smelters soon after commencing operations in the late 19th century (B, courtesy of Beatties Studio, Hobart, Tasmania), and the landscape of Queenstown 100 years later (C). From Hodgson et al. (2000); used with permission; www.schweizerbart.de.

to tree felling and burning associated with the mining activities, as well as acid rain killing small trees. Diatoms tracked the changes in the aquatic environment.

The paleolimnological data recorded two major biological responses to the smelting operations. First, there was a clear shift to more acidophilic diatoms with the development of the mining industry in the late 19th century. Perhaps not surprisingly, the diatom changes are different from those expected with similar pollution in North America or Europe, reinforcing the importance of developing regional calibration sets. This is especially true for places such as Tasmania, which are so distant from the surface-sediment training sets developed in Northern Hemisphere temperate regions.

Second, Hodgson et al. recorded a large number of deformed (i.e., teratological) diatom valve morphotypes, which coincided with the onset of acidic and heavy metal deposition. Although the link to metal pollution is still tenuous, other workers have recorded deformed diatom frustules from various types of contaminated and chemically stressed environments (e.g., Yang & Duthie 1993; Guilizzoni et al. 2001; Cataneo et al. 2004; Salonen et al. 2006).

The Owen Tarn sediments contain some evidence of post-acidification reversibility, as deformed diatom specimens are less common in the surface sediments, consistent with the termination of smelting in 1969 and the re-establishment of vegetation around Queenstown.

The extent of the influence from mining and smelting in other regions of Tasmania, and the Southern Hemisphere as a whole, awaits further paleolimnological study. Although acidic deposition may be decreasing in many European and North American regions, it is increasing in many developing nations, including many tropical and subtropical countries (Kuylenstierna et al. 2001). Paleolimnologists now have the tools to offer a suite of cost-effective approaches to assist in assessing and monitoring the extent of any problems, as well as providing data to assist future mitigation initiatives.

7.20 Acidification from the direct discharge of industrial effluents

Almost all the paleolimnological work dealing with acidification has dealt with environmental repercussions of atmospheric deposition of acids. However, lakes and rivers can be acidified by other industrial processes, such as direct discharges of chemicals causing marked acidification. In most of these cases, other pollutants, such as metals, are also discharged, and so it can at times be difficult to determine if the biological repercussions have more to do with pH change or the associated toxins. Acid strip-mine lakes would fall into this category, and paleolimnological approaches have been used to study their acidification and recovery processes (e.g., Brugam & Lusk 1986; Fritz & Carlson 1982). Another example of both metal pollution and marked acidification is Lake Orta in northern Italy, which experiences severe pollution and acidification from industrial discharges of copper and aluminum sulfate from a rayon factory, established in 1926 (Guilizzoni et al. 2001). This example is discussed in the following chapter, as most of the biotic responses were believed to have been caused by copper toxicity.

7.21 Some progress, but many lakes are still acidifying

Researchers working with the acid rain problem can certainly claim some successes. International, cross-boundary legislation has been enacted, and some lakes have begun to show signs of recovery. However, there is a growing consensus, which is confirmed by much of the paleolimnological data summarized in this chapter, that critical loads are still too high and that many sensitive lakes will continue to acidify. Scientists have also become increasingly sensitized to the importance of nitrogen deposition and its effects on the acidification process (and possible lack of recovery following sulfate reductions). Thousands of papers and scores of books have been written on the repercussions of lake acidification, but this problem has, in many respects, been dropped from many political and scientific agendas. This is partly due to new problems, such as enhanced global warming, contaminants, and so forth, which have shifted attention from the acidification issue. But acid precipitation is still a problem, and continued vigilance is required to remind policy makers that this is a major environmental concern.

7.22 Summary

The problem of lake acidification evoked a tremendous amount of political debate in the 1980s. In some respects, due to the trans-boundary nature of acidic emissions and transport, acidification was the first major international issue to involve many environmental scientists. Although many lakes in eastern North America and parts of Europe were obviously currently acidic, some challenges were put forward (primarily by those with strong industrial ties) that these lakes were naturally acidic and/or that factors aside from acidic deposition were responsible for the acidic nature of these lakes (e.g., land-use changes). The

pH metric was only developed in 1909 and so it was not possible to use water-quality monitoring data to determine the natural (i.e., pre-1850s) acidity levels. In fact, for the vast majority of lakes, little or no pH historical data were available even for the late 20th century. Consequently, only indirect proxy methods, such as those provided by paleolimnologists, could be used to infer long-term trajectories in lakewater pH and related variables. Diatom-based transfer functions took the lead in most of these investigations, although other indicators (e.g., chrysophytes) supplied additional information on limnological changes. A few naturally acidic lakes were identified, but the vast majority of these paleolimnological studies clearly demonstrated that acidic deposition was the root cause for the currently acidic state of many lakes. Fossilized mandibles of phantom midges (i.e., *Chaoborus* spp.) showed that many of the currently fishless lakes lost their fish populations as a result of acidification. Paleolimnological data were also used to determine if lakes were recovering as a result of subsequent cutbacks in emissions or other mitigation measures (e.g., liming), as well as to evaluate computer models. Many of the paleolimnological techniques honed during this period of intensive research were subsequently modified and transferred to the study other environmental problems.

8

Metals, technological development, and the environment

Everything is a poison, there is nothing which is not. Only the dose differentiates a poison from a remedy.

Paracelcus (1564)

8.1 Metals and the environment: a silent epidemic?

Metals enter lake systems naturally from a variety of sources, but mostly either from runoff and groundwater inflows that have scavenged metals from the catchment lithology via weathering and other processes, or because metals are emitted into the atmosphere. Although metals in our environment typically conjure negative images in most peoples' minds, some of these elements (e.g., iron, zinc) are required by biota in small quantities for physiological processes (Fig. 8.1A). Problems arise when concentrations exceed certain thresholds, and metals may harm organisms and even become toxic. Non-essential elements are tolerated to a certain concentration, and then they too can become toxic and lethal (Fig. 8.1B). In a limited way, paleolimnologists use these relationships by inferring trends in some metals from the biological responses they might elicit, such as changes in diatom (Dixit et al. 1991; Cunningham et al. 2005) or chironomid assemblages (Ilyashuk et al. 2003). As many human activities have greatly accelerated the supply of metals to freshwater systems, metal pollution is

an important environmental issue in many parts of the world.

Human technological advances have often been driven by our abilities to extract and use metals. In fact, archeologists often divide parts of our cultural history by the metal artifacts humans have learned to make (e.g., the Iron Age, the Bronze Age, etc.). The history of metallurgy can be traced back to about 6000 BC, with the scattered records of gold articles (mostly jewelry) found from that time. By 3600 BC, the first copper-smelted artifacts were found in the Nile Valley, and by 3000 BC copper weapons and tools were being made. The ancient Egyptians used lead sulfide as eye paint, and the Romans had developed a sophisticated water delivery system based on lead water pipes.

Most of the elements in the periodic table are considered to be metals, many of which we use directly or indirectly every day. As populations grew and technology progressed, the demand for larger quantities and different types of metals increased tremendously, to the point that many parts of our planet are now being drilled, excavated, and/or mined for metals. Between 1930 and 1985, production from mines of aluminum (Al), chromium (Cr), copper (Cu), nickel (Ni),

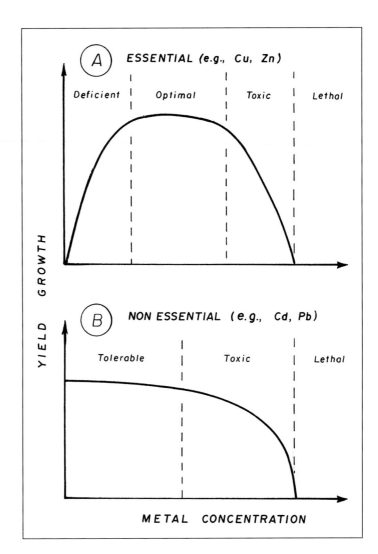

Fig. 8.1 A diagram showing the growth of organisms as a function of the supply of (A) essential and (B) non-essential metals. Modified from Förstner and Wittmann (1981).

and zinc (Zn) increased by 114-, 18-, 5-, 35-, and 4-fold, respectively (Nriagu 1988). Apsimon et al. (1990) estimated that the world's population consumes 50×10^9 tonnes of mineral resources each year; this exceeds the estimated 16.5×10^9 tonnes of sediment transported annually to the ocean by all the world's rivers! Humans have become the most important factors in the global biogeochemical cycling of trace metals, with millions of tonnes of metals being extracted annually from mines, and thereafter being redistributed into the biosphere (Nriagu & Pacyna 1988; Pacyna et al. 1995). Metals that were

once safely buried deep in the ground are now being emitted into our environment by industries (Nriagu 1996, 1998). Some are expelled directly from point sources into receiving waters, but many enter aquatic systems via atmospheric pathways. Dominant anthropogenic sources include the smelting of ores and the burning of fossil fuels, as well as different commercial and manufacturing processes. Some processes are more indirect, such as acidic precipitation falling on catchments, which thereafter mobilizes some metals that eventually are flushed into waterways.

The toxicological properties of some metals were recognized back in antiquity. For example, both the Romans and the Greeks knew that arsenic could be used to kill unwanted organisms, ranging from their political rivals to rodents. Metal pollution and its effects on human health have a long history. Lead pollution has been implicated in the fall of the Roman Empire, as well as contributing to the demise of Sir John Franklin's men on his last Arctic expedition. There is little doubt that high concentrations of many metals are acutely toxic to humans as well as many other taxa. However, on a subtler scale, metal pollution may result in sublethal/chronic effects, contributing to a suite of human ailments, leading Nriagu (1988) to question if we are falling prey to "a silent epidemic of environmental metal poisoning."

Not surprisingly, there is much interest in determining which metals have been increasing in which ecosystems, and tracking the timing and trajectories of these changes. An equally important issue has been to determine the natural metal loads to lakes, before human impacts, so that mitigation efforts can have realistic targets.

In this chapter, I summarize some representative studies that have tracked metal contamination using paleolimnological techniques. Because mercury (Hg) shares properties with both metals and persistent organic pollutants (Chapter 9), this element is discussed separately in Chapter 10.

8.2 Metals and lake sediments

Metals may be deposited on the water's surface via aerial transport, or directly supplied from the catchment via the weathering of rocks and other processes, as well as from industrial sources. Once in the water column, they are typically scavenged by particles and deposited in sedimentary profiles. Consequently, sediments often have metal concentrations several orders of magnitude higher than those in the overlying water (Rognerud & Fjeld 2001). As will be shown in this chapter, the veracity of the sedimentary record appears to be quite good for some metals, at least in qualitative terms (e.g., trends). Nonetheless, it has been well established that not all the metals in a water column will be archived in a lake's sediments; Nriagu and Wong (1986) estimated that about 40–60% of pollutant metals entering their study lake were actually retained by the sediments. Even once metals are deposited, diagenetic and post-deposition mobility can occur with some elements, and this is an area of active research (e.g., Belzile & Morris 1995; Boudreau 1999). The veracity of the sediment record will, of course, depend on the properties of the metal being studied and the environmental conditions in the water column and the sediments. A good example of post-depositional mobility is aluminum, whose solubility is partially pH dependent. Consequently, aluminum profiles can migrate in sediments when lakes acidify (Chapter 7), rendering their geochemical profiles problematic, at best, for interpretation. Downward diffusion of zinc is also problematic (Carignan & Tessier 1985), and other examples can be cited (e.g., Carignan & Nriagu 1985). Nonetheless, the stratigraphic profiles of several metals are often interpretable, at least in a qualitative sense. In addition to simply analyzing the concentrations of various metals in sedimentary profiles, paleolimnologists have also used sequential extractions (Boyle 2001) and stable isotope ratios (e.g., $^{206}Pb/^{207}Pb$ ratios) to glean additional information on, for example, the source of the pollution. Such data can be used to estimate natural, background metal concentrations, as well as to reconstruct past contamination trends. In fact, much of what we know about changes in long-term metal contamination has been deciphered from the paleolimnological record.

8.3 Regional-scale patterns of metal contamination in Norwegian lakes

Norway is blessed with a very high concentration of lakes, spanning a wide range of limnological characteristics and bioclimatic zones. Local land-use disturbances can often be quantified, but Norwegian lakes are also under significant threats of pollution via long-range transport. In the previous chapter, paleolimnological data were used to show how acidic deposition from other parts of Europe have affected many Scandinavian water bodies – but this is only part of the story, as wind currents also carry a suite of other pollutants, such as metals. With over 40,000 lakes greater than 0.04 km^2 in surface area, it is a daunting task for Norwegian lake managers, armed only with neolimnological techniques, to determine whether airborne pollution has affected these lakes or if elevated metal concentrations simply reflect natural levels, related to the local geology.

As described in other parts of this chapter, detailed geochemical analyses of sediment profiles can be used to pinpoint the timing and magnitude of past contamination for a wide range of metals. Stratigraphic studies of this nature are often the only ways to answer specific questions regarding the impact and timing of pollution in a particular lake (e.g., what happened and when). However, if the questions are more regional in nature (e.g., whether metal concentrations in Norwegian lakes have increased since pre-industrial times and, if so, what patterns can we discern from the data on a landscape scale), detailed, centimeter-by-centimeter analysis on hundreds of dated lake sediment cores is rarely a feasible option. Instead, the most cost- and time-effective approach to addressing these large-scale problems is to use "top/bottom sediment" analyses for a large number of lakes (Box 4.1). This "before and after" type of regional analysis was effectively used in acid rain work (Chapter 7), and has been applied to other lake management issues, such as eutrophication (Chapter 11). The following example shows how it can also be used on a regional scale to track atmospheric metal pollution.

As part of the national survey of Norwegian lakes, Rognerud and Fjeld (2001) sampled water chemistry and sediments from 210 lakes spanning the entire length of Norway. A number of criteria were used in the sampling, with the overall goal of obtaining a representative sampling of lake types, spanning a broad gradient of lake sizes and other limnological variables. However, because this survey focused on long-range transport of pollutants, lakes with significant local sources of pollution were excluded.

Gravity sediment cores were collected from each lake, and extruded and sectioned in the field. To document historical changes in metal deposition, geochemical analyses were performed on the surface 0.5 cm of sediments (representing present-day conditions; i.e., the "sediment core tops") and a 1-cm section from the deeper part of the core (usually within the 30–50 cm level in the core, representing the natural, background, or reference conditions in the lake; i.e., the "sediment core bottoms"). An enrichment in metal concentrations in the surface "top" sediments relative to the background "bottom" reference sediments was interpreted as indicating increased atmospheric deposition.

Rognerud and Fjeld (2001) found significant enrichment of several trace elements in the recently deposited sediments (Fig. 8.2). The relative increases, in descending order, were most pronounced in Sb (antimony), Pb (lead), Bi (bismuth), As (arsenic), Hg (mercury, which is treated in Chapter 10), and Cd (cadmium). Median concentrations in surface sediments were two to seven times as large as those recorded in the bottom, reference sediments. Smaller, but still significant, increases were found for Cu (copper), Zn (zinc), and Ni (nickel). Because the sampling design precluded lakes with local sources of contamination, these enrichments were most likely related to three factors: (i) deposition of atmospherically transported elements; (ii) chemical reactions and other diagenetic processes occurring in the sediment profiles; and (iii) accelerated release of these elements from the lakes' catchments due to increased mobilization from soils and bedrock. Regardless of the mechanisms, these data showed that metal concentrations were higher in the surface sediments of many Norwegian lakes.

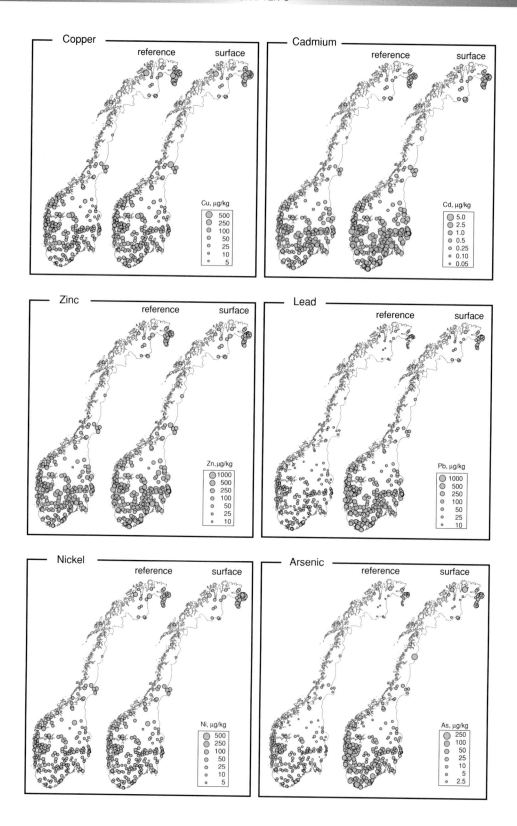

Principal components analysis (PCA) was used to determine which elements co-varied in their distributions. Hg, Pb, Si, As, and Bi formed a coherent group, reflecting a close co-variation of these elements in the surface sediments (but not in the pre-industrial reference sediments). Highest enrichments occurred in lowland areas of southern Norway (Fig. 8.2). The patterns in the recent sediments, which follow predictions based on estimated atmospheric Pb deposition, were also correlated with the loss-on-ignition (LOI) estimates for the sediments, suggesting that these elements are partly correlated with organic matter content. Cd and Zn concentrations were most enriched in lakes from southeastern Norway (Fig. 8.2), probably reflecting the greater influence of atmospheric deposition of these elements from Eastern Europe. Ni, Cr, and Cu concentrations were only slightly elevated relative to background concentrations (Fig. 8.2), with the most pronounced enrichment in the vicinity of smelters in southern Norway and close to the Russian smelters on the Kola Peninsula in the northeast. Co, V, and Mo did not record any significant elevations in recent concentrations.

This regional survey showed that many of the elements that health and environmental managers are concerned with have significantly increased in Norwegian lakes as a result of long-distance transport. Such data provide compelling evidence of the magnitude of change. Given the cost-effective nature of such "top/bottom analyses," a large number of lakes can be analyzed. However, to understand when concentrations increased, more detailed, contiguous sediment analyses are needed, as discussed below.

8.4 Aerial deposition of metal contaminants from point sources: The Sudbury (Ontario) nickel and copper smelters

Some of the most dramatic evidence of metal and acid pollution has been gathered from the Sudbury (Ontario, Canada) region, where some of the largest reserves of nickel were discovered in 1883 as the new Canadian Pacific Railway was routed through this region. The high-quality ore was initially sent to other regions for smelting, but in 1888 the first roast yards (open areas where the ore was roasted) were operational. Soon a number of local smelters were constructed. Copper was the first metal to be smelted, but by the 1890s the commercial value of nickel was also recognized, and so it too became a major commercial product for the region. The effects of this acidic deposition caused by these tremendous point sources were discussed in Chapter 7. However, in addition to the sulfuric acid that was being deposited in the region, the smelters were also emitting a suite of metals.

Because of the extent of deposition and the striking biotic repercussions, some of the earliest work on acid and metal pollution was undertaken in the Sudbury region. For example, Fig. 8.3 shows the results of some of earliest lake paleo-metal work done in Sudbury from the early 1980s. Nriagu and Wong (1983) showed that other metals associated with the smelting process had increased in recent sediments of Sudbury lakes. They documented three- to 18-fold enrichments in selenium in the post-colonial sediments. Subsequent geochemical analyses have recorded declines in some of these metals in the surficial

Fig. 8.2 (*opposite*) Maps showing the concentrations of copper, cadmium, zinc, lead, nickel, and arsenic in the surface (0.0–0.5 cm) sediments, representing recent limnological conditions, and bottom, reference (30–50 cm) sediments, representing pre-industrial conditions, in 210 Norwegian lakes. Marked enrichment of many elements is evident from these analyses. The areas of the symbols used are proportional to the square roots of the concentrations. Modified from Rognerud and Fjeld (2001).

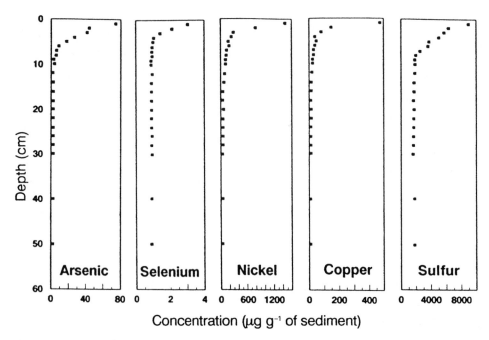

Fig. 8.3 Concentration profiles of a number of representative metals, as well as sulfur, from the sediments of McFarlane Lake in Sudbury. These data are amongst the earliest such studies done in the region, showing the increase in metals, coinciding with the local smelting operations. Later analyses of these sediments have also recorded declines in metal concentrations in the surficial sediments, coinciding with the striking declines in emissions following successful implementation of a number of mitigation strategies. The data for these profiles were originally published in Nriagu et al. (1982), Nriagu (1983), Nriagu and Coker (1983), and Nriagu and Wong (1983). From Belzile and Morris (1995); used with permission.

sediments (Nriagu & Rao 1987), coinciding with the successful mitigation procedures that have been implemented in Sudbury (Gunn 1995). However, as noted below, geochemical profiles should be interpreted cautiously, especially in very acid lakes, as the efficacy of metal retention is pH dependent (Dillon et al. 1988).

Because the Sudbury lakes were simultaneously impacted by acidic and metal deposition, and since some sedimentary metal profiles could be affected by pH declines (e.g., Carignan & Nriagu 1985; Carignan & Tessier 1985), it would be useful to have other types of paleolimnological data to reconstruct trends in metal pollution. Dixit et al. (1991) noted that the abundances of some diatom species in their surface-sediment calibration sets appeared to also be tracking the concentrations of some metals, such as Cu, Ni,

and Al, although with less statistical strength than the assemblages tracked lakewater pH. They reasoned that these transfer functions were sufficiently robust to at least track overall trends in some of these metals, and several examples have been published (e.g., Dixit et al. 1991). Similar approaches have been used in other environmental settings, such as tracking waste metal pollution to an Antarctic marine embayment (Cunningham et al. 2005).

8.5 Noril'sk, Siberia: limnological effects of smelter emissions on alkaline lakes

Noril'sk is situated approximately 300 km north of the Arctic Circle, on the southern portion of

Fig. 8.4 The metal smelting facilities at Noril'sk, Siberia. Photograph taken by B. Zeeb.

the Taymyr Peninsula in central-northern Siberia. The area is a major center for mining and smelting of high-sulfur nickel and copper ores, and is considered to be one of the most polluted regions on this planet. In fact, the Noril'sk Mining and Metallurgical Complex (Fig. 8.4) is the largest such industrial complex in Russia (Kunilov 1994), and is presently the largest point source of sulfur dioxide emissions in the world (Bouillon 1995). Since the mid-1930s, mining and smelting of the high-sulfur ores have resulted in emissions of heavy metals (e.g., nickel, copper, cobalt, zinc, cadmium, lead, mercury) and acidic compounds (e.g., oxides of sulfur, nitrogen, carbon; Dauvalter 1997a). Mining and smelting activities similar to those at Noril'sk have resulted in heavy metal contamination and acidification of thousands of lakes near Sudbury, Ontario (Keller & Gunn 1995), on the Kola Peninsula (Dauvalter 1992, 1994) and some, but not all lakes in Finnish Lapland (Dauvalter 1997b; Korhola et al. 1999). However, water chemistry analyses (Duff et al. 1999) showed that the Noril'sk-area lakes are

typically alkaline (range from pH 7.53 to pH 9.03) with low aqueous metal concentrations. Top/bottom geochemical analyses of Noril'sk area lake sediments indicated that the sediments were acting as sinks for most of the metal contaminants, as concentrations of metals (copper, nickel, cobalt, barium, zinc) were enriched in surface sediments relative to pre-industrial sediments (Blais et al. 1999). Sediment metal concentrations, particularly nickel, copper, barium, cobalt, and chromium, were distributed in a point-source pattern around the Noril'sk smelters, and were similar in concentration to those of other smelting regions (Blais et al. 1999).

In an attempt to further characterize any limnological changes that may have occurred in the Noril'sk study lakes, Michelutti et al. (2001a) analyzed the diatom assemblages preserved in the top/bottom samples of 17 lakes. They found that, relative to other regions of intense mining and smelting activities (such as the Sudbury examples noted earlier, and work in acid-sensitive mountain regions of Kola North, Russia; Moiseenko et al. 1997), the Noril'sk diatom assemblages have experienced relatively little change since pre-industrial times.

The Noril'sk lakes appear to be well protected against the effects of acidification due to their strong buffering capacities as a result of the surrounding bedrock and overlying glacial deposits. The alkaline nature of the lakes appears to have suppressed the environmental availability of metals, as most metals formed insoluble metallic complexes that were incorporated into the lake sediments. Although changes between recent and pre-industrial diatom assemblages were recorded for some lakes, the taxa contributing to the greatest amount of this dissimilarity were small, benthic *Fragilaria* species. Watershed disturbances resulting from mining activities, such as increased erosion, likely accounted for these species shifts over time. Thus, due to the high buffering capacity of this region, the water quality of the Noril'sk lakes has remained relatively unaffected, despite intense pollution from mining and smelting operations.

8.6 Copper toxicity and related pollution from industrial effluents

A common form of metal pollution is the direct discharge of untreated effluent from industrial sources into receiving waters. Lake Orta, in northern Italy, is an interesting example of such pollution. Following the establishment of a rayon factory in 1926, this lake has experienced high levels of metal pollution and subsequent acidification. Reports dating back to 1930 (summarized in Manca & Comoli 1995) attest to the large-scale ecological damage that occurred at this time, which was believed to be due primarily to copper toxicity, with much of the biota decimated or extirpated. In addition to the large amounts of copper that were discharged, ammonium sulfate in the effluent caused nitrate to accumulate in the lake, resulting in progressive acidification after 1960 (down to a pH of 4.0), and the depletion

of hypolimnetic dissolved oxygen through NH_4 oxidation. In 1958, a recovery plant was built that greatly decreased the amount of copper entering the lake by about an order of magnitude. However, new electroplating industries also began discharging into the lake in the late 1950s, and this continued until 1980, when new technologies were introduced that greatly reduced the loadings of these pollutants as well. In 1989–90 the lake was limed as a further attempt at reclamation. Lake Orta therefore provides an opportunity to show how paleolimnological techniques can assess biological responses to impacts from different industries, processes, and mitigation efforts.

Geochemical analyses of the lake sediments clearly showed the marked rise in copper concentrations in the late 1920s, followed by an equally striking decline after the installation of reclamation procedures at the rayon factory (Guilizzoni & Lami 1988). Fossil pigments tracked the marked changes in algae and cyanobacteria following pollution (Guilizzoni et al. 2001). For example, nuisance-causing colonial cyanobacteria increased in the recent sediments, as did carotenoids characteristic of green algae. In a study of the impact of metals on diatoms, Ruggiu et al. (1998) showed that, prior to the establishment of the rayon factory, the diatoms in Lake Orta were similar to those found in other deep, sub-alpine lakes. However, there was an almost 100% changeover in taxa with the release of industrial effluents in 1928. As the lake only acidified after 1960, these early changes were related to the acute copper pollution occurring at this time. Ruggiu et al. clustered the diatom responses into three groups: (i) taxa that were quickly extirpated by the discharge, such as many *Fragilaria* and *Cyclotella* species, many of which had been common in the lake for centuries; (ii) taxa that did not appear to be greatly affected by the high copper concentrations, such as some *Synedra* species; and (iii) taxa that appeared to thrive under the new environmental conditions, most notably *Achnanthes* species. A further observation was that, although taxa such as the

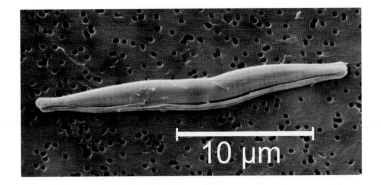

Fig. 8.5 A scanning electron micrograph of a morphologically deformed (teratological) frustule of the diatom *Synedra tenera* from the Lake Orta sediments. Photograph by Antonella Cattaneo.

Synedra species did not decline in numbers with copper pollution, many frustules (up to 45%) were deformed (i.e., teratological forms) after 1928 (Fig. 8.5). Interestingly, similar deformities in diatoms were recorded following metal contamination in Canada (Cattaneo et al. 2004) and Finland (Salonen et al. 2006). Although much more work remains to be done on the causes and range of deformities, such observations raise the possibility of eventually using morphological changes to track specific pollutants in lake systems.

Equally dramatic changes were recorded in cladoceran communities (Manca & Comoli 1995). The late 1920s in the sediment core were delineated by the disappearance of most Cladocera, even taxa believed to be tolerant of pollution. Other invertebrates, such as oligochaetes, represented in lake sediment by their cocoons, were also extirpated within a few years (Bonacina et al. 1986). Although cladoceran remains were still low, some recovery began around 1960 after the copper reclamation plant became operational. Early recolonization was dominated by one chydorid species (*Chydorus sphaericus*), which began around 1960, followed by increases in typically planktonic species by around the 1980s. These recent changes corresponded to the time that stricter standards were implemented on pollution abatement.

The Lake Orta sedimentary archive also provided an important data source for evaluating ecological theory. It has been suggested that, with ecosystem stress, the size distributions of

organisms should decrease, with a succession to short-lived, fast-reproducing, small organisms (Odum 1985). Using the size distributions of diatoms, cladocerans, and thecamoebans preserved in the Lake Orta sediments, Cattaneo et al. (1998) showed that the size distribution of the biota decreased markedly once pollution began in the late 1920s (Fig. 8.6). Not only did size classes across these broad taxonomic and trophic groups change, but the size of some individual species also decreased (e.g., the mean length of the diatom *Achanthes minutissima* decreased from 14 µm before pollution to a minimum of 9 µm between 1950 and 1970). Prior to this study, the link between size reduction and stress was based on short-term and laboratory experiments. Paleolimnological approaches provide much longer data sets, with many other potential applications for evaluating ecological theory.

Lake Orta provides a clear example of the magnitude of biological changes that can occur with copper pollution. The original invertebrate and diatom communities were extirpated. Although some recovery occurred in the recent paleolimnological record, coincident with mitigation efforts, the current assemblages are dramatically different from those that thrived in the lake before 1928. This study also raises the possibility that, in addition to simply describing species assemblage changes in lake sediments, it may eventually be possible to make inferences from the morphological variability of indicators (e.g., teratological forms, size), and so glean further

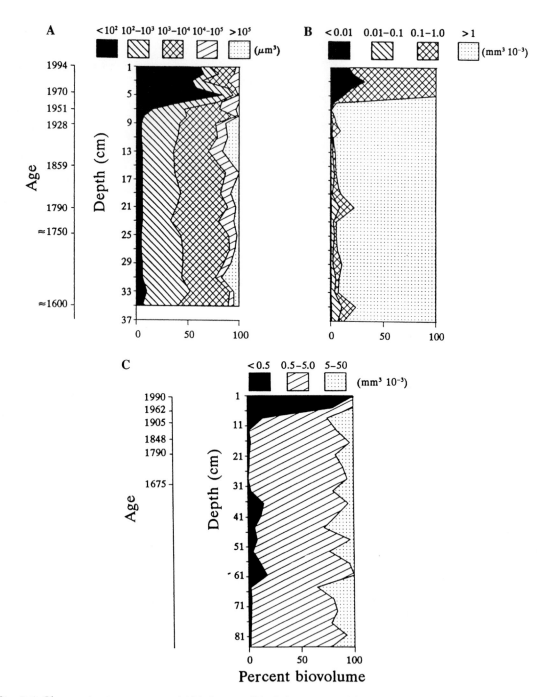

Fig. 8.6 Changes in size structure of (A) diatom, (B) cladoceran, and (C) thecamoebian assemblages in two sediment cores from Lake Orta, Italy. All assemblages record a marked shift to smaller organisms with the onset of pollution. Because both planktivorous and piscivorus fish were practically absent by this time, the succession to smaller organisms cannot be explained by size selective predation, but most likely to the resistance of smaller organisms to contaminants. From Cattaneo et al. (1998); used with permission of ASLO.

environmental information from the sedimentary record.

8.7 The medieval metal industry and atmospheric lead pollution in northern Europe

Paleolimnological approaches have also changed our historical perspectives on the relative magnitudes of past metal pollution. For example, Brännvall et al. (1999) used lead concentrations and stable lead isotopes ($^{206}Pb/^{207}Pb$ ratios) of annually laminated (varved) sediments in northern Sweden to document decadal level changes of atmospheric lead pollution over the past three millennia (Box 8.1, Fig. 8.7). A subsequent study (Brännvall et al. 2001) expanded this initial survey to 31 lakes covering most of Sweden, as well as one lake in northwest Russia, providing a comprehensive 4000-year inventory of atmospheric lead pollution for this region. As noted by the authors, the common opinion persists that large-scale atmospheric pollution is a problem that began during the Industrial Revolution of the

---

Box 8.1 The medieval metal industry and its effects on remote lakes: using stable lead isotope and total lead concentration data in paleoenvironmental assessments

Long-term trends in past deposition of certain metals can be inferred from total concentrations, as well as stable isotope ratios. The strongest interpretations are possible when these techniques are used in unison.

Lead is a suitable pollution indicator as it is comparably easy to analyze and is relatively non-mobile in lake sediment profiles. Furthermore, it is emitted from many sources, including the mining and metal industries, and from coal and other fossil fuel combustion. These pollution sources of course emit many other types of airborne pollution, but lead is easier to analyze as a proxy for this pollution.

Northern Scandinavia has always been sparsely populated, with few local emission sources of metals in pre-industrial times. The region also contains many lakes with annually laminated sediments, thus allowing for high-resolution studies of environmental change. Together, these factors make it an ideal study area for a paleoenvironmental assessment of long-term metal deposition.

The data presented in Fig. 8.7 are $^{206}Pb/^{207}Pb$ isotope ratios and "pollution lead" concentrations in sediment cores from annually laminated cores from northern Sweden. The results are plotted on calendar time scales, based on the annual varved couplets in the sediment profile. The authors used two approaches to track past metal deposition in these remote lakes: total lead concentration and the ratio of $^{206}Pb/^{207}Pb$ stable isotopes.

"Pollution lead" was distinguished from "natural lead" using the concentration and the stable isotope data. Stable lead isotope analyses can be used in environmental pollution studies to track or "fingerprint" past emission sources, as different sources of lead contain different isotope signatures. For example, in Sweden, recent lake sediments have $^{206}Pb/^{207}Pb$ isotope ratios typically between 1.1 and 1.2, similar to those found in aerosols, surface soils, and snow. These relatively low ratios distinguish atmospheric lead pollution from local, natural sources. In contrast, the natural, background signatures of $^{206}Pb/^{207}Pb$ isotope ratios in unpolluted sediments are higher, at about 1.5. These ratio differences allow paleolimnologists to use these data to distinguish between natural soil-derived lead and atmospherically derived lead pollution.

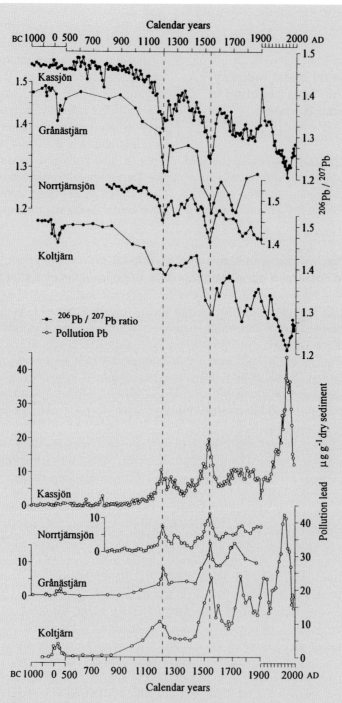

Fig. 8.7 The ^{206}Pb/^{207}Pb isotope ratios and pollution lead concentrations in annually laminated sediment cores from four northern Swedish lakes. From Brännvall et al. (1999); used with permission. Copyright 1999 American Chemical Society.

Brännvall et al. (1999) calculated "pollution lead" by applying a simple mixing model:

Pb pollution concentration = [(Pb ratio$_{sample}$ − Pb ratio$_{background}$)/Pb ratio$_{pollution}$
− Pb ratio$_{background}$)] × Pb$_{sample\ concentration}$

The background ratio for each lake is the mean isotope of unpolluted sediments (i.e., old sediments, where the $^{206}Pb/^{207}Pb$ isotope is high and stable).

The profiles clearly showed that the first signs of non-Swedish airborne lead date to about 3500–4000 years ago, which is probably related to increased long-range transport of soil dust from mainland Europe and the British Isles following forest clearance for agriculture. Airborne lead from Greek and Roman cultures occur about 2000 years ago; a pollution signature that is also evident in other lake sediment (Renberg et al. 1994) and peat core (Shotyk et al. 1998) studies.

Following the decline of the Roman Empire, there was a fairly long period (c. AD 400 to AD 900) of relatively "clean" conditions, and isotope ratios increase to approximately natural levels. However, with the onset of the medieval period, from about AD 900 to AD 1200, there was a large increase in lead pollution, as evidenced by a marked decline in isotope ratios. This period coincides with a time of rapid growth in Europe, which included expansion of mining and the discovery of new ore deposits, and the reopening of old mines. After AD 1200, coincident with economic declines, the trend was reversed and lead pollution was estimated to drop by about 50%, with a marked increase in stable lead isotope ratios. Part of this decline could also be related to the Black Death (c. AD 1350), when about 25% of the European population died. However, after c. AD 1400, the trends once again reversed as pollution lead increased with economic expansion, culminating in a minimum in $^{206}Pb/^{207}Pb$ and a maximum in lead concentrations at AD 1530. This corresponds to the peak in European metal mining. Thereafter, European mining and metal production declined with the expansion of American mining, and the pollution indicators in the core declined until the modern period (c. AD 1600 to the present). Following a minimum around AD 1600, there is a second peak in the 18th century, followed by less consistent trends, probably reflecting the use of more local sources of lead during recent times. Reductions in leaded fuel and other mandated declines in lead emissions are recorded by recent declines in pollution lead.

There is no doubt that atmospheric lead pollution increased in northern Sweden during the Industrial Revolution. However, when seen in this historical perspective, it was not as high as is often assumed. These data also provide important background data on past, natural lead concentrations, showing that one would have to use pre-impact lake sediments to estimate natural levels, as opposed to trying to simply use surface sediments, even in these seemingly remote regions of Sweden.

19th century, and that metal concentrations in remote localities, such as northern regions, represent background levels. Their data showed this is not the case. Brännvall et al. (1999, 2001) recorded an unambiguous signal in the sediments of airborne pollution from Greek and Roman cultures around 2000 years ago, with the first traces of this pollution beginning about 3500–3000 years ago. This was followed by a relatively "clean" period from AD 400 to AD 900. Thereafter, there was a clear and permanent increase in the fall-out of atmospheric lead, with peaks similar to current concentrations at AD 1200 and AD 1530, corresponding to peaks in metal production in Europe. Historically, lead deposition reflected trends in European economies.

The above studies show that metal pollution in northern Europe has its origins in antiquity, and that it was the medieval period when major metal pollution began, and not the later Industrial Revolution (although the latter clearly increased deposition). In fact, it was long before the Industrial Revolution that atmospherically deposited "pollution lead" was the dominant source of lead in northern Scandinavia (Brännvall et al. 1999, 2001), and undoubtedly also in places closer to the lead pollution sources. Furthermore, the paleolimnological data revealed a consistent geographical pattern, with higher deposition of atmospheric lead in the southern Swedish lakes, thus indicating that the main sources of pre-industrial atmospheric lead pollution were cultural areas in mainland Europe and the United Kingdom. These data are important for establishing realistic "background" levels, and for placing current pollution trends in context. Fortunately, lead deposition has been steadily declining since the 1970s as a result of stricter emission controls and the reduction in lead additives in fuels; the sediment profiles can be used to track this recovery as well (Fig. 8.7).

8.8 Assessing atmospheric transport and source of lead pollution in eastern Canada

Similar approaches to those just described have also been used in North America to track the deposition patterns and sources of lead pollution. For example, in a survey of 47 lake cores across Quebec and Ontario (Canada), Blais and Kalff (1993) estimated that surface Pb enrichment averaged 14 times the background Pb concentrations. They concluded that naturally derived lead constituted only about 7% of the lead in the

lakes' surface sediments. Blais (1996) expanded on these studies by incorporating $^{206}Pb/^{207}Pb$ isotope ratios, similar to the Swedish studies just described, in order to track the origin of the lead pollution. The isotopic $^{206}Pb/^{207}Pb$ ratios for Canadian industrial emissions are, on average, significantly less than the ratios from US emissions (Blais 1996). By using a simple isotope-mixing model, he determined the relative contributions of emissions from both countries using the isotopic signal preserved in 32 Ontario and Quebec sediment cores. These data suggested that over 50% of the total lead burden on the Canadian lakes was from US sources.

8.9 Intensive pre-Incan metallurgy and associated metal pollution in the New World

Just as metal pollution in Europe can be tracked back into antiquity (Brännvall et al. 1999, 2001; Box 8.1), paleolimnologists have also reconstructed the long-term environmental effects of metallurgy in the New World.

The importance of silver to the history of South America, and especially the Inca culture, is well known. Silver was used extensively in pre-Columbian South America in the 15th and 16th centuries, as the metal was highly prized by the Inca rulers for religious and symbolic rituals, jewelry, and other artifacts. It is also well known that, following Pizarro's conquest (AD 1545), silver was a major source of plunder that enriched Spanish expansionism. However an older chapter to the history of silver metallurgy was recently uncovered by Abbott and Wolfe (2003), who used paleolimnological techniques to track the history of metallurgy and its environmental impacts from the Bolivian Andes.

Cerro Rico de Potosi, which translates loosely to "the mountain of silver," is located in the southern Bolivian Andes and probably represented the richest silver deposit in the world, where initial ore grades contained over 20% silver by mass. Archeological artifacts have richly documented the Incas' use of silver from this deposit, but little was known about any possible pre-Incan use of metals or of any associated environmental impacts of long-term smelting in this region. These questions prompted Abbott and Wolfe (2003) to track the history of metallurgy in this culturally important region of the New World, using geochemical approaches from a high-resolution sediment core from Laguna Lobato, located 6 km east of the silver deposits at Cerro Rico de Potosi. Geochronology was established using ^{210}Pb, ^{137}Cs, and ^{14}C methods, and the sediments were analyzed for five metals (Fig. 8.8) associated with the smelting process: silver (Ag), bismuth (Bi), antimony (Sb), tin (Sn), and lead (Pb). The latter was a key element in their investigation, as a lead sulfide flux (galena, locally called *soroche*) was added to ores during smelting in wind-drafted clay kilns, probably to reduce volatile losses of precious silver. However, Pb would be released to the atmosphere during this process, which is well recorded in downwind lake sediment samples, as Pb concentrations were orders of magnitude higher than the other trace metals (Fig. 8.8).

Prior to AD 1000, the concentrations of all five trace metals associated with metallurgy were low in the Laguna Lobato sediment core, reflecting natural pre-anthropogenic conditions for metal accumulation in this region (Fig. 8.8). However, shortly after AD 1000, metal concentrations rose to well above background levels, with a distinct peak around AD 1130–1150; a profile that resembles the enrichment trends observed in Europe near medieval mining sites (e.g., Ek & Renberg 2001; see pp. 109–1). In the Andean highlands, this period coincides with the late stages of the Tiwanaku Empire, whose workers were apparently smelting the ore several hundred years before the Inca.

The demise of the Tiwanaku, and the associated decline in smelting activities (Fig. 8.8), has been linked to drought conditions in this region, based on proxy data gathered for this time period from the Quelccaya (Peru) ice core (Thompson et al. 1985) and low water stands in Lake Titicaca

(Abbott et al. 1997). The sediments accumulating between the Tiwanaku and the subsequent Inca regime (c. AD 1100–1400, the Altiplano Period) continue to record evidence of smelting, but at a much reduced rate, reflecting the political disorganization at this time, as well as repercussions from persistent drought. However, the rise in Inca metallurgy and the subsequent period of smelting following the Spanish conquest in AD 1545 is clearly recorded by the increases in Pb and Sb concentrations (Fig. 8.8).

By AD 1572, most silver-rich surface ores had been depleted, and silver production progressively declined until the abandonment of the deposits in AD 1930. However, large-scale tin production became important at Cerro Rico de Potosi in the late 19th and early 20th centuries, especially during the high demands for tin during World War I, but was then abandoned following the crash of the tin industry in AD 1950. Once again, the lake sediments faithfully recorded these changes (Fig. 8.8).

The Abbott and Wolfe (2003) study is important from several perspectives. First, it extends the record of metal smelting in the Andes back several centuries, to well before the Inca period, thus forging an important link to ongoing archeological work in the region, especially given the paucity of the surviving artifacts. Second, the researchers estimated that pre-Incan metallurgy had extracted several thousand tons of silver. This provides a historical perspective to the scale of environmental impacts exerted by a culture that flourished about 1000 years ago. Third, it clearly documents fluctuations in airborne pollutants, such as Pb, extending back many centuries. Clearly, not all pollution is a recent phenomenon, not even in the mountains of Bolivia.

8.10 Tracking past metal pollution from peat profiles

The focus of this book is lacustrine sediments, but we should not lose sight of the fact that

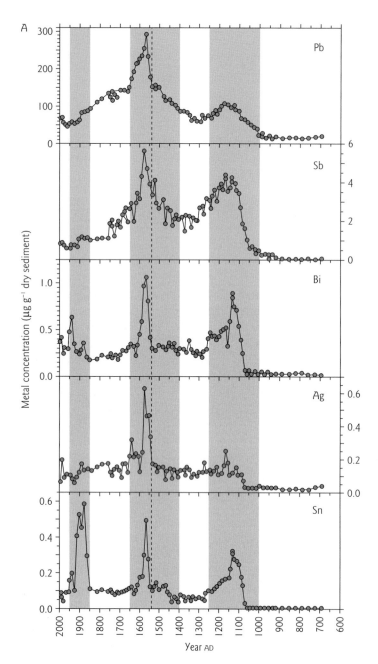

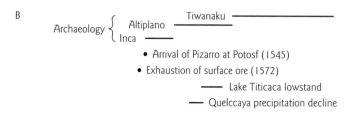

Fig. 8.8 The history of metal pollution from Laguna Lobato, Bolivia. (A) Geochemical concentrations of lead (Pb), antimony (Sb), bismuth (Bi), silver (Ag), and tin (Sn) for the past c. 1300 years of sediment accumulation. The three shaded zones identify three distinct metallurgical zones: Tiwanaku (AD 1000–1250), Inca – early Colonial (AD 1400–1650), and the rise and crash of tin mining (AD 1850–1950). The vertical dashed line separates Incan from Colonial periods. Reprinted with permission from Abbott, M.B. and Wolfe, A.P. (2003) Intensive pre-Incan metallurgy recorded by lake sediments from the Bolivian Andes. *Science* 301, 1893–5. Copyright 2003 American Association for the Advancement of Science. (B) Generalized archeology and historical events for the southern Andes region. The Lake Titicaca lowstand and the Quelccaya ice core precipitation decline indicate periods of extended droughts.

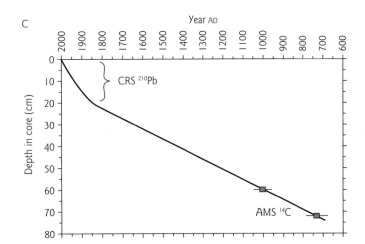

Fig. 8.8 (*continued*) The history of metal pollution from Laguna Lobato, Bolivia. (C) Geochronology for the sediment core was based on the constant rate of supply (CRS) ^{210}Pb model as well as two AMS ^{14}C dates.

many other natural archives contain very important information on pollution. A separate volume could easily be written on, for example, peatlands, and how they can be used to reconstruct environmental trends. Given the scope of this book, I will only show a few examples of how peat profiles can be used to track past pollution. Furthermore, there is no doubt that ice cores, such as those retrieved from Greenland, provide excellent, long-term, high-resolution proxies of the past deposition of metals, as well as other pollutants. However, ice cores are restricted to high-latitude regions. Peatlands occur throughout large parts of this planet wherever precipitation is sufficiently high.

Ombrotrophic peats (those receiving their water from precipitation, and not from runoff from a catchment) have proven to be excellent, high-resolution archives of some atmospheric contaminants (Bindler 2006), such as some of the metals (but not all; see Jones & Hao 1993) associated with mining and other activities (Shotyk et al. 1997; Koenig 1999). For example, Shotyk et al. (1998) analyzed nearly 15,000 years of atmospheric lead (Pb) deposition, representing the entire Holocene and part of the late glacial period, from a bog in the Swiss Jura Mountains to document a record of lead mining and smelting extending back to before the Roman Empire. Shotyk et al. used total Pb concentrations and

fluxes, as well as total scandium (Sc) concentrations and the ratio of Pb/Sc to track past deposition, as well as Pb isotopes to separate natural and anthropogenic sources. The reasoning behind analyzing for Sc is to calculate Pb enrichment factors (EF) in the peat. As summarized by Weiss et al. (1997) and Shotyk et al. (2001), titanium (Ti) bearing minerals are resistant to chemical weathering in acidic solutions, and so Ti can be used as a conservative reference element for calculating atmospheric soil dust inputs to bogs, as well as EFs of trace metals. However, in the Swiss peat core, Ti concentrations were near the detection limits obtained by X-ray fluorescence spectrometry (XRF), whereas Sc concentrations exceeded the detection limits obtained by instrumental neutron activation analysis (INAA). Because Sc also behaves conservatively during chemical weathering and could be measured more accurately in the peat profiles, it was selected over Ti as the reference element to indicate the concentration of mineral matter in the peat. An increase in the Pb/Sc ratio is assumed to reflect an enrichment of Pb in atmospheric aerosols, which is unrelated to chemical weathering, erosion, and other processes. The ratios of the stable isotopes ^{206}Pb/^{207}Pb were used to track the origin of the Pb.

Figure 8.9 summarizes the trends in Pb EFs (enrichment factors; i.e., the ratio of Pb/Sc in the peats, normalized to the background value) and the ratio of $^{206}Pb/^{207}Pb$, both of which were used to distinguish natural from anthropogenic sources of atmospheric Pb. Ten major periods of Pb deposition are clearly visible. Until 3000 ^{14}C yr BP, soil dust was the most important source of Pb. Higher fluxes reached their maxima during the Younger Dryas (a period of cooling around 10,590 ^{14}C yr BP) and 8230 ^{14}C yr BP, periods when dust deposition would have been high. Around 3000 ^{14}C yr BP, $^{206}Pb/^{207}Pb$ declined and

Pb EFs increased. From this point on, anthropogenic sources have dominated Pb emissions in Europe. The peat profile documents the deposition resulting from various types of mining activities throughout Europe, such as the early mining by the Phoenicians and Greeks, the Romans, and so on, as well as the introduction and eventual decrease in the use of leaded gasoline (Fig. 8.9). The highest Pb flux in the year AD 1979 was 1570 times the natural background value.

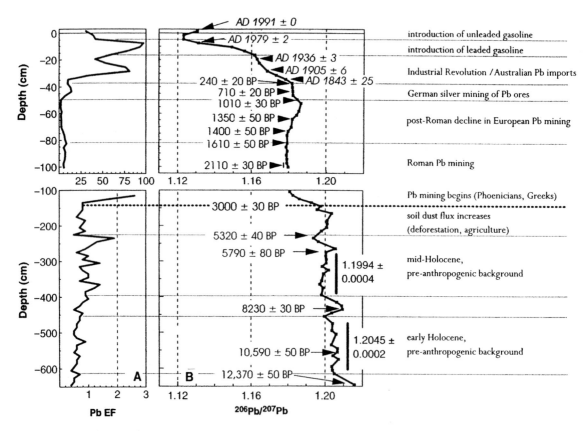

Fig. 8.9 (A) Lead enrichment factors (Pb EFs) calculated as the ratio of Pb/Sc in a Swiss mountain peat profile, normalized to the background value. An EF < 2 indicates that the sample is not enriched in Pb relative to background levels. (B) The isotopic composition of Pb, as $^{206}Pb/^{207}Pb$ graphed alongside the chronology of atmospheric Pb deposition since 12,370 ^{14}C yr BP. The heavy, horizontal dashed line at 3000 ^{14}C yr BP separates sections of the peat core where the dominant Pb source is soil dust, versus those where Pb is primarily from ore-dominated sources. Reprinted with permission from Shotyk et al. (1998) History and atmospheric lead deposition since 12,370 ^{14}C yr BP from a peat bog, Jura Mountains, Switzerland. *Science* 281, 1635–40. Copyright 1998 American Association for the Advancement of Science.

Data from peat profiles can also be combined with similar analyses from lake sediments to provide a clearer picture of past deposition. For example, Brännvall et al. (1997) compared Pb concentrations and stable Pb isotopes from three peat bogs and three lakes in Sweden. The concentration data provided similar patterns of past deposition, and closely followed the outline of historical Pb production over several millennia. Also, since there are large differences between the $^{206}Pb/^{207}Pb$ ratio between Fennoscandia and continental Europe, they were able to distinguish external sources of Pb to Sweden. In the three lake sediment cores, $^{206}Pb/^{207}Pb$ ratios and Pb concentrations followed similar but inverse patterns until about AD 1000 (Fig. 8.10). After about AD 1000, the ratio remains fairly steady at about 1.2, reflecting a dominance of pollutant Pb. A further decline to about 1.14 occurs in the 20th century as a result of the addition of alkyl-Pb to fuel. Although the patterns between lakes and peat profiles were similar for Pb concentrations, the isotopic data were different. Brännvall et al. suspected that, in ombrotophic peats, externally derived Pb from long-range transport of soil dust and atmospheric pollution has always been more important, as compared to lakes. The authors concluded that the long-term decline in the $^{206}Pb/^{207}Pb$ ratio in the peat cores was the result of the gradually increasing proportion of Pb from long-range transport of soil dust and atmospheric pollution from metal production.

8.11 Tracking declines in lead pollution following emission cutbacks: an example from suburban and urban reservoir sediments

As shown in the previous example, paleolimnology has been employed effectively to show the extent of metal pollution in a variety of settings. However, sedimentary techniques can also be used to determine if pollution abatement programs are working.

The Clean Air Act was passed in the United States in 1970 to set air quality standards and reduce emissions of various contaminants. One of the most striking declines was in Pb, where atmospheric emissions have declined 98% since 1972, the peak in Pb emissions. This has been achieved by eliminating Pb from gasoline and other industrial sources.

One of the first studies to determine if Pb concentrations were in fact decreasing in US waterways was a study by Trefry et al. (1985), who cored the sediments of the Mississippi Delta. The Mississippi River drains more than 40% of the contiguous United States, and carries more than half of the sediment and water of US rivers. Trefry et al. compared cores they collected in 1974 and 1975 (just following the peak in Pb emissions) with cores they collected and analyzed using identical procedures in 1982 and 1983. They reported that Pb concentrations began to increase in the late 1800s, which coincided with the increase in Pb mining in the Missouri Valley, and the industrial revolution in the United States. This trend increased steadily, likely fueled further by the introduction of Pb into gasoline during the 1920s. Their data, however, also suggested that Pb concentrations had declined by about 40% in the decade or so since the enactment of the 1970 Clean Air Act.

A subsequent study using sediment cores collected about a decade later extended the Pb trajectories even further. Taking advantage of the relatively fast sedimentation rates in many Midwest and Southeast US reservoirs, Callender and Van Metre (1997) used paleolimnological approaches to determine whether Pb concentrations have been decreasing in water supplies, concurrently with the known atmospheric reductions. If concentrations had decreased, they then hoped to determine how close contaminants have returned to pre-industrial levels. The sediments in all six sediment cores were characterized by uniform, fine-grained sediments, with no visible disturbances. Because the sediments appeared

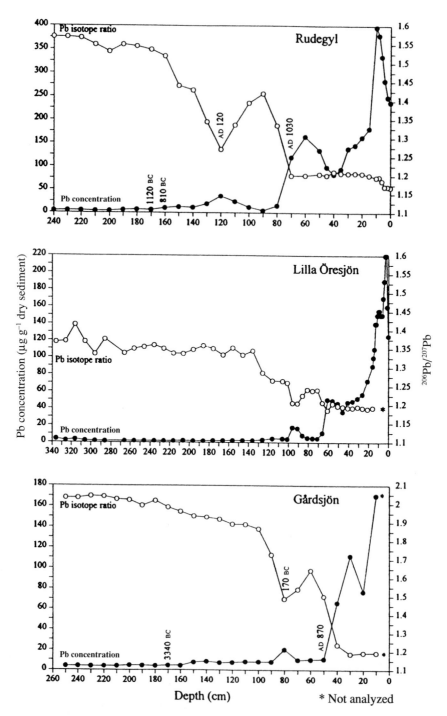

Fig. 8.10 Lead concentrations and ^{206}Pb/^{207}Pb isotope ratio data for three lake sediment cores from Sweden. From Brännvall et al. (1997); used with permission.

homogenous, Pb concentrations were not normalized and were reported simply as µg g^{-1} (Fig. 8.11). For the three urban and suburban sites (i.e., Lake Harding, downstream from Atlanta, Georgia; White Rock Lake in Dallas, Texas; and Lake Anne in Reston, Virginia, west of Washington, DC), there was a steep rise in Pb concentrations beginning in about 1950, 1960, and 1963, respectively, and equally marked declines after the peak values in 1972, 1977, and 1980, respectively. However, current Pb concentrations are still approximately double their values in pre-urbanization or early reservoir sediments. The earlier Pb peak in Lake Harding likely reflected local anthropogenic activities in Atlanta at this time. The Pb history for a rural reservoir was shown by the profile for Lake Sidney Lanier (Fig. 8.11), which is located in rural Georgia, about 90 km north of Atlanta. The trajectory of this curve is somewhat different from the urban and suburban sites discussed above, but it still forms a peak around 1970,

albeit a broad one. This profile may be expected, as rural reservoirs received substantially less airborne contaminants than urban ones, and these particulates mix with ambient, uncontaminated bedrock and soil, thus diluting the signal further. Hence, a smaller and broader peak would be expected.

The above data indicate that Pb concentrations in the water of these reservoirs has closely matched the recovery phase of Pb from atmospheric sources. For example, in the mid-1970s, Pb concentrations in ambient air were about 1.5 µg m^{-3}, but by 1990 were only about 0.07 µg m^{-3}. US leaded gas consumption peaked in 1972. However Pb concentrations in surface sediments are still about double what they were during pre-industrial times. These data provide positive support for the action taken with the Clean Air Act, but also show that significant concentrations of Pb still exist in soils and sediments, and that it will take more time to bring concentrations to baseline levels.

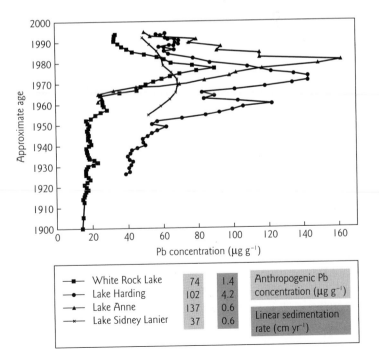

Fig. 8.11 Lead distributions in urban–suburban and rural reservoir sediments. Anthropogenic Pb in the reservoir cores represents the difference between the peak value and the baseline value before 1960. The cores were dated using ^{137}Cs and the depth of the pre-reservoir surface, defining the date of impoundment. Modified from Callender and Van Metre (1997). Copyright 1997 American Chemical Society.

8.12 Using biological indicators to help identify the source of anthropogenic metal inputs

Geochemical analyses of metals have thus far been conducted almost exclusively from bulk sediments. However, some recent work suggests that different components of the sediment matrix may be storing different types of pollution histories, with important management implications. In an exploratory study, Wyn et al. (2007) examined the historical trends in metal content in bulk sediment samples from a reservoir and two lakes in Saskatchewan (Canada), and compared these profiles to the metal concentrations of cladoceran ephippia (Fig. 5.9D) preserved in the same sediment sections. Their analyses showed that the metal content of ephippia differed substantially from the concentrations measured in the surrounding sediment matrix, with ephippia preferentially accumulating contaminants from urban or industrial sources (e.g., Cd, Cr, Mo). Meanwhile, elements associated with erosional and agricultural practices (e.g., Al, Ca, Fe, Mn) were present in higher concentrations in the sediment matrix, as opposed to concentrations measured in the ephippia. These data provide the exciting suggestion that chemical information stored in ephippia may provide paleo-ecotoxicological information that is unique from that available from bulk sediment analyses. Moreover, as *Daphnia* and other cladocerans are key components of many aquatic food webs, analyses such as these provide important information on food-web exposure to metals in the water column.

As a large number of biological indicators are preserved in lake sediments, approaches such as those used in the above example may eventually be applied to other organisms. For example, both the geochemical and isotope information stored in ostracod carapaces have provided paleolimnologists with additional environmental information (e.g., Holmes 2001), and recent work suggests that isotope analyses (and eventually possibly other physical and chemical techniques) can be used on chironomid head capsules (e.g., Fallu et al. 2004; Wooller et al. 2004) and diatom frustules (e.g., Leng et al. 2005). Yin et al. (2006) have shown that Antarctic seal hairs, which are preserved in some maritime lake sediments (Hodgson & Johnston 1997; Hodgson et al. 1998a; see Section 16.2), can also be used to reconstruct past concentrations of certain metals.

8.13 Resurrection ecology: combining descriptive and experimental paleolimnology to study impacts of mining discharge

The Keweenaw Waterway is a freshwater estuary on a peninsula jutting off the southern margin of Lake Superior. Early boat travel along the shore of Lake Superior was made difficult by the Keweenaw Peninsula, as it required a considerable detour around this land mass. However, an overland route, using Portage Lake on the peninsula, facilitated travel, as it allowed for a portage across this land barrier. Once copper mining began in the area, the southeastern and northwestern ends of the portage were channeled (first in 1859, and later improved from about 1862 to 1873), creating the Keweenaw Waterway. This 35.5 km long lake and canal system was open to navigation in 1890, and was expanded further in the 1920s, the late 1930s, and the early 1940s, resulting in an extensive freshwater estuary open to Lake Superior at both ends.

The major environmental impacts in the Keweenaw Peninsula were from copper and, to a lesser extent, silver mining as part of one of the great North American mining rushes, beginning in the late 1840s, peaking between 1890 and 1925, and ending in 1968. Between 1850 and 1929, it represented the second largest copper production region in the world. As might be expected, the mining process produced vast quantities of copper-rich stamp sands as a tailings product, with over half a billion metric tonnes of stamp sands discharged directly into the waterways from

1850 to 1968! Geochemical analyses of cores from Portage Lake, as well as cores from across eastern Lake Superior, tracked the widespread contributions of this copper to the sediments (Kerfoot et al. 1994).

Kerfoot et al. (1999) next posed the question: How did these mining effluents, as well as the eutrophication resulting from the expanding work force, affect the aquatic biota? A paleolimnological investigation focusing on cladoceran remains was initiated, supplemented by analyses of stratigraphic changes in sediment flux, copper and organic matter concentrations and fluxes, and biogenic silica concentrations, as well as changes in biotic remains such as diatoms and rhizopods.

The accumulation of biological indicators decreased moderately during the period of active stamp mill discharge, even though this represented a time of increased nutrient loading from the much-expanded local human population. Metal toxicity was likely a factor, as fluxes of the aquatic indicators only increased after 1947, when the discharge of active slime-clay ceased.

Up to this point, this case history would appear to be a standard paleolimnological investigation. Mining operations began, presumably toxic slime-clays were discharged, and the biotic community responded. However, Kerfoot et al. took the study one step further, into the realm of experimental paleolimnology. Copper toxicity was suspected to be the prime factor, based on the timing of the stratigraphic changes, but it was not yet proven. To illustrate the historical bioavailability and toxicity of copper, Kerfoot et al. tested the toxicity of sediment suspensions taken from before the mining operations began, during mining (containing quantities of the clay slimes), and after mining ceased. Bioassays were run using standard test protocols, with *Daphnia pulex*, a common cladoceran from the region, as a test organism. The results were striking (Fig. 8.12). Bioassays run with sediment from postmining (taken from a 7 cm sediment depth in Fig. 8.12) and from a pre-mining period (40 cm depth) were strikingly different from survivorship curves of bioassays using sediments from

a time when copper mining was at its height and slime-clays were being discharged and were accumulating in the sediments (33 cm). Quite clearly, the sediment layers containing the slime-clay layers were found to be highly toxic. An important factor affecting toxicity may have been the organic matter content of the sediments, as copper concentrations remained high in the sediments, even after the slime-clay discharge ceased. The authors suspected that the soluble copper interactions shifted from particle-bound dominance to more organically mediated (and less toxic) processes. As shown in Fig. 8.12, the organic-rich uppermost 7 cm were not at all toxic.

An additional component of this study was to use the fossil seed banks of past cladoceran ephippia preserved in the dated sediments to "resurrect" past populations. They found that *Daphnia* resting eggs could not be hatched from sediments older than 70 years in age, the period of most intense mining activity. These data suggest that the copper toxicity may have eliminated the seed bank of resting eggs due to copper toxicity.

The authors coined the term "resurrection ecology" for this combination of more traditional and descriptive paleolimnological approaches with experimental techniques. Using such combined approaches, cause-and-effect relationships can potentially be clarified for a wide range of environmental problems. It also illustrates the interaction (in this case, antagonistic interaction) between two different potential stressors; specifically, it showed how nearshore eutrophication and the resulting increased organic production enhanced copper complexation, and thus helped detoxify the sediments from the effects of copper pollution.

8.14 Other sources of proxy data for metal pollution

The discussion thus far has focused on sediments as archives of metal pollution. However, other

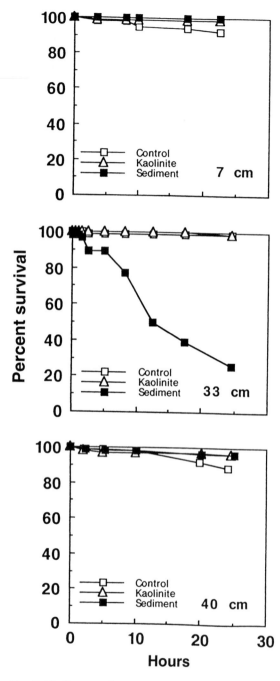

archival sources are being developed to reconstruct changes in metal concentrations of lakes and rivers. For example, museums contain collections of organisms (e.g., fish, birds, invertebrates), some of which were collected decades or even centuries ago, that can be used to measure past contaminant levels (e.g., Suarez & Tsutsui 2004), much the same way that forensic investigators use autopsies. The developing field of sclerochronology, which taps the environmental information stored in the growth rings of biological samples such as fish otoliths (e.g., Panfili et al. 2002) or mollusk shells (e.g., Ravera et al. 2005), may also provide chronologically constrained information on past contaminant loads. New approaches are continually being developed that can complement paleolimnological assessments.

8.15 Summary

Most of society's technical, industrial, and economic advances throughout history have been closely linked with our abilities to extract and work metals. However, there have also been some environmental costs, as elevated metal concentrations may result in a host of medical problems, as well as detrimental effects on a large number of other organisms. Geochemical analyses of lake, reservoir, and river sediments have been effectively used to track the trajectories of different metal pollution sources and to assess a large number of environmental problems. As in all paleolimnological studies, interpretations must be made cautiously, as not all metals entering an aquatic system will be scavenged onto sedimenting

Fig. 8.12 Survivorship patterns for bioassays using *Daphnia pulex* survival during 24-hour clay toxicity tests. The *Daphnia* clones were subjected to different inocula, using the same amount of lake sediment from post-mining (top), mining (middle, containing large amounts of clay-slime from the effluents), and pre-mining (bottom) periods. In addition, the assays were run on lake water alone (control, open squares), as well as with nontoxic clay (kaolinite, open triangles) to test for the possible toxic effects of clay alone. The marked toxicity shown in the sample using sediments from the mining period (33 cm) attests to the toxicity of the slime-clays. From Kerfoot et al. (1999); used with permission of ASLO.

particles and some sedimentary profiles may be altered by post-depositional mobility. Most studies have focused on tracking changes in metal concentrations, but other approaches, such as the use of stable lead isotopes, have been used to "fingerprint" the provenance of some pollution sources. Such studies have shown, for example, that the medieval mining industries of Europe contributed significant amounts of lead and other pollutants to the atmosphere. By combining biologically based paleolimnological approaches with geochemical studies, along with physiological investigations, the effects of past pollutants on lake biota can be assessed more effectively.

9

Persistent organic pollutants: industrially synthesized chemicals "hopping" across the planet

Scientists, physicians, and other technically informed people will also be shocked by Silent Spring – but for a different reason. They recognize Miss Carson's skill in building her frightening case; but they consider that case unfair, one-sided, and hysterically overemphatic. Many of the scary generalizations – and there are lots of them – are patently unsound. "It is not possible," says Miss Carson, "to add pesticides to water anywhere without threatening water everywhere." It takes only a moment of reflection to show that this is nonsense.
Book review, published in *Time* magazine (September 28, 1962, pp. 66–9),
of Rachel Carson's prophetic book on pesticides, *Silent Spring* (1962)

9.1 The threat from persistent organic pollutants (POPs)

Persistent organic pollutants (POPs), as their name implies, are organic substances that persist in the environment. "Persist" is, of course, a relative term, but in the case of POPs it is often defined as a pollutant with a half-life in water or sediments of greater than 8 weeks (and sometimes much longer). To say that there are many POPs that can potentially harm the environment is an understatement – some 500 to 1000 newly synthesized organic compounds are developed each year, adding to an already enormous pool of commercially developed POPs. Some of the first alarms about POPs such as insecticides were raised by the now classic book by Rachel Carson, *Silent Spring*. The above quotation from a 1962 review of this book in *Time* magazine is, in my view, a classic example of how society has a remarkable record of underestimating the

environmental seriousness of some of our actions. Yet little or no information is available on the toxic effects of the vast majority of the new compounds being synthesized and brought into production daily.

Most POPs are organochlorines (OCs), but they also include organobromines (e.g., hexabromobiphenyl, brominated flame retardants), perfluorinated acids, polycyclic aromatic hydrocarbons (PAHs), as well as other substances. POPs are characterized by their relative insolubility in water, persistence in the environment, and biomagnification in food chains. Many POPs have significant negative effects on human health, other organisms, and the environment as a whole. Moreover, it is becoming evident from paleolimnological data, as well as other types of investigations, that many POPs can be transported long distances and so they have become a global concern. The terms hydrophobic organic compounds (HOCs) and persistent bioaccumulative toxins (PBTs) are also sometimes used (although the latter includes mercury; Chapter 10), but I will continue to use the more common term POP in this chapter.

Most POPs are measured in concentrations of parts per billion or parts per trillion (also referred to as nanograms (ng) per liter, or 10^{-9} g l^{-1}). It is sometimes hard to visualize how small these concentrations are. If we compare these concentrations to time scales, one part per billion is equal to about 1 second in 32 years, and one part per trillion is equal to about 1 second in 32,000 years! To use a volumetric example, one part per billion is about one drop in a tanker truck, and one part

per trillion would be approximately equivalent to dropping one grain of salt into an Olympic-sized pool. Nonetheless, concentrations measured at these levels can severely affect ecosystems.

Four characteristics of POPs should be considered when assessing their use and potential effects:

1 *Persistence.* How long do they last, and in what form?
2 *Bioaccumulation and other biological processes.* How do the POPs change, concentrate, and magnify as they pass through the food chain (see Box 9.1)?

Box 9.1 Bioaccumulation and biomagnification of POPs

Some POPs may be present in water at such low dilutions that they cannot be measured even with our most sensitive instruments. However, many of these compounds accumulate in biological tissues (i.e., bioaccumulate) of aquatic biota (Fig. 9.1), usually in fatty tissue and certain organs. The longer an organism consumes or absorbs contaminants, the more concentrated the substance will be in its body will be (Fig. 9.1). For this reason, most health advisories warning people about the amount of fish they can consume from potentially contaminated waters often relate these guidelines to the size (which is also often related to the age) of the fish. Typically, the guidelines state that larger quantities of younger (hence less contaminated) fish can be eaten than older, larger animals, as the latter have higher contaminant loads due to the effects of bioaccumulation.

Using data from Lake Ontario (Environment Canada and the US Environmental Protection Agency 1995), PCBs are in very low concentration in the water but bioaccumulate

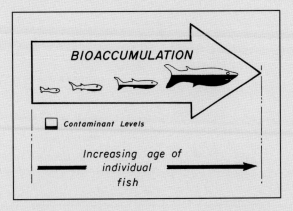

Fig. 9.1 Bioaccumulation of contaminants is the net result of more rapid accumulation than release of persistent contaminants by organisms. Biota take up contaminants from the water in which they live, and continue to bioaccumulate more contaminants as they grow. For this reason, older animals often will have higher loads of some contaminants.

in algae, where they can be measured at a concentration of about 0.025 ppm (Fig. 9.2). Small invertebrates, such as zooplankton, graze on the algae and biomagnify the concentration of PCBs further. Small fish, such as smelt, feed on the zooplankton, and are then preyed upon by piscivorous fish, and ultimately by a top predator and scavenger – the herring gull. The eggs of herring gull, which are high in fat, may have PCB concentrations about 10 million times higher than that of the lake water. Dead and malformed chicks (Fig. 9.3) are some of the indicators of POP contamination (Ludwig et al. 1996). Although most attention has focused on biomagnification of POPs in the food chain, other contaminants, such as mercury (Chapter 10), are also biomagnified.

Just as food advisories suggest that anglers preferentially eat younger, smaller fish, they also suggest that top carnivores should only be eaten sparingly. Humans have the option, in most cases, to vary their menus and avoid contaminated food. Herring gulls and other biota do not have this luxury.

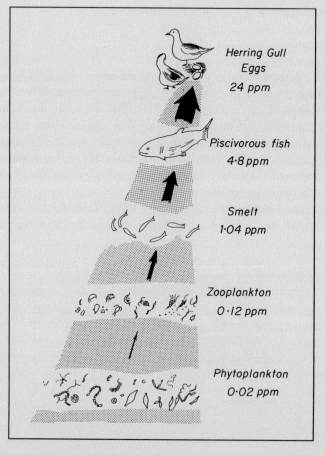

Fig. 9.2 A schematic representation of a Lake Ontario food web, showing the biomagnification of PCBs, starting with algae, to invertebrates, to smelt, to piscivorous fish, and finally to herring gulls and their eggs. Data from Environment Canada and the US Environmental Protection Agency (1995).

Fig. 9.3 Deformities, such as the "cross-billed syndrome" in this double-crested cormorant chick from the Great Lakes region, have been related to concentrations of persistent organic pollutants, such as PCBs. Photograph by J.P. Ludwig.

3   *Toxicity*. How much of the toxin, and in what form, will it take to kill organisms or cause serious damage? Toxicity can range from acute, which may result in almost immediate death, to sublethal and chronic effects, which may affect growth, breeding success, or future generations by, for example, increased mutagenicity.

4   *Volatility*. What vectors and pathways are available for the POP to be transported to other locations? As described more fully later, POPs may be deposited many thousands of kilometers away from where they were initially used.

There are 15 major groups of halogenated POPs, but these can be divided into more than 500 individual components. Common POPs that are of particular environmental concern include polychlorinated biphenyls (PCBs), which are mixtures of chlorinated hydrocarbons with a biphenyl backbone. PCBs have been used extensively since 1930 in a variety of industrial applications, such as dielectrics in transformers and large capacitors, heat exchange fluids, paint additives, and many other uses. DDT (dichlorodiphenyltrichloroethane), whose breakdown products, DDD and DDE, may persist for substantially longer than the parent compound, is also a well-known OC to environmental scientists. DDT was first synthesized in 1874 and has been used as an insecticide since 1939. Due to possible health and other environmental concerns (e.g., thinning of egg shells and hence declining reproductive success of birds of prey, such as eagles), it was banned in most industrialized countries in the 1970s, although it is still used by some nations today. POPs have obviously caused some serious environmental problems. However, for the sake of balance, it is also important to remember that insecticides such as DDT have probably saved millions of lives by helping to control insect-borne diseases, such as malaria.

The above POPs, such as DDT and PCBs, are now sometimes referred to as "legacy POPs": many of them were banned in the 1970s but still persist in the environment. However, new compounds are constantly being developed, and many of these chemicals are now also causing environmental concerns. These new POPs are often referred to as "emerging POPs," and include polybrominated diphenyl ethers (PBDEs), polychlorinated naphthalenes (PCNs), perfluorooctanesulfonate (PFOS), and perfluorinated compounds (PFCs). For example, PBDEs are brominated flame retardants, and in fact some governments passed legislation in the 1970s requiring the use of flame retardants on products such as carpets, aircraft interiors, and even children's sleepwear. Although the toxicological properties of PBDEs are not fully understood, they have now been linked to a variety of concerns, including reproductive, development, and carcinogenic effects. Like other POPs, these compounds have now been transported globally, including remote regions such as the High Arctic (e.g., de Wit et al. 2006).

Given the health and other environmental concerns of POPs, it is not surprising that considerable research efforts are under way to study the transport, fate, and accumulation of these anthropogenic compounds. As many POPs are effectively sorbed onto particles and accumulate in the sedimentary record, paleolimnological approaches have played many important roles in this field (Blais & Muir 2001). In fact, because there are few reliable quantitative data for POPs in water supplies prior to the late 1980s (as nearly all measurements before this time were conducted on water samples whose volumes were too small for the detection limits available at that time), sediment and other proxy records remain the only avenues to track past trajectories. Below, I summarize a few case studies showing some of the diverse applications that are possible.

9.2 Estimating the accumulation, inventory, and diagenesis of chlorinated hydrocarbons in Lake Ontario

I return to Lake Ontario, one of the Great Lakes straddling the border between Ontario (Canada) and several US states, as an opening example to the study of POPs because, similar to the eutrophication examples cited in Chapter 11, much of the early work on POPs was done on this system. The reasons are clear: Just as Lake Ontario was dramatically affected by nutrient fertilization from its highly developed catchment, it also received high doses of other pollutants, such as POPs. Given that Lake Ontario is the terminal lake in the Great Lakes system and that so many people depend on it for their drinking water, as well as many other uses and services (e.g., fisheries, wildlife, etc.), there are continued concerns regarding the levels of POPs (and many other pollutants) in this system.

Wong et al. (1995) used paleolimnological techniques to estimate the accumulation, inventory, and diagenesis of four major types of POPs (i.e., PCBs, DDT, mirex, hexachlorobenzene, or HCB) from three sedimentation basins in Lake Ontario, and compared these data to similar paleolimnological work done in the 1980s (Eisenreich et al. 1989). Figure 9.4 summarizes the accumulation data for the five ^{210}Pb-dated cores they analyzed, corrected for focus-normalized sediment accumulation rates (for details, see Wong et al. 1995), as well as production and sales data for the compounds of interest (the last

panel in each section of the figure). All profiles show a near-exponential increase in accumulation to a subsurface peak, followed by a decrease to the surface. The profiles show a relatively constant concentration of POPs at the surface of each core. Moreover, unlike profiles from cores collected in 1981 (Eisenreich et al. 1989), the peaks are now well below the surface mixing zone, preserved in the buried sediments, tracking the period of peak concentrations of POPs in the mid- to late 1960s, and the subsequent declines in concentrations following the North American bans in the early 1970s.

The second aspect of the Wong et al. study was to look at POP inventories in Lake Ontario, which they estimated by combining the total sediment abundance of POPs per unit area (i.e., by considering vertically integrated sediment inventories). Admittedly, using only five cores for a lake this size is an approximation when one attempts to make lake-wide calculations, but if presented with the required caveats, it provides some important observations. Similar to the accumulation rate profiles (Fig. 9.4), the inventories (the white bars in Fig. 9.5) vary significantly amongst the cores, but this variability is largely removed when the data are corrected for changing sedimentation rates and focusing (the black bars in Fig. 9.5). Given the similarity in the values after correction for sediment accumulation, one can conclude that the differences in the amount of POPs at any of the five studied sites is primarily due to the differences in the amount of accumulating sediment, as estimated by ^{210}Pb. This implies that either there is only one major source of POPs to Lake Ontario and/or the water column dynamics are sufficient to distribute POPs more or less evenly throughout the lake basin. Both of these processes appear to be occurring in Lake Ontario (Wong et al. 1995), as the Niagara River is the main source of POPs and circulation patterns are sufficient to mix the contaminants throughout Lake Ontario.

Such inventory data can be used for a variety of applications. For example, Wong et al. (1995) calculated the inventories of POPs in the top

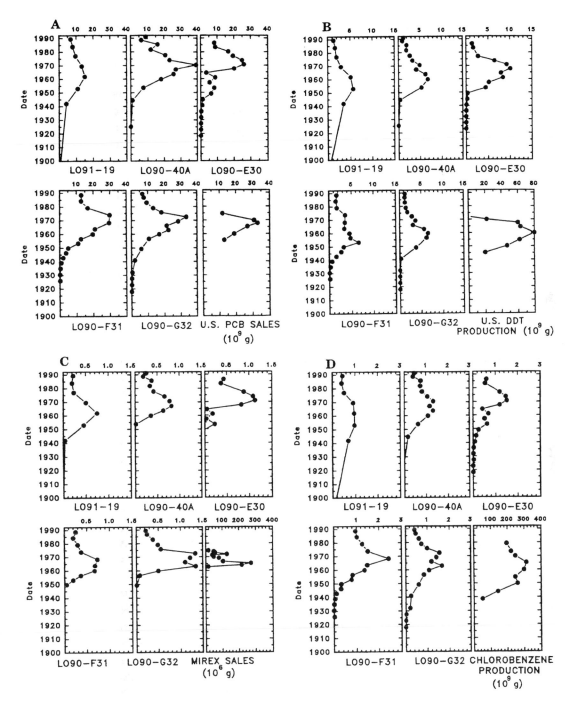

Fig. 9.4 Focus-normalized accumulation rates (ng cm^{-2} yr^{-1}) of total PCBs (A), DDT (B), mirex (C), and HCB (D) in five Lake Ontario sediment cores (i.e., LO91-19, LO90-40A, LO90-E30, LO90-F31, and LO90-G32) taken from the three major sedimentation basins of the lake. The final panel in each section summarizes the US production and sales data for the compounds. From Wong et al. (1995), used with permission. Copyright 1995 American Chemical Society.

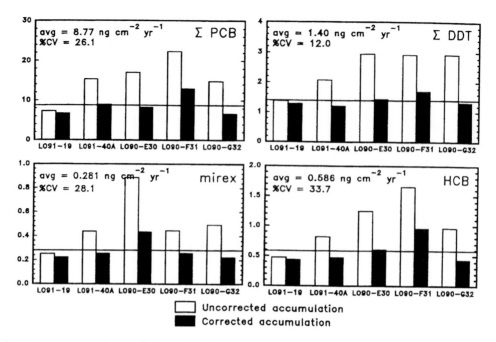

Fig. 9.5 POP inventories (ng cm⁻¹) for total PCBs, DDT, mirex, and HCB in the same five Lake Ontario cores described in Fig. 9.4. The white bars are the observed inventories, whereas the black bars are inventories corrected for ^{210}Pb focus-normalized rates. The horizontal lines represent the average focus-normalized inventories for each POP. Lake-wide totals are also estimated. From Wong et al. (1995); used with permission. Copyright 1995 American Chemical Society.

centimeter of sediment, as well as the ^{210}Pb-derived sediment-mixing zone, to estimate the amounts of different POPs that were potentially available for redistribution (i.e., are not permanently buried) into the water column by sediment–water exchange.

Finally, the topic of POP diagenesis was examined. Because sediment cores from two of the sites (at E30 and G-32) were collected and analyzed in a similar manner in 1981 (Eisenreich et al. 1989) to those collected in 1990 for the Wong et al. (1995) study, an important opportunity was available to study POP diagenesis in sediments over a 9-year interval. If POPs did not suffer from post-depositional diagenesis, then new sediment should have accumulated above the profiles described from the 1981 cores. These predictions were confirmed by their observations, with effectively no change in mass. The POP

peaks in the 1990 cores, however, had broadened somewhat compared to the 1981 cores, possibly due to diffusion.

9.3 Historical trends in organochlorines in river basins: reading the records of influent river water quality left in reservoir sediments

As has been discussed previously, river systems are often difficult settings for paleolimnological analyses, as finding sedimentation basins of sufficient quality may be challenging (but not impossible; e.g., Carignan et al. 1994). Nonetheless, rivers are of crucial importance to society, and river pollution is of prime concern to water managers and scientists, especially in regions that do not contain many lakes. As many rivers flow through agricultural areas, the study of OC pollution is often of prime interest in these areas. Fortunately, if suitable river profiles cannot be found, sediment accumulations in downstream reservoirs can often be used to track the past chemical patterns of the rivers feeding the reservoir.

Van Metre et al. (1997) used the above approach to study historical trends in water quality from six reservoirs in the central and southeastern United States. An advantage of reservoir sediment profiles over many lake cores is that the former typically have higher sedimentation rates, and so a finer temporal resolution is potentially attainable. Following ^{137}Cs dating, the cores were analyzed for PCBs, DDT and its derivatives DDD and DDE, and chlordane (Fig. 9.6). The oldest reservoir in this study was White Rock Lake in Texas, which was filled in 1912, whereas the youngest was Lake Seminole, where storage began in 1954.

The paleolimnological data appeared to accurately reconstruct the contaminant histories of these sites. For example, the profiles track the increase in DDT use until its ban in 1972. Chlordane, which is a mixture of over 140 compounds, was first produced commercially in 1947, and was then widely applied as an agricultural pesticide until 1974, although it continued to be used in urban areas as a termiticide and for the control of fire ants until at least 1990. The continued increase in chlordane in West Point Lake (Fig. 9.6) reflects the largely urban uses of this OC in its drainage (i.e., Dallas, Texas).

Van Metre et al. (1997) also compared their reservoir OC concentrations to those found in lake and peat bog profiles, where the primary focus was to find deposition sites that would estimate atmospheric fallout. They found that the reservoir concentrations were much higher. For example, DDT was from one to two orders of magnitude higher. These data showed that fluvial inputs of these contaminants were much more important than atmospheric sources, thus clarifying the pathway of these pollutants in this region.

9.4 Ecological effects of deliberate releases of toxaphene to lakes

Most work on POPs in freshwater systems deals with inadvertent contamination from either local or long-range (see below) sources. In a few cases, though, compounds that are now considered dangerous have been purposely applied to lake systems. One such example is toxaphene, which is a mixture of polychlorinated camphenes (PCC), and was once widely used as a piscicide and an agricultural pesticide. Toxaphene has been in use since 1949, and by 1975 was the most widely used insecticide in the United States. Although now banned in many industrialized nations, it is still used in other parts of the world. Many consider it to be a potential human carcinogen.

In the 1950s and 1960s, so-called "lake rehabilitation programs" were under way in many regions of North America. Although sounding like a very positive endeavor, "lake rehabilitation" was a euphemism for poisoning what were considered to be "undesirable" fish (e.g., yellow perch, northern pike), and then restocking them with "desirable" fish, such as rainbow trout. The most common piscicide used was rotenone, but during the 1950s and 1960s many agencies began considering toxaphene as a less expensive and more lethal alternative for extirpating undesirable fish. As summarized by Miskimmin and Schindler (1994), toxaphene applications had many direct and indirect unanticipated effects, in addition to killing the fish, including detrimental effects on non-target organisms, such as cladocerans, chaoborids, and other invertebrates. Little long-term monitoring data were available from the 1950s or 1960s, and so Miskimmin and Schindler (1994) used paleolimnological approaches to reconstruct the biological effects in treated lakes.

Miskimmin and Schindler (1994) chose three lakes in central Alberta (Canada) for this comparative study: two were treated with toxaphene at different concentrations in 1961–2 (Fig. 9.7) and one was a control. In the lake with the highest treatment (Chatwin Lake), planktonic

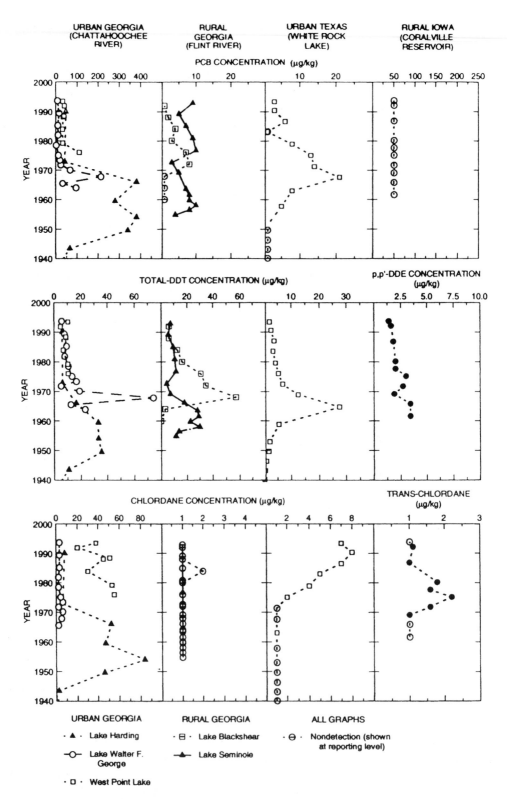

Fig. 9.6 The concentration of organochlorine compounds in reservoir sediment cores from the southwestern United States. From Van Metre et al. (1997), used with permission. Copyright 1997 American Chemical Society.

cladocerans decreased with treatment in 1962, and the assemblage shifted rapidly from small- to large-bodied taxa. For example, the accumulation of *Bosmina* body parts dropped precipitously in sediments dated to 1962, and were eliminated from the lake in the 1970s as invertebrate predators, such as *Chaoborus americanus*, increased in response to the poor survival of the stocked fish. As discussed in the chapter on acidification (Chapter 7), *C. americanus* can rarely coexist with fish and so the presence of its mandibles in lake sediments can be used to infer fish declines. *Chaoborus* is also a voracious predator on small-bodied cladocerans, such as *Bosmina*. Based on an assessment of fossil head capsules, chironomids appeared to be unaffected, as a group, by the toxaphene additions. In Peanut Lake, toxaphene was added at lower concentrations. Short-term toxic effects were not recorded, but large invertebrates also became abundant with the loss of the native fish population (Fig. 9.7).

The effects of losing top-level fish predators cascaded throughout the food web. Paleolimnology showed that toxaphene caused dramatic ecological changes in the lakes chosen for "rehabilitation." The changes were not identical in the two lakes, likely due to the differences in dosage. The study also emphasized an ecological side to contaminants work: It showed that it was the changes in fish populations that caused the most striking cascading changes in trophic structure. What were once natural, healthy lakes were now completely altered.

9.5 Determining the sources and pathways of PAH contamination of Lake Michigan

Polycyclic Aromatic Hydrocarbons (PAHs) are compounds that typically contain three to eight aromatic rings and are produced during the combustion of wood and fossil fuels. They therefore have both natural and anthropogenic origins. Ever since humans learned to use fire, they have

been enhancing the release of PAHs to the environment. However, more recent activities, such as high-temperature combustion of fossil fuels, are believed to have become major sources. There are considerable health and environmental concerns regarding elevated concentrations of PAHs, as many forms are considered to be carcinogenic, and they are very persistent in the environment. Neff (1985) estimated that nearly 230,000 metric tons of PAHs enter the aquatic environment annually. Because of the lack of long-term monitoring data, and because the source(s) of PAHs is (are) sometimes difficult to prove, as they can be distributed over long distances as gases and aerosols, paleolimnological approaches have aided much of this work.

Simcik et al. (1996) used sediment profiles to track the provenance of PAHs in Lake Michigan, one of the intermediate Great Lakes. The lake is situated in a primarily north–south orientation, with most industrial development in the southwestern corner, including the major urban/industrial complex of Chicago (Illinois). Five ^{210}Pb-dated sediment cores were collected along a north–south transect of the lake and analyzed for PAHs. Their working hypothesis was that Chicago and Gary (Indiana) were the two major sources of PAH pollution, and that atmospheric transport was the main vector for deposition. By reconstructing accumulation profiles and inventories for selected and total PAHs (similar to the POP data described in the previous Lake Ontario example; Wong et al. 1995), they showed that PAHs in the sediments increased dramatically around 1900, reached a plateau around 1930–50, and then decreased slightly (Fig. 9.8). The recent decreases were attributed to shifts from coal to oil and natural gas as a primary energy source, and to increased controls on emissions.

Similar to the Lake Ontario POP example described earlier, Simcik et al. (1996) then calculated PAH inventories (Fig. 9.9). However, since there are also natural sources of PAHs (in contrast to the previous example, where the POPs such as PCBs and DDT are all human-made), a background PAH accumulation rate from deep

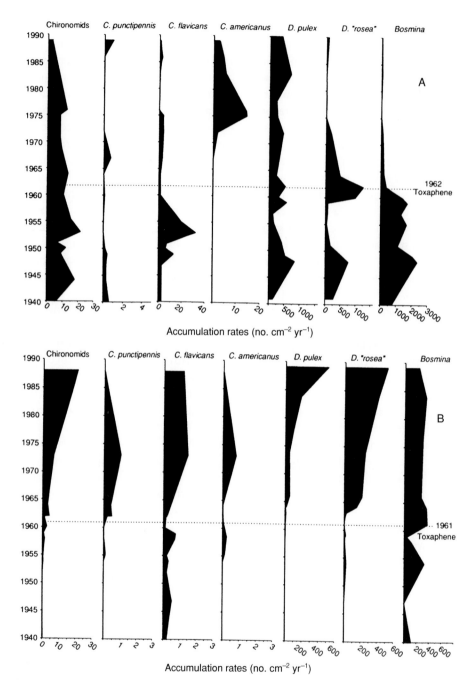

Fig. 9.7 Accumulation rates of chironomid head capsules, *Chaoborus* mandibles, and planktonic cladocerans from ^{210}Pb-dated sediment cores (last ~ 50 years shown) in (A) Chatwin Lake and (B) Peanut Lake. Both of these lakes were treated with toxaphene, in 1962 and 1961 respectively. An untreated basin of Peanut Lake, separated by an earth berm, was also used in the study as a control (not shown). From Miskimmin and Schindler (1994); used with permission of NRC Research Press.

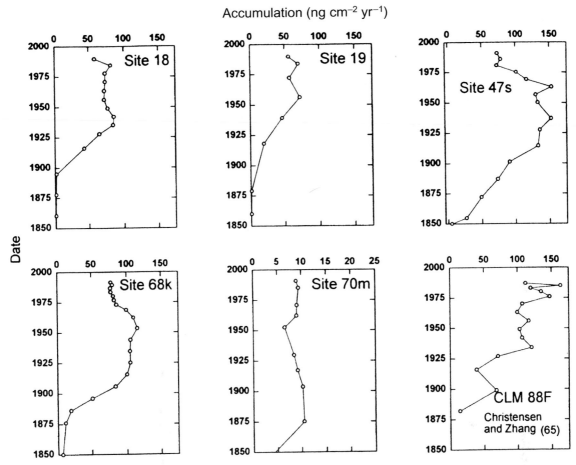

Fig. 9.8 Sedimentary profiles of total PAH accumulation rates for Lake Michigan sediment cores. The core from Site 70 was believed to have experienced significant sediment mixing. The final panel (CLM 88F) is from a core collected in 1988 near Milwaukee (Wisconsin) and analyzed by Christensen and Zhang (1993). From Simcik et al. (1996); used with permission. Copyright 1996 American Chemical Society.

sediments in each core was subtracted from the accumulation rates measured in the more recent sediments to estimate the anthropogenic contribution. If there was only one major source of PAHs to Lake Michigan, then focus-corrected inventories of PAHs should provide a common, lake-wide value. Such was the case with the Simcik et al. data, and so they concluded that there was one major source of PAHs and, similar to the POPs in Lake Ontario, the PAHs were well distributed in the lake (exceptions occurred in restricted embayments, such as Green Bay). Moreover, the PAH distribution over the last

approximately 70 years is statistically identical not only within each of the five cores, but also between the cores taken from over the whole lake, suggesting that the major source of PAHs to Lake Michigan have not changed over the past century. A comparison of individual PAH compounds recorded in the sediments with PAHs known to occur in Chicago air particles and from coke ovens (a major contributor of industrial emissions) suggest the latter as a likely source. A clear similarity between PAH accumulation in the sediment signatures and the known historical records of coal use in Illinois over the same time

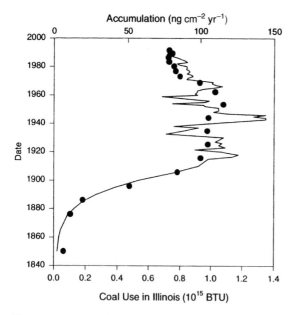

Fig. 9.9 Historical records of coal use in Illinois (solid line) and the accumulation of total PAHs from site 68K from the northern basin of Lake Michigan. From Simcik et al. (1996); used with permission. Copyright 1996 American Chemical Society.

period provides a convincing visual concordance that further supports their conclusions that particulate PAHs in the southern Lake Michigan atmosphere are primarily derived from coke and steel production. It also confirms that these emissions account for most of the PAHs in the Lake Michigan sediments, even those accumulating in the northern part of the lake, such as the core shown in Fig. 9.9.

9.6 Tracking long-term contamination of polycyclic aromatic hydrocarbon (PAHs): an example from a fluvial environment

Although most studies deal with lake sediments, PAHs can also be studied in fluvial systems, if the sampling site is carefully chosen. Smith and Levy (1990) documented the geochronology of

PAH contamination near the junction of the Saguenay River and Saguenay Fjord (Quebec), and related this contamination to the local aluminum operations. The economic production of aluminum requires a cheap and ready source of electricity. It is, therefore, not surprising that the industry was quick to recognize the economic benefits of locating in this region, given the hydroelectric potential of the Saguenay region, coupled with its ready transportation access to the Atlantic Ocean. The first hydroelectric station and aluminum smelter were operating by 1926, and by the 1960s the region was one of the world's largest producers of aluminum. The industry is also known to be a major producer of PAHs; however, as noted earlier, PAHs can be produced from many types of combustion, and so the source of contamination can be easily contested.

Smith and Levy (1990) used ^{210}Pb-dated sediment cores to track the type and amount of PAHs being deposited in the Saguenay sediments. Using capillary gas chromatograms, they found marked changes in PAH composition following the commissioning of the first smelter. By the late 1950s, however, marked changes had occurred in the composition as well as the concentration and accumulation rates (or flux) of PAHs (Fig. 9.10). For example, new components were identified in the PAH assemblage using gas chromatography/mass spectrometry. These changes coincided with developments in the local aluminum industry. Up until the 1940s, atmospheric deposition was the dominant mode of transport of PAHs from the smelters to the aquatic system. However, with increasing demands for aluminum, a new smelter was constructed with 15 new potlines between 1938 and 1943, and then yet another smelter with four new potlines went into production in 1943, 1951, 1952, and 1956. Technology had also changed, which produced approximately 1000 times more PAHs than the older "pre-bake" process for the same amount of aluminum. The PAH composition in the sediments dated to the 1950s was very similar to the PAH assemblages being released from the effluents at that time. Scrubbers

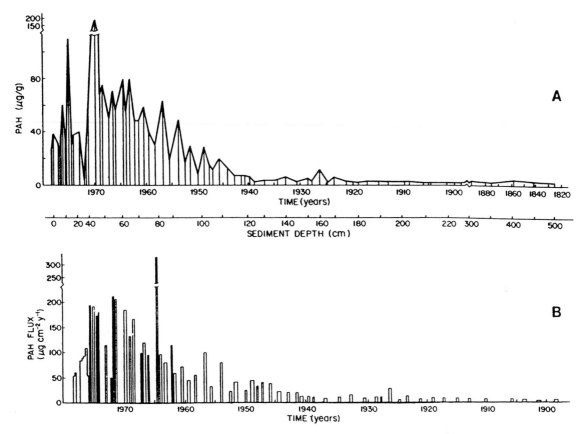

Fig. 9.10 PAH concentrations (A) and PAH flux (B) as a function of ^{210}Pb dates in a core from the Saguenay River/Fjord region. From Smith and Levy (1990); used with permission. Copyright 1990 American Chemical Society.

were subsequently installed to limit the amount of PAH contamination to the air, but the scrubber liquors were discharged directly into the Saguenay River between 1964 and 1976.

The paleolimnological data showed that the natural/baseline PAH concentration and flux to this system was only 2–3 µg g^{-1} and 5–8 µg cm^{-2} yr^{-1} (Fig. 9.10). The first smelter began operation in 1926, which coincided with the first sustained increase in PAHs. With the introduction of new technologies, and the continued expansion of the industry in the 1950s, PAHs increased further. In 1971, there was a massive landslide in the region, and the return to seemingly pre-disturbance PAH concentrations is a result of the dilution from the inwash of

this unaffected material. However, the subsequent decrease beginning in the 1970s mirrors the decline in the overall industry. The matching of the historical development of the local aluminum industry with the PAH records in the sediments leaves little doubt as to the provenance of the PAHs in this system.

9.7 Forensic paleolimnology: identifying pollution sources from POP signatures in river and lake sediments

One of the greatest challenges faced by environmental scientists and managers is to identify the sources of pollutants. Such information is of course crucial in determining how to manage water-quality problems, but it also has important implications for policy and enforcement (i.e., who is to blame – or perhaps, in more practical terms, who should pay for any mitigation or remedial action). As shown in the following two examples, sedimentary approaches hold considerable promise in identifying the sources of POP pollution in river and lake ecosystems.

The Fraser River system in British Columbia (Canada) is the most important salmon-bearing river system in the world. However, it is also subjected to multiple stressors, including threats from fishing, agricultural runoff, and industrial effluents. PAHs have emerged as a topic of great concern, as high concentrations have been recorded in the river. More alarmingly, river organisms have been observed with impairments that are typically associated with elevated exposure to PAHs. However, in order to correctly assess the significance of these observations, as well as to propose adequate and realistic pollution control strategies, it is first necessary to identify the composition and source(s) of the PAHs. As noted earlier, PAHs have both natural and anthropogenic sources, so it must also be clarified how anthropogenic PAH signatures have changed relative to natural or background conditions. Yunker et al. (2002) tackled this complex problem by analyzing PAH compounds from 345 suspended particulate and fluvial sediment samples from the Fraser River system. By examining PAH ratios and total concentration data, they showed that this ecosystem was only modestly impacted by a variety of PAH sources in its remote regions, where most of the anthropogenic PAHs were from nearby roads. However, the pattern was markedly different in urban areas, where elevated inputs of contaminants from vehicle emissions, as well as from stormwater and wastewater discharge, were clearly evident. They showed that the examination of all PAH ratios, where reliable concentration data are available, provide robust assessments of contaminant sources. As these POPs are archived in

sediment profiles, their changes over time can be correlated to the land-use, agricultural, and industrial development history, and so provide a forensic analysis of pollution sources.

Similar approaches can of course be used for lake systems and with other POPs. For example, Macdonald et al. (1998) used the polychlorinated dioxins and furans (PCDD/Fs) and other contaminants preserved in a British Columbia lake sediment core to track changes in pollution sources, ranging from atmospheric inputs of POPs to those originating from the watershed, to contaminants linked to pulp mill chlorination. From the sediment record, they were also able to identify various processing changes within the pulp mill (e.g., use of contaminated wood chips, switching to chlorine dioxide bleach) and to demonstrate that the chlorination of wood pulp was itself producing a number of PCB congeners.

Contaminant patterns recorded in sediments, especially for complex suites of compounds such as PCBs, PAHs, and PCDD/Fs, provide important forensic markers for identifying pollution sources. As shown with other examples in this book, sediments are always "on watch" and can often provide a better perspective on pollution trajectories of lake or river systems than the often patchy and incomplete data available from effluent monitoring.

9.8 The Global Distillation Hypothesis: long-range transport of organic contaminants to Arctic and alpine regions

One of the most remote areas on this planet is the High Arctic, where vast regions of the landscape appear to be totally untouched by humans and the only settlements are a few isolated Inuit villages, with populations of a few hundred people or less. It therefore came as quite a surprise when researchers began recording alarmingly high levels of toxic organic compounds such as organochlorines (OC) in Arctic wildlife and people. Although some of these compounds (such as some PCBs) had been used in, for example, local military installations, many other compounds were never used in or near the Arctic. The latter include chlorinated pesticides and herbicides (e.g., DDT and its derivatives), hexachlorocyclohexanes (HCHs), dieldrin, and polychlorinated camphenes (PCCs or toxaphene). Especially striking were the high levels of contamination, and that concentrations continued to remain high to the present, even though industrialized countries in the Northern Hemisphere had banned most of these substances several decades ago. Because many Inuit eat a diet high in animal fats (so-called "country foods," such as seals and whales), and since OCs are fat soluble and lipophilic, the health risks associated with the high OC concentrations in, for example, Inuit mothers' milk became an international concern.

The obvious question arose: What pathways could transport OCs, which were being used in places such as Central and South America, to regions such as the High Arctic? The global distillation and cold condensation hypothesis (sometimes colloquially referred to as the "grasshopper effect") appears to be involved (Box 9.2; Wania & Mackay 1993).

Box 9.2 The Global Distillation and Cold Condensation Hypothesis

Wania and Mackay (1993) hypothesized that the physical and chemical properties of OCs, as well as the environmental characteristics that typify cold regions, contribute significantly to the long-range transport of these compounds to polar regions. Critical factors are the volatility of a compound and the ambient temperatures. The process can be summarized as follows.

Vapor pressure and solubility are related to temperature. The high concentrations of OCs in polar regions are explained in part by the temperature-dependent partitioning of these low-volatility compounds. A key property is the vapor pressure, which varies greatly from compound to compound. Global fractionation may be occurring where OCs are being latitudinally fractionated and "condensing" at different temperatures, dependent on their volatility. As a result, OCs with vapor pressures in a certain low range may preferentially accumulate in polar regions.

For example, Fig. 9.11 shows the potential fractionation of three OC pesticides (top) and three congeners of PCBs. Compounds with high vapor pressures, such as benzo-a-pyrene (BaP), are deposited near their source of emission. Meanwhile, chloroflurohydrocarbons (CFCs), with very low vapor pressures, may not encounter temperatures sufficiently low to condense appreciably. Many compounds are of intermediate vapor pressures (e.g., DDT, HCH) and will be distributed, or fractionated, across the globe based on their volatility and local temperatures. Similarly, different congeners of PCBs will be distributed based on their respective vapor pressures (Fig. 9.11, bottom). The net result is that cold regions, such as the Arctic, may become "sinks" for many of these compounds, whereas many warm regions are "sources." The seriousness of the situation is compounded by the low temperatures, limited biological activity, and relatively small incidence of sunlight in polar regions, which can increase the persistence of OCs.

Because a compound may be volatilized in summer at one latitude, but then condensed and deposited farther north during one season, and then revolatilized the following summer and transported farther north, the process is sometimes referred to as the "grasshopper effect." These OCs "hop" farther north until the climate is too cold to revolatilize them further.

Organochlorine pesticides

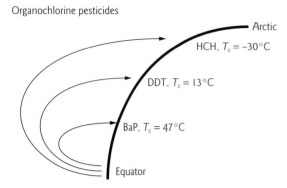

PCBs

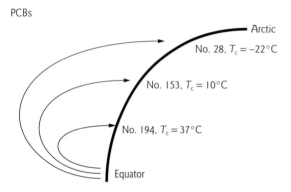

Fig. 9.11 The global distillation effect, showing how semi-volatile organic compounds tend to move toward cold climates as a result of their lower vapor pressure at colder temperatures. POPs with higher volatilities and lower temperatures of condensation (T_c) should be transported farthest. From Blais and Muir (2001), based on data from Wania and Mackay (1993, 1996); used with permission.

Long-term data can help evaluate the processes involved in potential global fractionation of OCs. For example, Muir et al. (1996) used paleolimnological approaches to test this hypothesis by studying PCB concentrations and fluxes in lake sediments from 11 remote lakes located between southern Canada at 49°N and the northern tip of Ellesmere Island at 82°N. The cores were dated using ^{210}Pb and ^{137}Cs. They found significant and exponential declines (about six-fold) in total PCB flux with latitude over 35°N. In the High Arctic sites, the onset of PCB inputs was much later (1950s–1960s) than in more

southerly sites (1930s–1940s). High Arctic sediment samples also contained higher proportions of lower-chlorinated congeners of PCBs. All these results are in agreement with the global fractionation model.

Similar mechanisms that are transporting organic contaminants to cold regions such as the High Arctic may also be transporting contaminants to alpine regions. Blais et al. (1998b) reasoned that if global distillation was operating on a latitudinal scale, it may also be operating on an altitudinal gradient, where similar mechanisms of fractionation and condensation may apply. High mountains are similar to polar regions in that they are both characterized by low temperatures. Blais et al. showed that OC deposition in mountain ranges of western Canada increased about 10- to 100-fold between 770 and 3100 m altitude. Cold condensation effects further enhanced the concentrations of more volatile OCs with altitude.

Paleolimnological studies appear to confirm these observations. Galassi et al. (1997), working on sediment cores from two lakes in the Himalayas (Nepal) at an elevation of 5050 m a.s.l., found relatively high concentrations of PCBs in the recent sediments of both lakes. In fact, concentrations were higher in these lakes than found in comparable studies in northern Italy.

9.9 Biological funnels: POPs swimming upstream and taking flight

The transport of POPs via cold condensation is alarming enough (Box 9.2); however, other mechanisms are also facilitating the dispersal of contaminants far from their original sources. One recently investigated vector is a biological one (reviewed by Blais et al. 2007), namely the movement of marine-derived contaminants to freshwater ecosystems via migrating fish, nesting seabirds, or other biota (Box 9.3, Fig. 9.12). As paleolimnological approaches can be used to track past changes in, for example, sockeye salmon (Finney et al. 2000, 2001; see pp. 290–5)

Box 9.3 Biologically mediated transport of contaminants: the "boomerang effect"

Although the transport of pollutants via cold condensation (Box 9.2) is now well recognized, biologically mediated movement of pollutants such as POPs, mercury, and other metals is less well known. Nonetheless, biovectors can significantly alter the contaminant loads of many freshwater ecosystems.

Gregarious animals that bioaccumulate and biomagnify certain contaminants (Figs 9.1 and 9.2), and then migrate and congregate (e.g., sockeye salmon returning to their nursery lakes or nesting seabirds), may represent prominent pathways for contaminant transport in certain circumstances. As reviewed by Blais et al. (2007), once a contaminant has been released to the environment via industrial or agricultural pollution, its pathway may then be intercepted and influenced by animal vectors.

For biologically mediated contaminant transport to occur, three crucial steps must be fulfilled by the biovector, namely: (i) collection of the contaminant; (ii) delivery of the contaminant; and (iii) deposition, release, or transfer of the contaminant at the receptor site (Fig. 9.12). To date, paleolimnological assessments of this phenomenon have been restricted to returning sockeye salmon populations (Krümmel et al. 2003, 2005) and nesting Arctic and Antarctic seabirds (e.g., Blais et al. 2005; Sun et al. 2005), but many other biovectors can significantly influence the transport of pollutants (Blais et al. 2007).

There is a depressing and insidious aspect to biologically mediated transport of pollutants, which is especially evident for biovectors that accumulate pollutants in the ocean and then return their contaminant load to terrestrial ecosystems, such as lakes and ponds. As most of the examples described thus far have shown, the main pollutant sources are those released from land (such as from industry or agriculture) that eventually make their way to the oceans, where it was assumed that the large volume and dilution effect of the ocean would render these pollutants less toxic, and that eventually they would effectively be removed from most contaminant cycles by sedimentation in deep-ocean sediments. However, the funneling effect of biovectors that occupy high trophic levels in the marine food web, and so bioaccumulate and biomagnify these contaminants in the ocean, and then either swim (e.g., anadromous salmon) or fly (e.g., seabirds) "uphill" toward land, represents an unwanted shunt returning pollutants to terrestrial ecosystems. Contaminants such as DDT and PCBs, long thought to have been removed from the terrestrial ecosystem by their eventual transport to oceans, are now "swimming" and "flying" back to land via biovectors such as migrating fish and birds. The return of these pollutants via a circuitous path from land, to the ocean, and then back to land-based ecosystems can be categorized as a "boomerang effect," and represents an important consideration for certain ecosystems and contaminant models (Blais et al. 2007).

and seabirds (Blais et al. 2005; see p. 295), so too can these same sediments be used to document the past contributions from biovectors to the contaminant loads of lakes and ponds.

One of the most spectacular examples of biological migrations is the returns of sockeye salmon (*Oncorhynchus nerka*) to their nursery lakes. After hatching in their nursery lakes, and spending 1–3 years in the freshwater environment, the young smolts leave the freshwater ecosystem, travel downstream to the Pacific Ocean, and then spend the next few years in the open-ocean

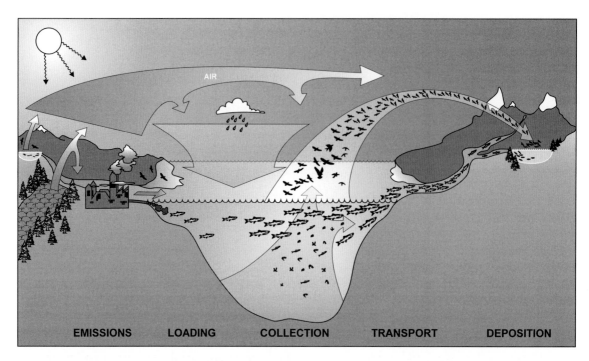

Fig. 9.12 The biologically mediated transport of pollutants to aquatic ecosystems. Moving from left to right, emissions release contaminants from industrial, municipal and agricultural regions, where they begin to cycle globally (see Box 9.2). Some of these contaminants are transported and loaded by air and water currents to the ocean, where they are diluted and become widely distributed in the upper waters. Biota occupying high trophic levels in the marine ecosystem (e.g., salmon, piscivorous seabirds) collect these contaminants by bioconcentration and bioaccumulation. Animals may then transport these contaminants, sometimes over long distances, to locations where they congregate in large numbers (e.g., spawning rivers, nursery lakes, nesting colonies) and subsequently deposit some or their entire contaminant burden through feces, molting or mortality. Modified from Blais et al. (2007). Copyright 2007 American Chemical Society.

ecosystem, where they accumulate over 95% of their body weight (paleolimnological applications for tracking Pacific salmon populations are discussed more fully later in this book, on pp. 290–5). Then, prompted by environmental cues that are still poorly understood, the salmon begin their arduous journey, with uncanny precision, back to their nursery lakes, undergoing upstream migrations for sometimes hundreds or in some cases over 1000 kilometers. Once the adult salmon reach their destination, they spawn and die; thus these lakes serve as both their nurseries and cemeteries. However, as Pacific sockeye salmon accumulate almost all of their biomass in the open ocean, where they occupy a

high trophic level in the marine food web, and because of their relatively high fat content, salmon accumulate contaminants such as DDT and PCBs. Although the contaminant load for each fish is quite low, what is critical to the overall calculations is the number of fish returning. In some nursery lakes, several million adult fish return each year. Several million multiplied by a relatively small amount in each adult fish is still a significant contaminant load!

Krümmel et al. (2003, 2005) used the sediments from Alaskan (USA) and British Columbian (Canada) nursery lakes, receiving a range in density of returning sockeye salmon, to show that anadromous salmon provided a more important vector for PCB transport than atmospheric deposition to some of the nursery lakes. Blais et al. (2005) used similar paleolimnological approaches to track the influence of Arctic seabirds on biovector transport. They showed that pesticides (such as HCB and DDT) and total mercury concentrations in Cape Vera (Nunavut, Canada) pond sediments could be linked to marine-derived contaminants released in guano and other biological debris from the northern fulmar (*Fulmarus glacialis*) bird colonies nesting above the study ponds. Evenset et al. (2004, 2005) similarly showed the effects of Arctic seabirds on Bjørnøya (Bear Island) lake ecosystems, including the biological transport of emerging POPs such as brominated flame retardants. In the Southern Hemisphere, Sun et al. (2005) used ornithogenic lake sediments and penguin bird droppings to track an approximately 50-year record of POP transport to King George Island, from the maritime Antarctic region. Similar approaches have been used to implicate avian biovectors in the transport of metals (e.g., Liu et al. 2005).

Although some of the examples cited above may represent relatively isolated ecosystems, a diverse array of biota can be involved in contaminant transport (Blais et al. 2007). Moreover, seabirds often represent the dominant form of wildlife for many coastal ecosystems, and their influences on contaminant transport have yet to be addressed fully. It should also be remembered that nursery rivers and lakes are the home to resident biota (e.g., freshwater fish) and supply food for terrestrial animals (e.g., bears, minks, otters, birds, insects). Accordingly, the same biovectors that focus contaminants within anadromous fish or migratory bird life cycles may also result in "collateral damage" when biotransported pollutants are ingested by predators and scavengers, thus further contaminating the terrestrial food web (e.g., Christensen et al. 2005; Gregory-Eaves et al. 2007). As paleolimnological approaches can be used to track long-term changes in some biovector population sizes (see pp. 290–5) as well as the contaminants

themselves, these long-term data provide important information for ecosystem managers and for the development and evaluation of contaminant models.

9.10 Determining the distribution of PCB contamination from a local Arctic source

Although most OC research in polar regions has focused on long-range transport of pollutants, there are also some significant point sources of contaminants in certain areas of the Arctic. Because lake sediments are effective traps for many OCs, and because lakes are a ubiquitous feature of many Arctic regions, paleolimnological analyses can be used to track the distribution of contaminants from a known point source.

Point sources of PCB contamination in the Canadian Arctic are often associated with former military installations, such as the Distant Early Warning (DEW) Line and the Polevault Line radar facilities, which have now been mostly decommissioned. These sites typically housed equipment containing large concentrations of PCBs. Since there are few other (if any) local contaminant sources in these regions, it is often possible to map out the distribution of contamination from a single source. One such example is the former Polevault Line Long-Range Radar (LRR) station, now modernized and incorporated into the new North Warning System (NWS), at Saglek (LAB-2), at the northeast tip of Labrador, Canada. The past use and careless disposal of PCBs and other contaminants at northern military installations is well known. What is not well known, however, is whether past PCB contamination was limited to the area immediately around the radar station, or was this contamination dispersed to other surrounding regions. This is an important health and cultural issue, as the local Labrador Inuit live, hunt, and fish in the area.

An analysis of PCBs in the surface sediments of 19 lakes, selected in a radial pattern around the Saglek station, helped document the dispersal of contaminants from this point source. Betts-Piper (2001) found that although PCB concentrations generally decreased with increasing distance from the site, detectable levels of PCBs were recorded from sediments as far as 30 km away from the radar facility, which was about 10 km farther than previously suspected.

9.11 Linking land-use practices to POP dynamics using delta sediments from tropical southern China

As shown in the previous two sections, tropical and subtropical regions are important sources for POPs that may then be transported long distances via ocean and air currents (e.g., via cold condensation distillation; see Box 9.2), and, in some cases, biological vectors (Box 9.3). However, only a few contaminant paleolimnological studies have been conducted from these warm water regions (e.g., Lipiatou et al. 1996; Peters et al. 2001). The paucity of long-term monitoring data is especially troublesome, as pesticide and other POP applications have been relatively high in tropical and subtropical regions. Furthermore, many POPs, such as DDT, which faced production bans in most temperate regions in the 1970s, continued to be used in many tropical and subtropical countries, and in fact are still used in some developing countries today (Li & Macdonald 2005).

The lack of long-term monitoring data for two important pesticides (DDT and hexachlorocyclohexane or HCH) commonly used in tropical regions led Zhang et al. (2002) to determine the contaminant inventories from eight Pearl River delta sediment cores in southern China. China is the second largest producer of pesticides in the world, and the Pearl River Delta is one of the most prosperous regions in China, with the dubious distinction of also holding the record for the highest rate of pesticide application in that country. Not surprisingly, Zhang et al. (2002)

recorded increases in DDT and HCH inventories beginning c. 1963 in the ^{210}Pb-dated deltaic sediment cores, which coincided with the time of extensive pesticide use in China. However, although China banned the use of DDT and HCH in 1983, the expected declines in pesticide fluxes were not evident in the sediment cores, and instead tracked increasing trends and even strong rebounds in DDT profiles, as well as changes in the ratios of DDT to its derivatives DDE and DDD, in many of the sediment sections dated from the 1990s to the present. This initially puzzling finding prompted Zhang et al. (2002) to investigate the land-use changes occurring in the Pearl River basin during this period. As documented by the authors, the last two decades coincided with rapid economic development and land-use changes in the Pearl River basin, which resulted in increased erosion and soil runoff (see Chapter 12), as well as regional flooding. The authors concluded that the transfer of POPs from the soil to the river (and thereafter to the delta sediments) was linked to the increased erosion and subsequent transfer of soils to the Pearl River, thus linking two seemingly separate environmental problems, and thereby exacerbating the environmental consequences.

9.12 Perfume paleolimnology: documenting the discharge of synthetic musk fragrances

Synthetic musk fragrances, which are much cheaper than their natural counterparts, are now commonly used in a wide variety of products, including cosmetics, detergents, and soaps. In 2000, the United States alone manufactured over 6500 tons of synthetic musks. As noted by Pelley (2006), synthetic fragrances share many features with POPs, such as persistence in the environment, accumulation in the food chain, and an affinity for fat tissue. These characteristics, coupled with the large volume of fragrances that are being used, have alerted researchers to the potential environmental impacts of these compounds once they enter waterways from sewage and other discharges. The lack of long-term data on these compounds in the Great Lakes region prompted Peck et al. (2006) to undertake a paleolimnological analysis of seven synthetic fragrances from Lake Erie and Lake Ontario. Using fragrance concentrations in ^{210}Pb-dated sediment cores, Peck et al. (2006) estimated that the burden of synthetic musk fragrances to Lake Erie was an astonishing 1900 kg, and to Lake Ontario 1800 kg, and that the input of these compounds was still increasing. The ecological and environmental effects of synthetic musks are still unclear, although some recent studies suggest sub-lethal effects on fish (Pelley 2006).

9.13 Summary

Industrial chemists have been very successful in producing a large number of organic compounds that have helped society with a number of problems and applications (e.g., insect and fungal infestations, other pesticides, heat exchange fluids, chemical additives, etc.). Environmental scientists, however, slowly began to demonstrate that many of these synthesized organic compounds had potentially serious side-effects on non-target organisms, such as birds and fish, as well as people. The problems were compounded by other properties of these persistent organic pollutants, such as their propensities to be bio-magnified through the food chain. In addition, due to the volatility of many compounds, they can be transported long distances via atmospheric processes, and then deposited far from their points of application, such as polar regions (e.g., de Wit et al. 2006) or in remote mountain lakes (e.g., Blais et al. 1998b; Fernández et al. 2000). Biological vectors, such as migrating salmon and seabirds, may also affect contaminant pathways. Although paleolimnological approaches have as yet rarely been applied to these emerging issues (especially given the potential magnitude of these problems), sedimentary profiles potentially allow us to track the trajectories and patterns of deposition of many of these pollutants back into time, long before the technology was even available to measure these chemicals in the water. Given that new compounds are being synthesized and applied daily, the problem of persistent organic pollutants in our ecosystems is likely to escalate further in the coming years.

10

Mercury – "the metal that slipped away"

Human history becomes more and more a race between education and catastrophe.
H.G. Wells, *The Outline of History*, volume 2 (1920)

10.1 A metal on the move

In the February 1998 issue of the *Wisconsin Natural Resources Magazine*, Katherine Esposito aptly referred to mercury (Hg) as "the metal that slipped away," alluding to this element's capacity to be transformed and transported far from its source. As such, the placement of mercury in a book such as this one is a bit problematic. On one hand, it is a metal (Chapter 8), but one that shares many features with persistent organic pollutants (POPs, Chapter 9), such as its propensity to be biomagnified through the food chain (Box 9.1), to be aerially transported long distances (Box 9.2), and to be influenced by biovector transport (Box 9.3). I have therefore chosen to devote this separate short chapter to mercury. Given the importance and consequences of mercury pollution, as well as recent research activity in this area, a separate chapter is justified.

The toxic properties of mercury have been known for some time. The mineral cinnabar has been refined for its mercury content since the 15th or 16th century BC. Its health hazards have been known at least since the time of the Roman conquest of Spain, as prisoners sentenced to work

in the mines had a life expectancy of only 3 years. The most celebrated case of mercury poisoning was from Minamata Bay (Japan) in 1956, when hundreds of people died and many more were affected when a local industry released methylmercury into the bay, resulting in the steady poisoning of residents from nearby fishing villages as a result of eating the contaminated fish and seafood. (Interestingly, Tomiyasu et al. (2006) have used Minamata Bay sediments to document the release of mercury into this ecosystem). In 1971, more than 400 people died in Iraq from eating bread made from wheat treated with a mercury-containing fungicide, which was initially intended as seed for planting. Most cases of mercury exposure are not lethal, but may include damage to the central nervous system, resulting in reductions of motor skills and other sensory problems, and alteration of genetic and enzyme systems. Developing embryos are especially vulnerable.

Mercury has a number of sources, both natural and anthropogenic. Significant natural sources include local geological deposits, volcanic eruptions, and volatilization from the oceans. Human activities, however, have greatly accelerated the passage of mercury throughout our planet. Mercury released in the 20th century due to human activities is almost an order of magnitude greater than the amount released by weathering (Moore & Ramamoorthay 1984). Major sources include the alkali and metal-processing industries, burning of coal and other fossil fuels, incineration of medical and other waste, the pulp and paper industry, and the mining of gold and mercury. However, what is more ominous from a global environmental perspective is the volatilization and atmospheric dispersion of mercury. The total emissions of mercury to the atmosphere was already more than 6000 tonnes in 1983, of which more than 60% was from the combustion of fossil fuels (Arctic Monitoring and Assessment Programme 1998).

Mercury comes in several forms. Elemental mercury (Hg^0), the silver metal in thermometers, is rarely the focus of limnological studies, although it can be found in high concentrations around areas such as gold mines that use mercury as part of the extraction process. The form that is of most concern is organic or methylmercury (CH_3Hg^+), whose formation is often mediated by bacteria that transform elemental mercury to this organic and much more toxic form (ranked by the International Chemical Safety Program of the United Nations amongst the top six most serious pollution threats on the planet). Methylmercury is readily absorbed by organisms and thence effectively biomagnified in the food chain, just as POPs are (Box 9.1). For example, mercury can be biomagnified 100-fold for each step in a food web; even a relatively short food chain can therefore result in a 10^6 magnification of Hg concentrations. It can also be released back into the atmosphere by volatilization, following reduction to Hg^0. Because Hg^0 re-oxidizes relatively slowly, its residence time in the atmosphere is on the order of 1 year (Morel et al. 1998). Therefore, water bodies far from any local sources of pollution can still have serious mercury problems, especially in top carnivores. Lakes with high concentrations of dissolved organic carbon (DOC) appear to have higher rates of mercury methylation, thus increasing its mobility. Meanwhile, exposure to high ultra-

violet (UV) radiation appears to have a detoxifying effect, as it causes photoreduction to Hg^0, although data suggest that UV radiation may also result in the photooxidation of Hg^0 to Hg^{2+}.

10.2 The sedimentary record of mercury pollution

It has long been known that diagenetic processes can reshape elemental profiles in sediments (Lockhart et al. 2000), and certainly considerable attention has focused on mercury, with an at times polarized debate trying to determine whether mercury enrichment in surface sediment reflects digenesis (Rasmussen 1994), or if these profiles are actually tracking anthropogenic sources (Fitzgerald et al. 1998). These discussions are still ongoing; however, data have been gathered which suggest that, at least in some sedimentary sequences, the majority of mercury released into water bodies eventually becomes associated with sediments (Rada et al. 1989), is tightly bound to organic particles, and thereafter shows little diffusion (Fitzgerald et al. 1998; Lockhart et al. 2000). Paleolimnology therefore may provide powerful tools to track the trajectories of these pollutants. Because mercury can be transported long distances, the range of applications has been quite diversified. Below, I summarize a few studies in which mercury was tracked via point and diffuse sources in rivers, lakes, as well as other sedimentary environments.

10.3 Assessing the fidelity of lake sediment cores to track known point sources of mercury contamination

As noted in the previous paragraph, some heated discussions have debated whether sedimentary profiles faithfully track mercury fluxes. I have chosen the Lockhart et al. (2000) study to introduce mercury contamination, as it provides an important investigation showing how point sources of mercury can be tracked in lake sediments. It also provides support that paleolimnological approaches can faithfully track mercury fluxes. Three case studies were explored from very different regions of Canada where a known (but different) point source of mercury contamination had occurred over the past century. The three local sources of pollution were a chlor-alkali plant, a gold mine, and a mercury mine.

Clay Lake, Ontario, is the first large lake downstream of a former chlor-alkali plant that discharged mercury (estimated to have been 9000–11,000 kg) from 1962 to 1970. Cores were collected in 1971 (just after the contamination stopped), in 1978, and more recently in 1995. The latter core was dated using ^{210}Pb and ^{137}Cs geochronology.

Figure 10.1 shows the history of mercury concentrations in the 1995 Clay Lake sediment core. The paleolimnological record nicely mirrored the known contamination history of the lake. These data alone would provide strong evidence that the mercury profile was not greatly altered by diagenetic processes. However, further evidence was collected by measuring mercury in the cores collected in 1971 and 1978, and then tracking the temporal sequence of changes in these profiles, which are superimposed on Fig. 10.2. The mercury peak is essentially identical in all three cores, but it is much nearer the surface in the 1978 core, and is at the surface in the 1971 core, just as one would expect if no or negligible mobility or diagenesis were occurring.

In addition to the Clay Lake data, Lockhart et al. investigated two other mercury-contaminated lakes. Giauque Lake is in the Canadian Sub-arctic, part of the McCrea River system, about 84 km north of the city of Yellowknife in the Northwest Territories. Underground mining for gold began in 1946 and continued until 1962, over which time about 1 million tons of

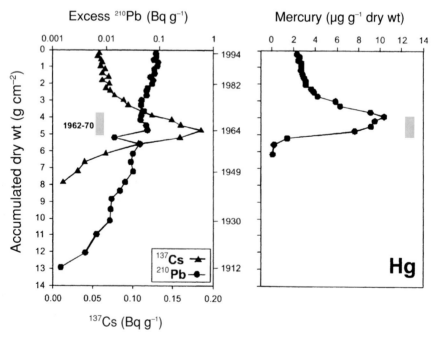

Fig. 10.1 ^{210}Pb and ^{137}Cs dating profiles (left panel) and mercury concentrations (right panel) from the 1995 sediment core from Clay Lake, Ontario. The gray bars indicate the period of industrial discharge from the chlor-alkali plant. From Lockhart et al. (2000); used with permission of Elsevier.

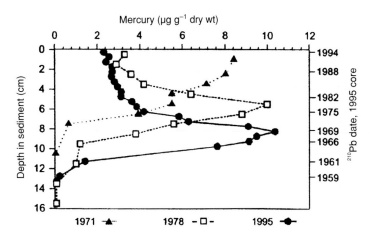

Fig. 10.2 Sedimentary mercury profiles from three cores collected from Clay Lake, Ontario, in 1971, 1978, and 1995. These profiles faithfully track the peaks in mercury contamination from the chlor-alkali plant, and provide further evidence that mercury was not mobilized in these sediments. From Lockhart et al. (2000); used with permission.

ore were produced. During this time, approximately 2.5 tons of mercury were used for the amalgamation and cyanidation processes. Tailings were dumped directly into the lake from 1965 to 1968, but even since that time, leakage from the accumulated tailings has been a problem. The history of mercury contamination, as reflected in the dated sediment core from this site, was consistent with the known history of mining operations and tailings management.

Finally, Lockhart et al. studied mercury concentrations in cores from Stuart Lake, the first one downstream from Pinichi Lake, British Columbia. The Pinichi Lake Cominco mine had two distinct periods of mercury mining: one between 1940 and 1944 and a second from 1968 to 1975. However, the mining and tailing management strategies were quite different during these two periods. In the 1940s, tailings were dumped directly into Pinichi Lake. During its second period of operation, much more stringent environmental controls were in place, and no waste was disposed into the lake, as the mine operated as a "no-discharge facility." The sediment cores faithfully tracked the 1940s mining operation, with expected peaks in mercury concentrations. In contrast, the sediment cores did not record any additional mercury contamination from 1968 to 1975, attesting to the efficacy of the environmental measures put into place for its second period of operation.

Collectively, these case studies confirm that, at least in these three lakes, mercury is not sufficiently mobile in sediments to generate profiles that would misrepresent historical records. These data also show how point sources can be faithfully tracked using paleolimnological techniques, and can also be used to demonstrate the efficacy of environmental controls used in some modern mining operations (as shown by the Stuart Lake example). However, not every site will have the potential to contain such clear and unambiguous archives. For example, if sedimentation rates were slower or if sediment mixing was a more serious problem, then such high-resolution records would not be possible.

10.4 The contributions of mercury and other metals from domestic sewage: a case study from an Arctic Inuit community

Sewage disposal is a continuing problem in Arctic communities, as the cold temperatures, long periods of ice cover, and permafrost conditions pose challenges to engineers and municipal officials to devise effective (and economically feasible) ways to remove domestic sewage (Heinke & Deans 1973). The situation is further complicated by the remoteness and the small population sizes of many Inuit hamlets (often less than 1000 and sometimes less than 200 inhabitants). Options are quite limited in these polar regions. As a result, some communities have relied on piped (utilidor) systems of sewage removal (discussed more fully on pp. 224–6, dealing with Arctic eutrophication) or small tanker trucks that collect domestic sewage (much like garbage is collected on a regular basis in more southerly communities). Although some communities use land-based dump sites for sewage disposal, others have designated a lake to act as a sewage lagoon for their domestic waste. One such example is Sanikiluaq, situated on the Belcher Islands in Hudson Bay (Canada). This small Inuit hamlet was established in the late 1960s, when the Canadian government began enforcing a policy to concentrate native populations into communities, as opposed to their more traditional nomadic lifestyles.

Annak Lake was chosen as a sewage oxidation pond for the Sanikiluaq community (*annak* is the Inuktitut word for excrement). Human sewage was trucked from the homes and dumped into the lake, beginning in the late 1960s. In 1983, this sewage stream was augmented by the addition of all liquid domestic waste. Clearly, the dumping of raw sewage into a naturally ultra-oligotrophic lake would lead to eutrophication (Chapter 11) and enhanced algal production was clearly evident in this lake. However, human sewage is also an excellent indicator of total human exposure from ingestion and inhalation of metals (such as mercury, lead, and cadmium), as a high percentage of each metal is excreted. This prompted researchers (Hermanson 1998; Hermanson & Brozowski 2005) to collect a high-resolution sediment core from Annak Lake and document the contributions of this human point source of metal pollution (and thereby they also tracked the history of community exposure to these metals). Paleolimnological studies of a nearby lake (Imitavik) were used as control samples.

Geochemical analyses of the control Imitavik Lake showed that anthropogenic mercury (Hermanson 1998), as well as lead and cadmium

(Hermanson & Brozowski 2005), began to increase modestly in the mid-1800s, reflecting aerial inputs of these elements (see Chapter 8). Annak Lake's geochemical history closely matched the control lake's profiles during the pre-1960s period. However, following the establishment of Sanikiluaq and the resulting sewage inputs, the geochemical data differed dramatically, with Annak Lake sediments tracking large increases in mercury, cadmium, and lead. In fact, mercury inputs to Annak Lake by c. 1990 were about 15 times higher than in nearby Imitavik Lake (Hermanson 1998). Furthermore, from 1970 to 1990, metal inputs to Annak Lake exceeded the rate of population growth in the hamlet, thus suggesting that the per capita contributions of metals were increasing as well during this time (Hermanson & Brozowski 2005).

Studies such as those conducted at Sanikiluaq provide compelling evidence of the bioconcentration and bioaccumulation of contaminants by humans, and how metals contained in sewage can be tracked in sewage lakes using paleolimnological approaches. Similar sedimentary techniques can be used to study the effects of other organisms, such as the contributions of mercury and other contaminants from the guano from large populations of nesting seabirds (Blais et al. 2005; see p. 295).

10.5 Tracking mercury pollution from a point source down a river system

One of the most efficient ways to transport pollutants is via moving waters, and so it is no surprise that rivers have become dumping grounds for toxic substances such as mercury. However, given mercury's affinity for fine-grained organic sediments, it might not travel as far as the polluters may have initially hoped! An example from the Sudbury River (Massachusetts, USA) illustrates this point, and shows how paleolimnological techniques can be used to follow a pollution source through a riverine system.

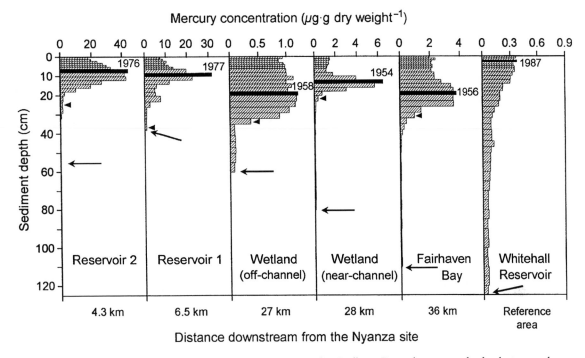

Fig. 10.3 Mercury concentrations from sediment cores in the Sudbury River basin, graphed relative to the distance from the Nyanza industrial site. Reservoirs 1 and 2 were filled in 1879 to provide drinking water for Boston. The last panel (Whitehall Reservoir) is upstream of the putative pollution source, and was used as a reference site. It was impounded in 1900. Arrows mark the bottom of each sediment core. The black histogram denotes the highest mercury concentration recorded in each core, with the estimated ^{210}Pb date noted. The black triangle denotes the last sediment layer deposited before 1917, the year the Nyanza site began operations. From Frazier et al. (2000); used with permission of NRC Research Press.

The Sudbury River system had been contaminated by mercury. A chemical dump site from a former industrial complex (Nyanza), which operated from 1917 to 1978, was a likely suspect for the source. However, due to the absence of long-term monitoring data, the extent, duration, and magnitude of mercury pollution were not known. Frazier et al. (2000) recorded the historical distribution of mercury accumulation in five sedimentary basins of the Sudbury River system; four were progressively downstream of the Nyanza site (i.e., two reservoirs, a wetland area, and a bay), whilst the fifth site (Whitehall Reservoir) was outside the contamination flow and was used as a reference site. An off-channel wetland site was also investigated.

Figure 10.3 summarizes the Frazier et al. mercury data, which confirm that the Nyanza site was the primary source of mercury pollution to this fluvial ecosystem. The timing of the rise in mercury accumulation in the two sites closest to the putative source (i.e., Reservoirs 1 and 2) coincided with the onset of operations at the complex in 1917. Concentrations peaked in the late 1970s and early 1980s, and then declined a few years after closure of the plant in 1978, and decreased further following excavation and capping of the contaminated soil in 1991. Concentrations were highest in Reservoir 2 (44 μg (g dry weight)$^{-1}$), the impoundment closest to Nyanza. Their estimates of baseline (i.e., pre-impact) mercury concentrations (0.038–0.049 μg g^{-1}) were similar to those from other fine-grained sediments in North America, indicating no elevated natural sources of mercury in this system.

The mercury concentrations in the recent sediments (postdating the closure of the plant) from the two downstream reservoirs (0.93–17.7 $\mu g\ g^{-1}$) still far exceeded natural baseline concentrations. Given the high organic content of the sediments (and hence the high binding affinity of mercury to the sediment matrix), upward diffusion of mercury in the sediment profile was considered highly unlikely, and so the continued elevated concentrations in the surface sediments were attributed to the continued input and recycling of some of the industrial mercury.

The wetland and bay sites, further downstream from the industrial complex, provided additional insights into the overall pollution history of this area. The off-channel wetland and the bay sites did not record any marked declines in mercury in the most recent sediments. It is highly unlikely that this recently deposited mercury could be transported from the upstream reservoirs, based on modeling and observations of the system. An alternate source, such as remobilization and recycling of contaminated sediments from wetland areas or the river channel upstream, is more likely. Ancillary data suggest that water draining from adjoining wetlands may be the source.

As expected, mercury levels were much lower in the reference site, Whitehall Reservoir (Fig. 10.3). Nonetheless, just as many other cores taken from relatively pristine areas have recorded, there was a clear enrichment in mercury in the recent sediments. This increase, however, was much smaller than the contaminated sites explored in this study, and reflects overall anthropogenic sources (e.g., burning of fossil fuels) deposited primarily by atmospheric transport.

The Frazier et al. study illustrates some important lessons that can be gleaned from the sedimentary record. First, these approaches can be used on a landscape scale to provide forensic data on the source of pollution. The historically

elevated mercury concentrations in the two downstream reservoirs can be linked to industrial activities at the Nyanza site. As this material is buried by new sediment, it is made less available for recycling and the likelihood of it becoming a future contamination source is steadily decreasing. Second, by calculating concentrations in the pre-impact sediments, the natural baseline concentrations for the region were estimated. Mercury levels in the bay and wetland sites illustrate that mercury pollution is not always straight forward, and can have a variety of sources and pathways. Finally, the study of the reference site provided information on ambient anthropogenic mercury loading from more diffuse sources, such as the atmosphere.

10.6 Evaluating the regional dimensions of mercury deposition in Arctic and boreal ecosystems using lake sediment mercury flux ratios

The examples cited so far in this chapter have focused primarily on local inputs of mercury contamination. Mercury pollution, however, is clearly far more pervasive and complex, with considerable potential for long-distance transport by air currents similar to the problems described with POPs (Chapter 9). For example, regions far removed from point sources, such as Arctic and boreal areas, may have elevated mercury concentrations, sometimes surprisingly so. Questions that are critical for human and environmental health concerns are often heatedly debated, such as the likely source(s) of the mercury, whether it is natural or anthropogenic, and the patterns and magnitudes of change. To put these questions more plainly: Where is it coming from? Who is at fault? What can be done about it?

Given that many northerners (e.g., Inuit) include a significant proportion of native foods in their diets, such as marine mammals, and that mercury can biomagnify similar to POPs

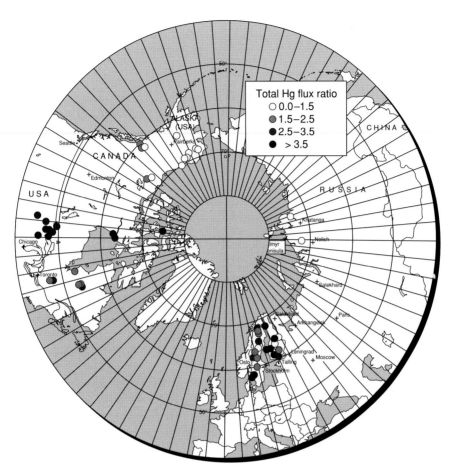

Fig. 10.4 A polar projection map of the Northern Hemisphere, showing the location of the 51 lakes used in the mercury flux ratio database. Each lake is grouped into one of four categories, based on its total mercury flux ratio (i.e., post-industrial to pre-industrial mercury flux). From Landers et al. (1998); used with permission of Elsevier.

(Chapter 9), the potential health effects are worrisome. Challenges associated with the detection and the study of mercury pollution are compounded by the broad geographical ranges that these problems encompass. The absence of long-term monitoring data, which is especially evident in remote high-latitude regions, necessitates the use of indirect monitoring approaches, such as those that can be provided by the paleolimnologist.

Landers et al. (1998) addressed the complex issue of anthropogenic mercury contamination in the Northern Hemisphere's vast Arctic and boreal regions by estimating mercury flux ratios from a database of 51 lake sediment cores, which included paleolimnological data from five of the eight Arctic countries (Finland, Sweden, Canada, the United States, Russia; Fig. 10.4). Mercury profiles collected using similar approaches were also included from seven additional mid-continental North America lakes (Swain et al. 1992). Flux ratios,[1] defined as the ratio (or

[1] Intuitively, flux rates appear to provide the most realistic estimates of past deposition, and they certainly can provide important information, as shown from these examples. However, they may also vary greatly within a lake basin due to sediment focusing, so ideally whole lake averages should be worked out, using multiple cores.

enrichment factor) of mercury flux rates in sediments deposited before c. 1880 (defined as the pre-industrial period) with flux rates in the post-1880 sediments, were chosen as the primary method of data presentation, as this approach tends to minimize some of the biases inherent in these types of measurements. For example, the sediment cores were collected and analyzed by different researchers and laboratories, and sampled at different levels of resolution. Moreover, different geographical regions have dissimilar background (pre-anthropogenic) mercury concentrations, and so flux ratios decrease this variability when sedimentary data are compared across broad spatial scales. Increases in mercury flux can generally be attributed to anthropogenic emissions of mercury. By examining flux ratios in sediment cores from this broad landscape of lake sites, Landers et al. (1998) evaluated the spatial and temporal characteristics of the enrichment of mercury to determine if deposition had increased as a result of industrialization, even in these remote sites, and if so, then what was the magnitude and pattern of this enrichment.

One of the most obvious features of this database was that background levels can be strikingly different between regions, and even within lakes in a particular region. For example, the lowest pre-industrial flux rate was from Hawk Lake (Canada), at $0.7\ \mu g\ m^{-2}\ yr^{-1}$, whilst the greatest flux, about 77 times higher, was from Feniak Lake (Alaska, USA), with a pre-industrial flux rate of $54.3\ \mu g\ m^{-2}\ yr^{-1}$. The serpentine geology characterizing the latter lake explains its elevated, natural levels. Nevertheless, pre-industrial flux rates as low as $2.5\ \mu g\ m^{-2}\ yr^{-1}$ were recorded for some Alaskan lakes, demonstrating the variability in background mercury levels even within a geographical region.

Just as background levels differ amongst regions, changes in post-industrial mercury accumulation rates were markedly different across the spatial gradient. A summary of the changes in post-industrial to pre-industrial mercury flux ratios (i.e., mercury enrichment factors) calculated

for the 51 lakes is summarized in Fig. 10.4. Landers et al. selected four classification categories: enrichment factors less than or equal to 1.5; greater than 1.5, but less than or equal to 2.5; greater than 2.5, but less than or equal to 3.5; and greater than 3.5. Background mercury levels can affect the flux ratio. For example, a site such as Feniak Lake, noted above, would require a proportionately larger post-industrial increases in mercury flux to give the same ratio as one receiving the same elevated mercury loading but having a lower pre-industrial mercury influx (e.g., Hawk Lake).

Certain trends in flux ratios are clearly evident. High flux ratios were recorded in sites associated with strong regional sources of atmospheric mercury emissions, such as central and eastern North America and Central Europe (Fig. 10.4). For example, the two Finnish lakes with the highest increases in flux ratios (up to about nine-fold increases) were those located nearest the two largest urban/industrial centers in Finland. Meanwhile, remote lakes from Alaska and the two sites from the Taimyr Peninsula of Russia showed only moderate (~ 30%) enrichment. At first examination, the relatively low mercury increases in the two Russian lakes may appear surprising, as the large metal smelting facilities at Noril'sk (see pp. 126–8) are only 375 and 900 km away. The relatively modest mercury increases in the Russian cores can be attributed to the fuel source at Noril'sk, which does not use coal or oil, but natural gas, which releases much less mercury. One lake from the Canadian High Arctic (Amituk Lake), on Cornwallis Island, had a surprisingly high flux ratio of 4.1 for such a remote site. This, however, is consistent with previous work on Arctic atmospheric chemistry, which suggests that this area may be a high-deposition zone for atmospherically derived contaminants (Barrie et al. 1992).

Data sets such as these have only rarely been compiled, but they have considerable potential for lake management studies. For example, we now know that lakes in boreal and Arctic regions have had dramatically different pre-industrial mercury fluxes. Increases in mercury accumulation have also been strikingly different, with some lakes only showing modest enrichment, whilst others recording increases by almost an order of magnitude. Sites located far from industrial complexes generally have relatively low flux ratios, whilst those located within about 2000 km from major urban/industrial sources have higher flux ratios. Even the remote sites, however, document about 30% enrichment in post-industrial mercury deposition, suggesting that this value may be a circum-Arctic background enrichment factor. Finally, this study showed that, especially in remote regions where long-term monitoring programs are rarely logistically feasible, lake sediments provide a convincing substitute for more conventional sampling programs, and that these approaches should be extended to other regions and problems.

10.7 Regional patterns of mercury deposition in Norwegian lakes

The above example showed how mercury analyses can be done using lake sediments on a broad spatial scale to track overall patterns of deposition. Similar studies can be conducted on country-wide scales. As introduced in the discussion dealing with metal contamination, Norwegian scientists had used "top/bottom" sediment analyses from 210 Norwegian reference lakes, spanning the length of the country, to track regional depositional patterns for a suite of metals related to industrial activities (see pp. 122–5). Rognerud and Fjeld (2001) also included mercury in these analyses. As shown in Fig. 10.5, mercury concentrations were markedly enriched in the surface sediments of many lakes, but especially those in lowland areas in the south, closest to

Fig. 10.5 A map showing the concentrations of mercury in the surface (0.0–0.5 cm) sediments, representing recent limnological conditions, and bottom, reference (30–50 cm) sediments, representing pre-industrial conditions, in 210 Norwegian lakes. Other metals used in these analyses are shown in Fig. 8.2. The areas of the symbols used are proportional to the square root of the concentrations. Modified from Rognerud and Fjeld (2001).

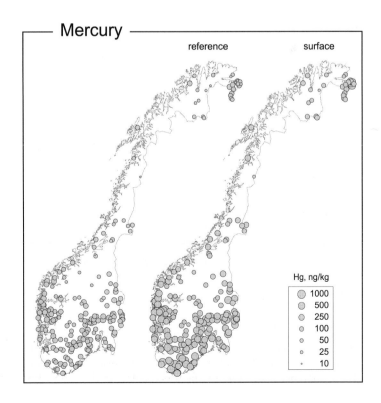

industrial sources in other parts of Europe. An exception was the elevated surface mercury concentrations in the far northeastern corner of Norway (Fig. 10.5), an area that is likely affected by the smelters in the Kola Peninsula of northern Russia.

10.8 Four thousand years of atmospheric mercury deposition recorded in peat profiles

Just as lead isotopes were used effectively to track long-term metallurgical activities in peat cores (see pp. 137–9), similar approaches can be used to track past mercury contamination. Martínez-Cortizas et al. (1999) used similar approaches to those used by Shotyk et al. (1997), described in Chapter 8, to derive a record of atmospheric mercury (Hg) levels since about 4000 yr BP in a peat core from northwestern Spain. The study area is about 600 km northwest from Almadén, the site of the largest Hg mine in the world.

In fact, the area has been mined at least since the time of the Romans, and possibly earlier. Martínez-Cortizas and colleagues showed how Hg deposition has changed in the past, including both natural sources (such as from volcanoes) as well as anthropogenic sources, such as mining (Fig. 10.6). Hg levels started to increase in the eighth century AD, which coincides with increased mining activity for cinnabar (a Hg ore) during Spain's Islamic period, when demand for metals was high. Perhaps more surprisingly, they also found an earlier, gradual increase in Hg around 2500 years ago (Fig. 10.6). This may reflect a period when the Celts were using the cinnabar, but probably more for its colorful pigment than as a metal.

The long-term record of Hg concentrations gathered by Martínez-Cortizas et al. (1999) had other interesting conclusions. By using about 500 years of historical records of past mining in the region, they found that not all the changes in the peat profiles could be attributed to human effects. For example, there was a significant increase in concentrations between about 300 and

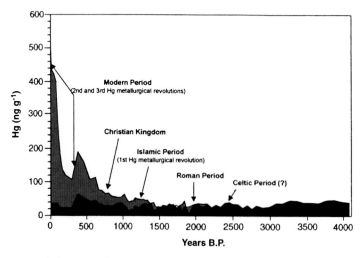

Fig. 10.6 Total mercury and the natural component of mercury accumulation (black area) recorded in a Spanish peat core. The shaded area represents estimates of the anthropogenic component of accumulated mercury, calculated as the difference between total mercury and the natural component. The major phases of mercury exploitation in Spain are marked on the figure. Reprinted with permission from Martínez-Cortizas et al. (1999) Mercury in a Spanish peat bog: Archive of climate change and atmospheric deposition. *Science* 284, 939–42. Copyright 1999 American Association for the Advancement of Science.

600 years ago, a period predating the Industrial Revolution. This period approximately coincides with the so-called Little Ice Age, a period of general cooling in the region. The authors suggested that during colder periods, the bog retained more Hg than during warmer periods. The reasoning here is that Hg, in its elemental form, is a volatile element and is more prone to evaporate from the bog surface during warmer periods (and, conversely, is more prone to be left in the bog during cooler periods). This relationship may potentially have applications in the reconstruction of past climatic trends.

10.9 Summary

Controversy still exists about whether mercury levels in sedimentary profiles archive an accurate record of past accumulation rates or are greatly influenced by diagenetic processes. Many sedimentary profiles, however, appear to be faithfully tracking known changes in mercury deposition. These studies have shown how both local and distant sources can affect aquatic resources. Given the toxicological properties of mercury, coupled with its ability to be transported long distances and to be biomagnified in the food chain, there is certainly much interest in its distribution and trajectories of past changes. What is also evident is that almost all studies completed thus far are geographically focused on temperate and Arctic regions, yet mercury pollution is a global phenomenon. However, recent paleolimnological studies on, for example, Lake Victoria (East Africa), the world's largest tropical freshwater lake, show how sediment analyses can provide important information on pollution trajectories in these regions as well (Campbell et al. 2003). There is little doubt that this area of paleolimnology will see considerable new research efforts in the years to come.

11

Eutrophication: the environmental consequences of over-fertilization

Ecology should be to environmental science as physics is to engineering.
Charles J. Krebs (1994)

11.1 Eutrophication: definition and scope of the problem

Eutrophication refers to the problem of nutrient enrichment of water bodies. One might first think that too few nutrients would be a more serious ecological issue, but eutrophication continues to be ranked as the most common water-quality problem in the world, and remains an active area of scientific research (Schindler 2006). The reasons for this are related to the myriad of symptoms that eutrophic lakes, reservoirs, and rivers often exhibit, such as unsightly algal blooms, large growths of aquatic macrophytes, excessive accumulations of decaying organic matter, taste and odor problems, decreased deep-water oxygen levels, marked shifts in food web structure – including possible extirpations of some fish species and other organisms – as well as many other problems (Fig. 11.1).

Once algal blooms reach high concentrations, other problems may develop. For example, cyanobacteria (blue-green algae) are often associated with eutrophic waters, and are well known for their negative impacts on water quality, including the production of taste and odor prob-

lems, and interfering with certain water treatment processes. When certain cyanobacteria populations (e.g., *Microcystis*) reach very high proportions, they can also produce toxins that can render water unsafe for consumption. As summarized by Carmichael (1994, 1997), these toxic blooms have been reported in the scientific literature for over a century (Francis 1878) and can pose serious problems, especially to livestock that are ingesting large quantities of these infested waters (Fig. 11.2).

As with most environmental issues, there was considerable controversy regarding the causes of eutrophication, especially in the 1960s and 1970s, when the problem began to be more fully recognized. Interest heightened in the 1960s partially due to an increased awareness of environmental issues, but also because it was becoming clear that a large numbers of water bodies were being choked by excessive algal and macrophyte growth from increased effluents. Few would deny that nutrient enrichment was involved – but which nutrients? This had important economic implications, as whichever industry was responsible for adding the putative nutrients, it would also be responsible for addressing the problem.

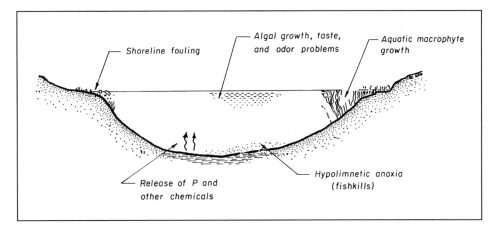

Fig. 11.1 Eutrophication may result in a large number of water-quality problems. The most obvious ones include excessive phytoplankton blooms (often with a shift to less desirable species, such as cyanobacteria, that are not readily grazed by many invertebrates), and sometimes thick growths of aquatic macrophytes and periphyton. Taste and odor problems may also occur. Once this increased biomass dies and settles to the bottom of the water body where decomposition occurs, oxygen levels may become depleted, thus killing off fish and other animals, releasing phosphorus from the sediments (i.e., internal loading), and causing a suite of other problems.

Fig. 11.2 Nerve toxins from cyanobacteria blooming in a pond used for watering livestock killed 16 cows on a farm near Baldur, Manitoba (Canada) in 1996. Photograph courtesy of Manitoba Environment.

Vallentyne (1974) provides a popularized summary of the eutrophication debates, written from the perspective of a scientist keenly involved in the major issues of the time. Vallentyne warned that, if action was not taken to limit phosphorus inputs, inland waters would soon become "algal bowls," similar to the "dust bowl" that characterized large parts of North America during the droughts of the 1930s. Sometimes referred to as the "detergent wars," one can see many similarities to the acid rain debates that would dominate much of the 1980s (Chapter 7). Industries, and in particular detergent companies, whose products then contained large concentrations of phosphorus, fought hard (and some would say mean) to protect their profits. They lost (Box 11.1).

Box 11.1 Liebig's Law of the Minimum, limiting nutrients, and whole lake manipulations

In order for an organism to grow and reproduce, it must have sufficient raw materials to live. These materials include water, certain chemical constituents (such as nutrients, the specific requirements for which will vary between species), and other variables (e.g., for most higher organisms this includes oxygen and light for plants, etc.). The German botanist Justus von Liebig's experimental work on the concept of limiting factors in 1840 showed that if any one element required by an organism is not available in sufficient quantity, the organism will no longer be able to survive. *Liebig's Law of the Minimum*, as it is often referred to, states that under steady state conditions, the growth of an organism is dependent on the amount of essential material that is available in least supply. Whatever element is in shortest supply with respect to demand, then it will limit growth. For example, if you provide an alga with 10 times more nitrogen and carbon than it needs, but not enough phosphorus (another essential nutrient), then growth will stop. Stated more plainly: "You are only as strong as your weakest link."

There can, of course, also be too much of a factor. For example, too much oxygen can harm or kill organisms, too much light can cause photo-damage, and elevated concentrations of many chemicals can poison life forms. These sensitivities vary from species to species, and in fact form the ecological basis of surface-sediment calibration sets and transfer functions (see Chapter 6). Some of the earliest work on the subject of tolerance was done in 1913 by V.E. Shelford, who incorporated these ideas in what is now often referred to as *Shelford's Law of Tolerance*. It states that different organisms can survive within a range of environmental conditions, but have an ecological minimum and maximum, which represents their limits of tolerance. This was shown graphically in Fig. 6.2, where environmental optima and tolerances were discussed.

By the 1960s and 1970s, the role of limiting nutrients in lake systems had become a subject of considerable debate. By the mid-20th century, many lakes, rivers, and reservoirs were exhibiting the typical symptoms of cultural eutrophication. But what was causing this excessive plant growth? Nutrients were almost certainly involved, but which one(s)? There are many nutrients that can potentially limit plant growth. By the 1970s, the relative roles of three elements (carbon, nitrogen, and phosphorus) in causing eutrophication were being hotly debated. Many detergents at that time contained high concentrations of phosphates, and so these industries typically led the charge of the "anything but phosphorus camp"! But studies were beginning to mount showing the critical role of phosphorus in limiting algal growth, at least in most temperate freshwater systems. For example, empirical studies (e.g., Sakamoto 1966; Vollenweider 1968; Dillon & Rigler 1974) were reporting strong correlations between lake-water phosphorus concentrations and chlorophyll *a*, a pigment common to all algal forms and typically used as an estimate of overall phytoplankton abundance. Other studies were also accumulating. But perhaps the most compelling evidence, at least from a visual perspective, came from the manipulation of Lake 226 in the Experimental lakes Area (ELA) of northwestern Ontario. As summarized by Schindler (1974), Lake 226 was a typical, oligotrophic, undisturbed, Precambrian Shield lake, with a constriction in the middle part of the lake giving it an hour-glass shape. By partitioning the lake into its two basins using a sea curtain at this constriction, David Schindler and his colleagues fertilized one basin with phosphorus, nitrogen, and carbon, whilst treating the other basin with identical concentrations of carbon and nitrogen, but no additional phosphorus. As shown in Fig. 11.3, the results were striking: The basin that received the fertilizing effects of phosphorus additions bloomed with cyanobacteria, but the control basin remained in its pre-fertilized state. Other manipulations were done on nearby Lake 227. There could be little doubt that phosphorus was the limiting nutrient. Shortly thereafter, legislation was adopted limiting the phosphorus levels in detergents and other effluents.

Nutrients, such as phosphorus and nitrogen, naturally enter water bodies from sources such as overland drainage and through groundwater. However, human activities can greatly accelerate the rate of nutrient input, and this is often referred to as cultural eutrophication. For example, typical anthropogenic sources include untreated sewage, other municipal and industrial effluents (i.e., point sources), as well as more diffuse or non-point sources (e.g., fertilizer run off from agriculture). The potential of significant additional nutrient enrichment from an aerial component has recently been identified (e.g., see pp. 219–21).

It is important to keep in mind that when one is referring to limnic nutrient concentrations, one is often dealing with very small amounts. Total phosphorus concentrations are typically expressed as $\mu g \, l^{-1}$, which is equivalent to parts per billion (ppb). It is sometimes hard to visualize how small these concentrations are and how dramatically they can affect aquatic systems. For example, 4 ppb is equivalent to one teaspoon of a liquid dropped into a volume of water equivalent in size to an Olympic swimming pool. Many pristine lakes often have TP concentrations below $10 \, \mu g \, l^{-1}$ (i.e., oligotrophic lakes). An increase in TP of $4 \, \mu g \, l^{-1}$ would certainly be considered a significant nutrient addition to such a lake.

Given the critical role of nutrients in determining the characteristics of lake systems, such as their overall levels of production, it is not surprising that limnologists categorize lakes according to their lakewater nutrient concentrations (typically phosphorus, but nitrogen and other elements are also used) and algal biomass (often estimated from chlorophyll *a* concentrations). For example, lakes with low concentrations of limiting nutrients and primary production are referred to as oligotrophic lakes. Lakes at the other end of the extreme are referred to as eutrophic, and mesotrophic systems occupy intermediate levels. These terms, however, can become confusing when one begins to compare lakes between regions, as the boundaries between trophic categories vary from country to country. For example, in Canada, a country blessed with large numbers of nutrient-poor lakes, a lake with a total phosphorus concentration of, say, $32 \, \mu g \, l^{-1}$ would

Fig. 11.3 Lake 226, in the Experimental Lakes Area of northwestern Ontario, was the subject of a classic eutrophication study in 1973. The far, northeastern basin was fertilized with phosphorus, nitrogen, and carbon. The near basin received the same nutrient fertilization, except that no phosphorus was added. Algal blooms dominated the phosphorus-enriched basin within 2 months (Schindler 1974). Photograph supplied by D.W. Schindler.

already be considered eutrophic. Meanwhile, in Denmark, where many water bodies have total phosphorus concentrations greatly exceeding $100\,\mu g\,l^{-1}$, such a lake would still be considered oligotrophic by many managers. One must be careful, then, to avoid misunderstandings when water bodies from different regions are being described and compared (e.g., "eutrophic indicators" described in North American lakes may be common in some European "oligotrophic lakes," simply because lake trophic states are being categorized differently).

Much of the early (pre-1985) applied paleolimnological work dealt with cultural eutrophication, but with the heightened interest in acidification in the 1980s, many paleolimnologists shifted their efforts to these new problems. Although controls on effluents have been implemented, the eutrophication problem is still a serious one and many important management questions remain to be answered. Over the past few years, many new paleolimnological programs have been initiated to tackle some of these issues. This chapter summarizes some of these research directions.

11.2 Inferring past nutrient levels

An increased supply of limiting nutrients such as phosphorus and nitrogen is the root cause of eutrophication, and so it is not surprising that a considerable amount of attention has focused on developing methods to reconstruct past nutrient levels using the sedimentary record. One might first think that since sediments archive a geochemical record, then simply analyzing sediment profiles for the total concentrations of nutrients such as phosphorus and nitrogen would provide a history of past loading. This simple approach, unfortunately, is fraught with difficulties. As summarized by Engstrom and Wright (1984), there are many problems and challenges in using sedimentary phosphorus concentrations, for example, as a proxy of past lakewater nutrient concentrations. Phosphorus retention in sediments is strongly controlled by sorption onto iron oxides, and so variations in both iron content and redox conditions might change phosphorus accumulation in sediments independent of its concentration in the overlying water. Under anoxic conditions, post-depositional mobility of phosphorus can occur and phosphate is returned to the water column (so-called "internal loading"). Determining past nitrogen levels directly from geochemical data is also difficult. However, if interpreted cautiously, and the proper methods are used, sedimentary nutrient concentrations can provide important information on past processes (Boyle 2001). Nonetheless, as described before, more indirect, often biological-based proxy methods are primarily used to infer past nutrient levels.

Fortunately, the distributions of many limnic organisms are related, at least partly, to nutrient concentrations. In a qualitative sense, limnologists have long known that many taxa are characteristic of certain trophic states. For example, it was established early in the 20th century that certain diatoms were more common in nutrient-enriched environments. Such qualitative assessments are still useful but, similar to other aspects of paleolimnology, there has been a trend toward more precise quantification of the ecological optima and tolerances of taxa. Using similar surface-sediment calibration approaches (Section 6.2) developed and refined in the acidification studies, paleolimnologists have steadily been constructing and improving transfer functions to infer past phosphorus concentrations, as well as other nutrients (e.g., nitrogen). Although the nutrient inference functions are not as strong as those typically developed for pH, they are already sufficiently robust to infer long-term trends in the directions and magnitude of eutrophication. Diatoms have again taken the lead (reviewed in Hall & Smol 1999), but other indicators such as chrysophytes (Wilkinson et al. 1999), chironomids (Lotter et al. 1998; Brooks et al. 2001) and Cladocera (Brodersen et al. 1998; Bos and Cumming 2003) have also been used. In the few cases where long-term monitoring data were available, comparisons between diatom-inferred and lakewater nutrient levels have shown good correspondences (e.g., Bennion et al. 1995; Lotter 1998; Bradshaw & Anderson 2001).

11.3 Tracking the major symptoms of eutrophication

The above discussion showed how nutrient levels can be inferred from biological indicators, and will be exemplified further in the case studies that make up the bulk of this chapter. Eutrophication results in many symptoms (Fig. 11.1) and it is these conditions that lake managers are often most concerned with. For example, have blue-green algal blooms increased in intensity and frequency? Have benthic oxygen levels declined to below levels of natural variability? Has biotic integrity changed? Paleolimnologists have steadily been developing methods to track these different symptoms using information preserved in the sedimentary record.

11.3.1 Algal and cyanobacterial (blue-green algal) blooms

One of the most common symptoms of lake eutrophication is the development of large blue-green algal (cyanobacteria) blooms. As noted earlier, cyanobacteria blooms in particular can cause serious lake management problems, including the production of toxins that may harm or even kill mammals. Although paleolimnological evidence, such as the examples presented in this chapter, has shown that these blooms have often been exacerbated by human-induced eutrophication, there is also paleolimnological evidence recording large cyanobacterial blooms prior to significant anthropogenic impacts. For example, Züllig (1989) used fossil pigment analyses to show increases in cyanobacteria beginning thousands of years ago, which persisted to the mid-Holocene in Soppensee, Switzerland. Fritz (1989) recorded an increase in *Oscillatoria* taxa in about 6000 yr BP in a small lake from the United Kingdom, and fossil pigment analyses on Canadian prairie lakes have revealed long periods of blue-green algal populations (Hickman & Schweger 1991). Meanwhile, McGowan et al. (1999) provides compelling evidence for a long history of cyanobacterial blooms in Whitemere (UK), possibly related to prolonged periods of high thermal stability. They conclude that blue-green algal blooms may be a normal feature of these lakes, and are not necessarily a pathology to be controlled.

Studies such as those described above can also be applied to brackish and more saline environments. For example, Bianchi et al. (2000) tracked past cyanobacterial blooms in the Baltic Sea using fossil pigments. Massive blooms of nitrogen-fixing cyanobacteria have been recorded there since the 19th century, and are believed to have increased in frequency and intensity as a result of anthropogenic eutrophication. However, using a combination of fossil pigment and diatom analyses, as well as. ^{15}N and geochemical data, they showed that large cyanobacterial blooms have been a common phenomenon since about 7000 years ago, soon after the salinity in the Baltic Sea increased. The pigment zeaxanthin was used primarily to track past cyanobacterial abundance and the depleted. ^{15}N data suggested that nitrogen fixation was common. As these blooms appear to be a natural feature of the Baltic Sea, the overall goals of current restoration and nutrient reduction strategies should be re-evaluated.

In addition to pigment analyses, past algal production has been inferred using other approaches. For example, biologically produced calcite ($CaCO_3$) in lake sediment has been used as a proxy of past production, since this calcite production is induced as a function of epilimnetic carbon dioxide (CO_2) removal by photosynthesis. For example, Lotter et al. (1997b), working on the annually laminated sediments of Baldeggersee (Switzerland), showed that average calcite size can be used as a proxy indicator of past phosphorus (P) loading to the lake. Isotope analyses also hold considerable potential (see pp. 190–1), as do different types of organic geochemical analyses (e.g., Meyers & Ishiwatari 1993), such as geolipid analyses (Bourbonniere & Meyers 1996).

11.3.2 Excessive aquatic macrophyte growth

Other forms of primary production, in addition to algae and cyanobacteria, can be stimulated by increased nutrient levels. The littoral zones of many nutrient-enriched water bodies are often choked with excessive growths of aquatic macrophytes, which can impair recreational and industrial activities, as well as altering the food web structure. Although macrophyte growth has often been regarded as a lake management problem (witnessed, for example, by the large number of chemical, biological, and mechanical treatments advertised in lake management magazines for removing macrophytes), as discussed later in this chapter, macrophytes may play a key role in regulating the trophic status of a lake.

Past macrophyte abundance and assemblage composition can be estimated using macrofossils (Birks 2001) and/or more rarely pollen (Sayer et al. 1999; Zhao et al. 2006) of the plants themselves. Quite often, though, past macrophyte populations are inferred using more indirect approaches, such as using the abundance of epiphytic diatoms (Moss 1978), chironomids (Brodersen et al. 2001), Cladocera (Thoms et al. 1999; Jeppesen et al. 2001), and/or other indicators (e.g., Odgaard & Rasmussen 2001). These and other approaches are discussed more fully on pages 213–18.

11.3.3 Deepwater oxygen depletions

Oxygen is required for all life forms on this planet, with the exception of some bacteria. Dissolved oxygen measurements, along with temperature, pH, and some water chemistry variables, represent the standard metrics limnologists use to classify lakes. As G.E. Hutchinson (1957) wrote in his classic treatise on limnology: "To classify or to understand a lake, one must know its oxygen regime." It is no surprise, then, that oxygen depletion is considered to be a serious lake management problem, which is typically associated with eutrophication. Hypolimnetic anoxia in eutrophic lakes is typically driven by the biological oxygen demand (BOD) from primarily bacterial respiration and decomposition of the sedimenting organic matter (e.g., algae). Quite simply, with eutrophication and the subsequent increased organic matter production, more material is sedimenting down into the profundal waters, and oxygen is being used up. Dissolved oxygen can also be removed from the water via chemical reactions (chemical oxygen demand, COD). Deepwater oxygen depletion is especially a problem in late summer in stratified lakes, as well as under ice cover (which can lead to so-called "winter fish kills").

Hypolimnetic oxygen depletions can cause many limnological problems. As all eukaryotic organisms need oxygen for metabolism, prolonged oxygen depletions will preclude fish and other biota from inhabiting deepwater regions of anoxic lakes. For example, valuable salmonid and other cold-water fish species cannot tolerate prolonged exposure to dissolved oxygen levels below approximately 4 mg l^{-1}. The distributions of other biota, such as benthic invertebrates, are also closely linked to oxygen levels, and these relationships can be exploited by the paleolimnologist to infer past oxygen levels (see examples below). Determining and monitoring benthic oxygen levels, however, are also critical for many reasons. For example, under reducing conditions, sediments begin to release orthophosphate that had been held in sediments by, for example, iron. With reducing conditions, this phosphorus is released back into the water column (i.e., "internal loading" of phosphorus), adding further to the eutrophication problem. Certain anoxic microorganisms are also capable of releasing orthophosphate during their metabolism under eutrophic conditions. Moreover, maintaining oxic conditions in a lake can help decrease certain types of contamination, such as mercury release.

Dissolved oxygen (DO) concentrations are difficult to compare and track, as they can be very variable, both within and amongst lakes. Not surprisingly, attempts at tracking changes in DO using paleolimnological techniques are especially challenging. Both chemical and biological approaches have been used.

Historically, some of the earliest attempts at inferring past DO in lakes came from geochemical principles. In a classic study, Mackereth (1966) explored the use of a suite of geochemical techniques in paleoenvironmental assessments in the English Lake District. One aspect of his study focused on iron (Fe) and manganese (Mn) concentrations. The supply of these elements to a lake and its sediments could be controlled by many factors, including their concentrations in the drainage basin, erosion intensity, and the limnological conditions overlying the sediments where they accumulate. It is this latter factor that Mackereth believed could be used to reconstruct past deepwater oxygen levels.

Under oxidizing conditions, both Fe and Mn are highly insoluble in lake sediments. However, with lower oxygen levels and reducing conditions, both elements will become more soluble, and would be expected to be mobilized into the overlying water. The key factor is that Mn would be more soluble than Fe. Since the mobility of Fe and Mn in lake sediments will be closely related to the redox potential of the overlying waters, and since Mn will be mobilized more easily than Fe, Mackereth hypothesized that the ratios of these two elements could be used to reconstruct paleo-redox levels, and hence estimate hypolimnetic DO. Should the redox potential continue to decrease to very low levels, FeS_2 (pyrite) may form and precipitate, thus further increasing the Fe : Mn ratio.

The use of Fe : Mn ratios to infer past oxygen levels has found support in a number of studies, but, as reviewed by Engstrom and Wright (1984), many other factors can affect the concentrations and hence ratios of these two elements in lake sediments.

Since pyrite may form in lake sediments from the accumulation of iron sulfides under conditions of anoxia, this may also provide the paleolimnologist with information on past anoxia. For example, Raiswell et al. (1988) used the degree of pyritization (DOP) of Fe to infer overlying oxygen conditions at the time of burial of organic sediments. Cooper and Brush (1991) used sedimentary iron sulfide measurements to help infer the anoxia and eutrophication history of Chesapeake Bay, near the mouth of the Choptank River (USA).

Although chemical inferences of anoxia continue to be used, paleolimnologists are now often using biological approaches to track past anoxia. Benthic invertebrates, and especially the chitinized head capsules of Chironomidae larvae (Brodersen & Quinlan 2006), have been the most frequently used indicators. It has long been known that the distribution of chironomid larvae was at least partially related to dissolved oxygen levels. For example, Thienemann (1921) first classified the trophic statuses of European lakes according to their larval chironomid assemblages, recognizing that some taxa (e.g., *Chironomus* spp.) could survive eutrophic conditions with prolonged periods of low oxygen concentrations, whilst other taxa required high dissolved oxygen

levels (for an introduction to these indicators and their use in reconstructing oxygen levels, see p. 67). Qualitative paleolimnological assessments using chironomid assemblage changes have been used for some time (reviewed in Walker 2001), but several transfer functions have been developed, including those to infer the duration of anoxia (Quinlan et al. 1998) and late-summer hypolimnetic oxygen levels (Little & Smol 2001; Quinlan & Smol 2001).

Other biotic indicators have also been proposed to track past oxygen levels. For example, Delorme (1982) and Belis (1997) used ostracodes to infer the prehistoric oxygen levels in Lake Erie (one of the Great Lakes) and Lago di Albano (Italy), respectively.

11.4 Tracking the causes and symptoms of eutrophication

Below, I summarize several case studies showing how paleolimnological techniques have been used to track various aspects of eutrophication on several temporal and spatial scales. There are many different applications and examples available, and I have focused on relatively recent human-related changes, occurring over the past two centuries or so. Certainly, though, paleolimnological approaches can be used on longer time scales, ranging from the effects of Neolithic and Bronze age disturbances on water quality (e.g., Fritz 1989; Bradshaw et al. 2005), to eutrophication from early agricultural activities of native North Americans (Ekdahl et al. 2007), to the effects of prehistoric Thule Inuit whalers in the High Arctic (Douglas et al. 2004), to the limnological consequences of the medieval beer industry (de Wolf & Cleveringa 1999)!

11.4.1 The Great Lakes: large-scale ecosystem disruptions in inland freshwater seas

The St Lawrence River – Great Lakes system is the largest freshwater system in the world,

Fig. 11.4 Growths of the macroalga *Cladophora* along the shores of the Great Lakes – St Lawrence River system. In the 1960s and 1970s, these growths were far more extensive. Photographs taken by D. Selbie and N. Michelutti.

draining over 25% of the world's supply of fresh water. Not surprisingly, this waterway has had a profound impact on the history of North America (e.g., early exploration and transportation), and has played and continues to play a key role for industry, fisheries, transportation, energy generation, recreation, and society in general. However, this ecosystem has paid a high environmental price for its contributions. Early in the 1900s, raw sewage discharge led to typhoid and cholera epidemics. Overfishing resulted in significant fisheries losses from the 1930s to the 1960s, and water-quality deterioration, habitat destruction, and introduction of exotic species (see Chapter 13) have led to an almost complete alteration of the original ecosystem. The Great Lakes' watershed is currently highly developed, supporting over 35 million residents in its catchment.

Paleolimnological techniques have been used to track a number of environmental problems in this system (e.g., Chapters 9 and 13), but only issues primarily related to cultural eutrophication will be dealt with in this chapter. Due to space limitations, I will further focus this overview primarily on Lake Ontario, the terminal lake in the Great Lakes system before it empties into the St Lawrence River. The area around Lake Ontario was also the first to be influenced by European settlers.

Effluents have plagued the Great Lakes since the expansion of European-style agriculture and towns in the early 19th century, but it was especially in the 1960s, which coincided with increased urbanization, sewering, and the use of phosphate detergents, that the symptoms of eutrophication began to be clearly recognized. For example, algal blooms began to form, lake shores became covered with rotting masses of the green filamentous alga *Cladophora* (Fig. 11.4), deepwater anoxia became a problem, and fish spawning shoals were deteriorating.

Due to the lack of consistent monitoring data, the ontogeny and development of these problems were not documented or fully understood. One of the first detailed paleolimnological studies using species assemblage data from the Great Lakes was the work of Stoermer et al. (1985a), who recorded shifts in diatom species composition and concentration, as well as chrysophyte cysts and biogenic silica changes over the past approximately 300 years in a ^{210}Pb-dated sediment core from offshore Lake Ontario. Pollen analyses for ragweed (*Ambrosia*) were also used to help pinpoint the time of European arrival. The overall goal of this study was to determine the history of modifications to this system (i.e., by knowing where the system has gone, one can better plan for where it may be going), and interpret these changes in terms of the most likely causative mechanisms.

Stoermer et al. showed striking modifications to the diatom flora, which coincided with human impacts. Certain taxa were either extirpated or reduced to such low abundances that they were below detection limits. Meanwhile, several taxa, likely exotics, were introduced into the system, and some became abundant and caused water-quality problems (see Chapter 13). The biggest changes, though, were related to nutrient enrichment.

The period between c. 1700 and 1770 represents the diatom flora that existed prior to large-scale human intervention, and was dominated by taxa characteristic of low to moderate nutrient levels, representing a Lake Ontario ecosystem that was in quasi-equilibrium with its natural environment. The diatom valves were excellently preserved, including many small, thinly silicified forms, indicating that silica levels at this time were in excess of nutrient needs. In the sediment interval dated to approximately 1769, a shift in diatom species composition and a decline in chrysophyte cysts suggest a modest eutrophication event. After c. 1815, though, changes become more apparent. This coincides with periods of major deforestation and the introduction of large-scale agriculture. There is a modest reversal of this trend from c. 1860 to the turn of the century, which coincides with the period of re-vegetation and stabilization of the landscape following large-scale clearances that occurred earlier in the 19th century. A noticeable feature of this period, though, is the decline in absolute siliceous microfossil accumulation, which the authors interpreted as the commencement of silica depletions in the lake.

It is in the 20th-century sediments, however, coinciding with increased residential, agricultural, and industrial activity in the Great Lakes' catchment, that dramatic ecosystem changes were recorded. For example, all the oligotrophic *Cyclotella* diatoms previously present in the system were extirpated, and replaced by taxa characteristic of highly eutrophic waters or by eurytopic species. The lake was now strikingly different than its "natural" state.

The situation continued to deteriorate in the post-1945 sediments, which coincided with increased population growth in the drainage, increased sewage inputs, and the introduction of phosphate-based detergents. Production and/ or preservation of siliceous microfossils further declined, likely due to increased silica limitation with the phosphorus enrichments. Moreover, Stoermer et al. (1985b) showed progressive changes in the frustule morphology of important diatom taxa, indicative of silica limitation. With declines in planktonic diatoms, another symptom of silica depletion, benthic taxa were more common.

These paleolimnological data showed the controlling influence of phosphorus enrichment in the eutrophication process, but also emphasized the importance of secondary nutrients, such as silica. Maximum phosphorus loading, and corresponding silica limitation, occurred during the 1960s and early 1970s.

Diatoms are clearly important indicators for tracking past eutrophication trends in the Great Lakes, but other proxies can supply additional information. For example, there is considerable potential in using sedimentary isotopic data to infer past lakewater production levels. Schelske and Hodell (1991), Hodell et al. (1998), and Hodell and Schelske (1998) used the carbon and nitrogen isotopic signatures in Lake Ontario sediments to track past eutrophication patterns. The overall concepts are as follows. During photosynthesis, carbon and nitrogen isotopes are used differentially (fractionated) by algae and cyanobacteria, because the lighter isotopes (i.e., ^{12}C before ^{13}C in CO_2, and ^{14}N before ^{15}N in NO_3^-) are removed preferentially from the dissolved inorganic carbon (DIC) and dissolved inorganic nitrogen pool, leaving the remaining carbon and nitrogen enriched in ^{13}C and ^{15}N. A change in the $\delta^{13}C$ and $\delta^{15}N$ of organic matter produced

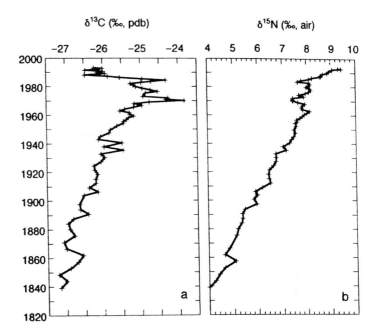

Fig. 11.5 Carbon (A) and nitrogen (B) isotopic composition of bulk sedimentary organic matter in core LO-G32-93 from eastern Lake Ontario. From Hodell and Schelske (1998); used with permission of ASLO.

in the surface waters should reflect changes in productivity in a stratified lake.

All the Lake Ontario cores they analyzed for C isotopes displayed a progressive increase in the 1900s, peaking in the early to mid-1970s, and then decreasing to the present (Hodell & Schelske 1998). These reproducible patterns matched the historical trends of phosphorus loading, suggesting that $\delta^{13}C$ of organic carbon is a reliable proxy for paleo-productivity and responds to spring phosphorus supplies (Hodell & Schelske 1998; Fig. 11.5A). Meanwhile, the $\delta^{15}N$ of sedimentary organic matter increased linearly from c. 1840 to 1960, and then remained relatively constant except for an increase in the upper few centimeters of sediment (Fig. 11.5B). Hodell and Schelske (1998) interpreted the nitrogen data to reflect a combination of factors, including early forest clearance by Europeans, increased sewering by municipalities after about 1940, increased nitrate utilization as productivity increased in the lower Great Lakes, and possibly increased rates of denitrification in upstream Lake Erie from about the 1930s to the early 1970s. The carbon isotope signal in calcite was more complicated, and may also reflect climatic

factors (Schelske & Hodell 1991). Similar to the calcite patterns recorded by Lotter et al. (1997b) on the annually laminated sediments of Baldeggersee (Switzerland), Hodell et al. (1998) showed an exponential rise in carbonate accumulation after 1930 in cores from Lake Ontario, culminating in peak values in the early 1980s. These patterns matched the known eutrophication history of the lake.

11.4.2 Exploring the relative roles of phosphorus and silica in limiting diatom growth in the Great Lakes

Some of the earliest paleolimnological data tracking eutrophication in the lower Great Lakes was derived from sedimentary biogenic silica analyses (a measure of past siliceous algal production, primarily diatoms). In one of the first paleolimnological applications of this technique, Schelske et al. (1983) showed that diatom production peaked from c. 1820 to 1850 in Lake Ontario, at about 1880 in Lake Erie, but much later (c. 1970) in Lake Michigan. They postulated that silica depletion resulted from a small phosphorus enrichment of < 10 µg phosphorus per liter, a level experienced in Lake Ontario and Lake Erie in the 1800s, but not present in Lake Michigan until the mid-1950s. Schelske et al. reasoned that phosphorus enrichment would stimulate diatom growth, but only as long as silica was available in the system. Once diatom production increased to the point that lakewater silica reservoirs were depleted, diatom populations would become silica limited, decline, and be replaced by other algal and cyanobacterial populations.

Schelske (1999) summarized the management implications of this pioneering study, with further context supplied by subsequent paleolimnological data from the Great Lakes. Biogenic silica accumulation increased in Lake Superior and Lake Huron coincident with enhanced phosphorus enrichment following European settlement in their catchments, but no peak occurred as eutrophication was relatively minor (total phosphorus concentrations in these lakes never exceeded 4–5 μg l^{-1}) and silica was never limiting. However, in systems such as Lake Ontario, a peak and subsequent decline in biogenic silica in the late 1800s signaled a relatively early depletion of the epilimnetic silica reservoir, as diatom production (and subsequent sedimentation) increased. Silica depletion was also indicated in cores from Lake Erie and Lake Michigan, although depletion occurred much later in the latter system. Comparisons of profiles collected from all five Great Lakes supported the hypothesis that once phosphorus loading was sufficient to increase diatom growth past a certain threshold, then depletion of silica reservoirs began to limit diatom production.

Paleolimnological inferences on resource limitations have important lake management implications. As water becomes progressively more silica-limited, diatom populations decline and are replaced by primarily cyanobacteria and green algae, two groups that are often associated with many of the problems discussed in this chapter. Replenishment of silica in these large lakes with long residence times is controlled by geochemical factors, so response times to

phosphorus reduction will be slower than those measured by reduction in standing crops of algal biomass (e.g., chlorophyll a). But replenishment of silica supplies will be an important tool in assessing long-term responses to phosphorus remediation. Data such as those gathered by Schelske (1999) can be applied to other lake systems to help determine which nutrients are limiting algal populations, and so assist mangers in setting water-quality guidelines.

11.4.3 Have phosphorus reductions had any effect on eutrophication in the Great Lakes?

By the late 1960s it had become clear to residents, and then to politicians, and eventually (often requiring legislation) to industries that it was time to take serious action on water quality. One of the major mitigation efforts instigated was reductions in the amounts of nutrients entering the lakes. These efforts have generally been seen as a success story. Loading from point sources has been drastically reduced since 1972, primarily through sewage treatment plants (over $9 billion dollars has been spent by the Canadian and US governments over the past three decades). Moreover, the phosphate content of detergents was dramatically reduced. Nutrient from diffuse sources, however, such as from agricultural lands, is still a significant problem. Paleolimnological data have been used to track and evaluate this recovery.

Continuing with the Lake Ontario example described above, Wolin et al. (1991) sampled the recent sediments of a core collected in 1987 sectioned into 0.5 cm intervals, and studied the diatoms preserved within. They showed that, by the 1980s, small changes in diatom assemblages had occurred, which appeared to be indicative of reduced phosphorus loading, and some indications of reduced silica depletion. However, there was no reappearance of taxa that characterized the Lake Ontario flora prior to c. 1935.

Much more rapid trophic state changes have been described from the recent sediments of Lake Erie's central basin (Stoermer et al. 1996), the most eutrophied lake in the Great Lakes

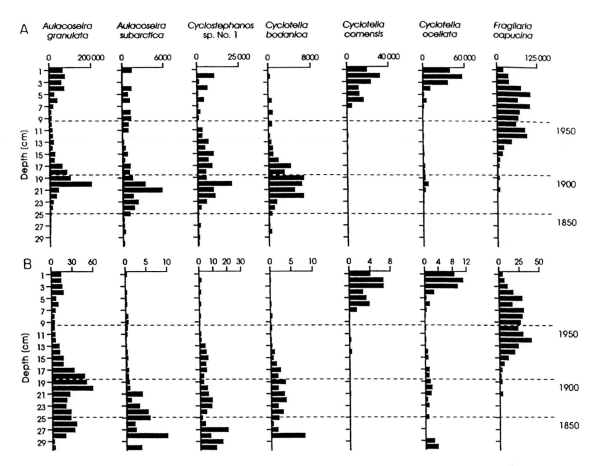

Fig. 11.6 Concentrations (valves/mg; A) and relative frequency (% total diatom sum; B) of the most common diatom taxa versus sediment depth, dated using ^{210}Pb geochronology, from the central basin of Lake Erie. From Stoermer et al. (1996); used with permission of NCR Research Press.

system. Diatoms preserved in sediments deposited after c. 1972 contain increasing numbers of taxa associated with oligotrophic conditions, and especially *Cyclotella* species, which grow in the summer, deepwater chlorophyll maximum layer in the lake (Fig. 11.6). Further indication of more oligotrophic conditions since the instigation of controls on nutrient inputs is that the closest analogues to the assemblages deposited after c. 1986 are assemblages that were deposited c. 1900. However, there are notable exceptions. For example, several taxa common in material predating European impact have not recovered (e.g., *Cyclotella americana*), and now may be extirpated (see Chapter 13). Furthermore, some

of the recent species changes were unusual, in that they did not have analogues in the lake's earlier assemblages, and may be related to the higher nitrate levels.

What was perhaps most striking in the profiles was the rate and magnitude of changes in the recent sediments. Abiotic factors related to nutrient reductions were clearly important. In addition, the authors surmised that biotic factors, namely the limnological repercussions of the establishment and incredible colonization success in the late-1980s of the exotic invader (see Chapter 13) the zebra mussel (*Dreissena polymorpha*), may have altered the limnological environment to such an extent as to influence diatom assemblages. The voracious filtering abilities of this invader, coupled with the decreased nutrient load to the lake, resulted in increased water clarity and the establishment of deepwater diatom populations.

11.4.4 How closely do eutrophication patterns in local bays reflect trends recorded in open-water regions?

More work has been completed on the open-water regions of the Great Lakes rather than the nearshore environments. For many residents, however, it is the inshore bays that are most important. Embayments may also be characterized by more intense water-quality problems, due to the concentration of local inputs and less dilution in the open water.

One of the most studied bays in the Great Lakes with respect to cultural eutrophication has been the Bay of Quinte, a long and relatively narrow embayment on the Canadian side of Lake Ontario. This region was one of the earliest centers of European colonization in Ontario, and it supported a productive (now defunct) fishery. Paleolimnological analyses of a core from the Bay, sampled for both chironomids (Warwick 1980) and diatoms and chrysophytes (Stoermer et al. 1985c), recorded marked changes in assemblages that closely matched known historical disturbances. Prior to the arrival of European settlers, the area was inhabited by indigenous peoples, who appeared to have some minor impact on the bay, as the high relative abundance of benthic diatoms and chrysophytes during this period suggests an ultra-oligotrophic system. With the arrival of European settlers (first the French in the late 1700s, and then primarily the English and Loyalists from the United States in the 1800s), eutrophication of the bay became severe, and silica depletions, similar to the open-water example discussed above, were evident. The modern eutrophic diatom flora was established by c. 1928.

An even more dramatic example of water-quality degradation in an embayment is from Hamilton Harbour, near the head of Lake Ontario. Following the arrival of French explorers in the late 1600s, the area gradually became a major industrial region (currently the location of the cities of Hamilton and Burlington, Ontario). By the early 1970s, the harbor was identified as one of the most polluted areas in North America. However, historical records of water quality before the 1940s are absent, and so paleolimnological studies were instigated.

Yang et al. (1993) provided a historical context for this bay's degradation by studying the distribution of diatoms, chrysophytes, and sponge spicules in a ^{210}Pb-dated core from the harbor. Prior to the arrival of European settlers, the bay was oligotrophic. The assemblages changed rapidly with deforestation and land clearance, and indicated increased eutrophication. The eutrophication signal, however, was complicated further by other forms of pollution, such as industrial waste. The greatest shift in assemblage composition occurred after c. 1970, with marked diatom species changes, and the extirpation of chrysophytes and sponges.

These data showed how local disturbances can be tracked in cores taken from embayments and, in some cases, the inferred declines in water quality can be markedly more severe in these restricted water bodies than in the open-water pelagic regions, where most monitoring data are often available.

11.4.5 Has eutrophication resulted in increased nearshore bio-fouling in the St Lawrence River?

A natural extension of the Lake Ontario studies was an analysis of long-term environmental changes in the St Lawrence River, which is the conduit between the outflow of the Great Lakes at Kingston (Ontario) to the Atlantic Ocean, a distance of about 1600 km. The St Lawrence Seaway was opened in 1959 and became the world's largest inland navigable waterway. In addition to receiving water from the Great Lakes, with all the water-quality problems associated with those systems, additional pollution sources occur throughout the river proper. For example, major industries line the river, as do dams and power generation stations, as well as densely populated cities and agricultural lands. Similar to the Great Lakes, the river has been assaulted by a variety of chemical (see Chapters 8–10) and biological (see Chapter 13) pollutants and stressors. In this chapter, I will focus on problems related to nutrient enrichment, and specifically one of the most obvious symptoms of eutrophication along the river: excess growth of macrophytes and macroscopic algae. Similar to the lakes that feed into the St Lawrence River, the nearshore habitats have been choked with excessive plant growth, which was especially noticeable in the 1960s and 1970s.

There has been considerable interest in restoring the St Lawrence ecosystem, but important questions remain. For example, what conditions should the river be restored to? What were pre-impact conditions like? Were macrophytes and macroalgae always present in high abundance in the river system, or has nearshore bio-fouling only occurred with increased nutrient loading? What were the causes, timing, and extent of deterioration? Have recent declines in nutrient levels resulted in similar biological responses to those recorded in paleolimnological studies from the Great Lakes? What was the impact of the construction of the seaway and the changes in hydrology? How similar were successional changes occurring in the river to those occurring in the upstream lakes?

It would seem that paleolimnological approaches could effectively answer these questions. However, as noted in Chapter 3, river systems are notoriously challenging for paleolimnological work, as it is difficult to find areas of the river system where continuous and uninterrupted sedimentation occurs. Rivers, by definition, are flowing, and so sediments (and the important information they contain) may be continuously or episodically swept downstream. Although few paleolimnological studies with a water-quality focus have been attempted,

careful surveying of river systems can often yield restricted areas where sedimentation can match the record archived in most lentic systems, such as lakes. This can take the form of protected river bays – but these may, in some cases, not totally reflect river systems, as some bays are sufficiently isolated to be at least partially independent of the mother river system. This is, of course, ideal if the paleolimnologist wishes to track changes in a particular bay, rather than the river itself (just as the Lake Ontario bay examples discussed above). More direct repositories of past river environments are sediments from "fluvial lakes," which are wider areas of the river where flow slows down considerably, and sediment profiles can accumulate (Fig. 3.2). Paleolimnologists have also used floodplain sediments to track past changes in river environments.

The general lack of accumulating sediments throughout large stretches of some river systems results in interesting calibration challenges. As discussed throughout this book, surface sediments often form the mainstay of our estimates of a taxon's environmental optima and tolerances. Without reliable accumulating sediments, other approaches must be attempted. In our diatom-based paleolimnological study of fluvial lakes in the St Lawrence River, we sampled a variety of microhabitats to calibrate diatom indicators to environmental variables to develop a habitat inference model (Reavie & Smol 1997). This involved samples from 50 sites along the river, reflecting a gradient of pollution, and included diatoms attached to *Cladophora* (O'Connell et al. 1997), which is the major macroalga associated with nearshore bio-fouling along the river, as well as other aquatic macrophytes (i.e., epiphyton; Reavie & Smol 1998), as well as diatoms attached to rocks (i.e., epilithon; Reavie & Smol 1999). In general, similar statistical techniques to those used for surface-sediment training sets were then applied to the data. Several water chemistry variables often associated with eutrophication were shown to be significantly related to the diatom assemblages. However, habitat type (i.e., substrate) was clearly important (Fig. 11.7),

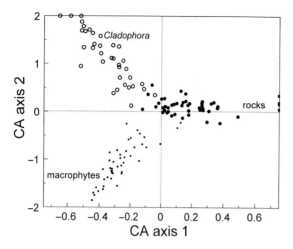

Fig. 11.7 Correspondence analysis of the diatom assemblages (samples) in the three main periphytic habitats present in the St Lawrence River (i.e., *Cladophora*, macrophytes, and rocks). The grouping of diatom assemblages according to habitat was clearly shown, indicating the potential for the development of a paleolimnological inference model to track past changes in the relative importance of these three substrates. Most attention focused on the history of *Cladophora* proliferation and the development of aquatic macrophytes (as indicators of cultural eutrophication) as opposed to epilithic habitats, which indicated relatively clean waters. From Reavie and Smol (1997); used with permission from the International Association of Great Lakes Research.

and so a diatom-based model to infer the relative importance of *Cladophora*, aquatic macrophytes, or rocks as substrates for diatom growth was developed (Reavie & Smol 1997).

As noted above, one of the biggest problems noted in the St Lawrence River was the proliferation of macrophytes and *Cladophora*, which were especially noticeable in the 1960s and 1970s, before the introduction of phosphate restrictions in detergents and other nutrient mitigation efforts. There was no doubt that algal and macrophyte growth was extensive once environmental issues began to be raised; there was little information, however, on the extent of river

eutrophication and associated bio-fouling before the media and other groups began to bring these issues to the forefront.

For this discussion, I will focus on attempts to reconstruct past limnological and shoreline habitat characteristics in two fluvial lakes: Lake St François, near the city of Cornwall, Ontario; and Lac St Louis, near the large metropolis of Montréal, Québec. Previous work (Carignan et al. 1994) showed that useful and datable sediment cores could be retrieved from the fluvial lakes and be used to track past contaminant histories of the region. However, many river management questions remained. For example, little information was available on trends in other limnological variables, nor were there data on how nearshore habitats had been altered since the 19th century. Reavie et al. (1998) used the diatom calibration data they had gathered from the 50 sampling sites along the river, as discussed above, and used these inference models to reconstruct past trends from the diatoms preserved in ^{210}Pb-dated sediment cores from the two fluvial lakes. Their data showed that intense agricultural practices, increased population growth, and canal construction coincided with increased growths of littoral macrophytes and macroalgae. In addition, diatoms characteristic of highly eutrophic environments dominated the assemblages from about the 1940s until the late 1960s. The inferred eutrophication trends coincided with those documented for the upstream Great Lakes, as discussed in the previous section. However, although the overall trends were similar, the actual assemblages were often strikingly different. This, of course, is not surprising, as the fluvial lake cores were more closely tracking riverine changes based primarily on lotic, periphytic diatoms, and not taxa from the upstream lakes. Similar to the Great Lakes studies, diatom shifts in the recent sediments of the fluvial lake cores indicated some recovery, coinciding with nutrient abatement programs. Not surprisingly, less improvement was noted in the cores collected closer to Montréal, a major urban center.

The above study attempted to track past *Cladophora* abundances using an indirect approach, namely the diatoms known to be characteristically associated with this macroalga in this system (O'Connell et al. 1997). In some limnological settings, it is also possible to estimate past macroalga abundance by counting individual *Cladophora* fragments on microscope slides (Einarsson et al. 1993).

11.4.6 A summary of the important management questions answered concerning the St Lawrence River – Great Lakes ecosystem

Many other paleolimnological studies were conducted on the Great Lakes system, and many important insights were gained. E.F. Stoermer, who has led much of the diatom-based work, concluded that "It would be virtually impossible to obtain a spatially and temporally integrated sample of any of the lakes in a given year of the quality routinely available through paleolimnological techniques by any other method" (Stoermer 1998, p. 523). Looking back at his work, Stoermer summarized the critical information that paleolimnology offered on the following management questions:

1   To what extent has the chemistry and hence the food chains of the lakes been altered by human activities? The work of Stoermer and colleagues showed that nutrient additions have greatly altered the quality of food sources for organisms higher in the food web. Limnological conditions appeared to be similar in all the lakes prior to European settlement, but then changed dramatically (Stoermer et al. 1993).

2   How early in the course of settlement and industrialization did these changes occur? Large-scale ecological change in the Great Lakes was primarily thought to be a recent phenomenon (c. 1930), with most of the problems occurring in the second half of the 20th century. Paleolimnological data, however, showed that changes began to occur as soon as substantial European populations were established, and that, for example, Lake Ontario was severely impacted by as early as c. 1830. This has important implications for issues dealing with critical loads and assessing the sensitivity of systems.

3   Were chemical factors (e.g., nutrient concentrations) more important in determining phytoplankton populations (i.e., bottom-up controls; see p. 253), than were biotic factors (i.e., top-down controls, such as grazing). In contrast to some smaller lake systems, paleolimnological studies could not easily discern any top-down controls, suggesting that chemical factors (e.g., nutrients) were more important in the Great Lakes. For example, large-scale changes in diatom assemblages typically predated changes in fish populations by 30–50 years (a more recent exception, though, has been the introduction of zebra mussels; see below). This has important management implications.

4   How important were inputs from outside the drainage basin? Due to the very large surface areas of the Great Lakes, a significant source of pollutants can be the airshed. Paleolimnological data showed that the loadings of nitrogen and some trace metals potentially affected biotic communities to a degree previously unrecognized (Stoermer 1998).

5   How important were exotic invaders to the Great Lakes system? This is discussed in more detail in Chapter 13, but clearly the effects of this "biological pollution" cannot be divorced from eutrophication effects (e.g., zebra mussels filtering out a significant part of the phytoplankton, thus decreasing biomass, even without concomitant declines in nutrients).

6   How permanent were some of these changes? The paleolimnological data showed that some recovery had occurred in systems with nutrient abatement, but that severely perturbed biotic systems will likely never completely return to their original states. There is also an inherent lack of predictability in such large, complex systems.

11.5 Tracking past nutrient changes in rivers by using offshore sediments

The St Lawrence River example showed that cores taken from river systems can, in some cases, be used directly to undertake paleolimnological assessments of river eutrophication and other environmental changes. Nonetheless, finding reliable stratigraphic sequences in flowing systems is rarely an easy task for the sedimentological reasons noted earlier.

An alternate approach is to use sediment cores from the water body where the river eventually discharges, and tracking river-related changes that are archived in the receiving water body (Fig. 3.2). For example, if a river is flowing into a lake, then riverine material and sediments, as well as indicators (such as lotic diatoms and other biota), are also carried into the inflow lake and become incorporated into the lake's sediments. Ludlam et al. (1996) used this approach to reconstruct past river inflow (and its paleoclimatic significance) for an Ellesmere Island drainage system by tracking the proportion of river diatoms in a High Arctic lake's sediments. If the river water was also a major source of, for example, nutrients or acidity, then a change in the lake flora could also be used to track these changes, which could ultimately be related back to the river. The same approach can be used for rivers flowing into the ocean, as was shown by a study of the eutrophication history of the Mississippi River.

The Mississippi River is the largest river in North America (and the third largest in the world), draining 40% of the contiguous United States. Much of its catchment has undergone dramatic land-use changes in the past, and nutrient loading has consequently increased. However, once again, historical monitoring records are not adequate to gauge these changes (nutrient data were only available for four years before the 1950s, when more regular monitoring began) or to determine their ecological consequences. Turner and Rabalais (1994) posed the question: If nutrient input has supposedly been increasing in the Mississippi River, can one track these changes by using the paleoenvironmental record stored in the continental shelf records influenced by the river?

Sediment cores collected from the river's delta bight were analyzed for biogenic silica (BiSi),

which was used to estimate past diatom abundance. The general pattern that emerged was one where an equilibrium accumulation of BiSi occurred from c. 1800 to c. 1900, which was followed by a modest increase, which was then followed by striking increases since the 1980s. The recent changes in BiSi closely paralleled the available monitoring data for Mississippi River nitrogen loading, as measured since the 1950s. This eutrophication would significantly affect food webs in the coastal waters, and Turner and Rabalais (1994) surmised that the resulting increased primary production would also influence the frequency and duration of hypoxic (< 2 mg l^{-1} dissolved oxygen) events at the bottom of the water column, a significant water-quality problem that has been related to seasonal declines in benthic marine biota.

The question of decreased deepwater oxygen levels was addressed more explicitly in a subsequent study by Gupta et al. (1996), who studied benthic foraminifers from four of the cores collected as part of the Turner and Rabelais study. Foraminifera are marine protozoa that have a calcareous shell (or test) and are widely used as bioindicators by paleoceanographers. By using the ratio of two dominant taxa (which differ in their oxygen requirements), they concluded that oxygen stress had increased in the continental shelf waters of Louisiana, and this increase in hypoxia was especially evident over approximately the past 100 years, coincident with the progressive increase in river-borne nutrients.

11.6 Eutrophication of estuaries: a long-term history of Chesapeake Bay anoxia

Estuaries are defined as semi-enclosed bodies of often brackish water that have a free connection to the sea. Human populations have used these often rich and productive systems to harvest food throughout history. As estuaries are the mouths of river systems, they are often repositories of human, agricultural, and industrial wastes. Not surprisingly, eutrophication is a major environmental concern.

Paleolimnological studies on estuaries have rarely been attempted, as they typically have complex and variable sedimentation patterns in space and time, and, similar to rivers, resuspension of sediments can be a serious problem (Cooper 1999). Nonetheless, reliable sediment deposition zones have been identified in several investigations, and some successes should encourage future endeavors. For example, Cooper (1995, 1999) and Cooper and Brush (1991, 1993) have shown how multi-proxy analyses of sediment cores from Chesapeake Bay, at the mouth of the Choptank River (Maryland, USA), could be used to track cultural eutrophication patterns in this important estuary. Since the first recorded occurrence of anoxia in the bay in the 1930s, there has been considerable debate on the relative importance of natural (i.e., climate) versus anthropogenic (i.e., nutrient fertilization) factors in causing these serious problems. Regular monitoring began in 1984, and so the only archives available to determine the causes of anoxia were those preserved in the sedimentary record. By combining analyses of sedimentation rates, pollen, diatoms, biogenic silica, geochemistry, and the degree of pyritization of iron (DOP), Cooper and Brush were able to convincingly show that European settlement and land-use changes were responsible for the poor water quality affecting the estuary. Although human impacts were already discernable from the 17th-century sediments, the paleolimnological data showed that eutrophication accelerated greatly after the mid-1800s, when the estuary was being affected by the higher concentrations of nutrients from fertilizers that were being applied in large concentrations on farmers' fields. Increasing discharges of sewage and subsequent industrial effluents further exacerbated the problems, with the most marked changes in anoxia occurring in the post-1940 sediments. These data provided strong evidence to managers that anoxia and other symptoms of eutrophication were the result of human activities, and

not within the realm of natural variability for the bay. Similar trends have been observed in estuaries on the North Carolina coast (Cooper 2000).

11.7 Separating the influence of nutrients from climate on lake development

Lake sediments store information from a variety of sources, and track many different environmental signals. Not surprisingly, it may be difficult to pinpoint which variables are the most important in influencing limnological conditions, as typically many variables are shifting simultaneously. For example, as significant climatic changes have often coincided with periods of recent lake eutrophication, how can one determine if the changes are due to climatic variability or to nutrient enrichment? Lotter (1998) tackled this problem by taking advantage of the precise temporal resolution made possible from an annually laminated (varved) lake sediment core from hypereutrophic Baldeggersee (Switzerland). Diatom assemblages were enumerated at an annual level (i.e., varve by varve) from AD 1885 to 1993 to address four major goals: (i) to investigate the dynamics and inter-annual variability of assemblages and compare these data to the historical phytoplankton collections available for this lake; (ii) because lakewater total phosphorus (TP) data were available for about 40 years, to compare the diatom-inferred TP data to measured values, in order to evaluate the accuracy of the transfer function; (iii) to use these paleolimnological data to track the eutrophication history of this site; and (iv) to use variance partitioning analyses (VPA) to separate the influence of climate and nutrient enrichment, and so determine which variable is most important in determining diatom species composition. VPA is a powerful statistical technique that allows one to partition out the effects of different stressors (Borcard et al. 1992).

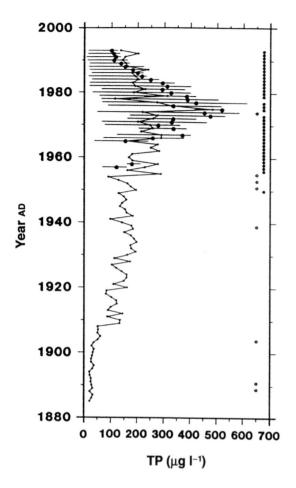

Fig. 11.8 Measured water column (the top 15 m) total phosphorus (TP) during spring circulation in Baldeggersee (solid circles) compared to diatom-inferred TP in a varve-by-varve analysis from this annually laminated sediment core. The horizontal lines show the annual range in measured TP. The dots to the right represent samples with close (filled dot; second percentile) and good (open dots; fifth percentile) modern analogues in the Swiss training set used for the TP inferences. From Lotter (1998); used with permission of Sage Publications.

The diatom-inferred TP closely matched the available measured spring water chemistry data (Fig. 11.8), demonstrating the strength of the transfer function. Furthermore, the fossil assemblages were similar to the available archived phytoplankton data, further confirming that the preserved diatoms were a true reflection of lake communities. Although inter-annual changes in diatom assemblages were low, three major steps in eutrophication were recorded at 1909, the mid-1950s, and the mid-1970s (Fig. 11.8).

But how does one disentangle these changes from concurrent climatic changes? To answer this question, Lotter also collated the yearly (1885–1993) values for mean spring air temperature. Using VPA, he partitioned the total variance in the diatom data into four components: (i) variance due to lake trophic status only; (ii) variance due to climate only; (iii) covariance between trophic state and climate; and (iv) unexplained variance. He found that trophic status was the best explanatory variable, and so diatoms assemblages were most closely tracking changes in variables such as lakewater phosphorus in this lake.

Not all paleolimnological studies have such outstanding, annually resolvable sediments, such as the core from Baldeggersee afforded. However, similar approaches can be used in ^{210}Pb-dated sediments, if researchers are prepared to acknowledge the additional errors that are inherent in these analyses (see the following example).

11.8 Examining the effects of multiple stressors on prairie lakes: a landscape analysis of long-term limnological change

Paleolimnological analyses have shown that lakes in different regions may have startlingly different limnological trajectories. For example, due to the combined effects of climate, vegetation, soils, hydrology, and other variables, some lakes will have very different background, pre-impact conditions. This information is critical for effective lake management strategies, such as setting realistic mitigation goals. For example, Brenner et al. (1993) showed that some Florida lakes have been eutrophic for millennia, long before human populations could have significantly accelerated the eutrophication process. A

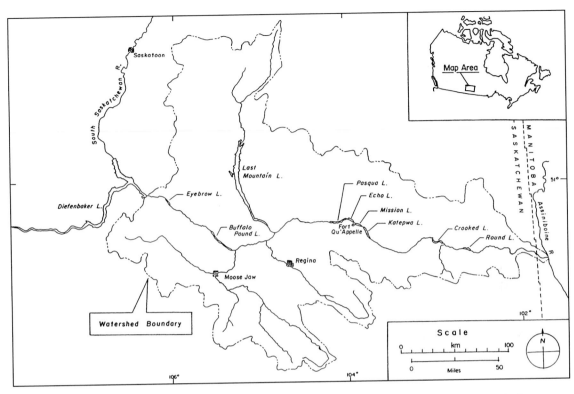

Fig. 11.9 A map of the Qu'Appelle Valley (Saskatchewan) drainage system. Water flows through the system from west to east. From A.S. Dixit et al. (2000); used with permission of Blackwell Publishing.

series of hyper-eutrophic lakes in the Canadian Prairies (Saskatchewan), which will be discussed in more detail here, provide another example of naturally eutrophic lakes with important management implications.

The prairie lakes of the Qu'Appelle Valley (QV) drainage system (Fig. 11.9) provide water to about a third of the population of the Canadian Great Plains. Due to the high production of grain, this area (along with the adjacent US prairies) has been referred to as the "breadbasket of the world," and is therefore of great economic and social importance for Canada and its trading partners. Water quality, however, is considered to be poor and may be deteriorating further, although long-term monitoring data are not available. Currently, the lakes are plagued by excessive algal blooms and macrophyte growth,

fish kills, and other problems typically associated with excessive eutrophication. Nutrient concentrations are exceptionally high. For example, total phosphorus and total nitrogen levels can exceed $600 \, \mu g \, l^{-1}$ and $4000 \, \mu g \, l^{-1}$, respectively. Management initiatives are hampered due to the inability to identify and therefore regulate agents causing the eutrophication.

Before any effective mitigation efforts can be initiated, the natural, pre-impact conditions of the lakes should be assessed (i.e., to set realistic targets). Paleolimnological work in many other temperate regions has shown that, prior to the arrival of European-style agriculture and other activities, lakes were generally quite oligotrophic (e.g., see other examples used in this chapter). This is certainly the perception of the general public: Lakes should not have massive growths

A

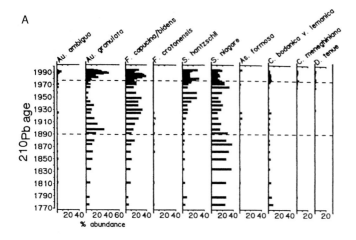

B

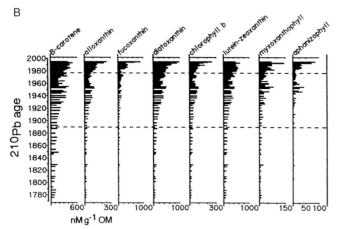

C

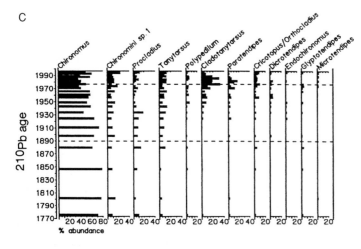

Fig. 11.10 Temporal changes in biological communities from a ^{210}Pb-dated sediment core from Pasqua Lake, Saskatchewan. Diatom assemblage changes (A) are expressed as relative percentages of total number of valves counted; pigment concentrations of the most common chlorophyll and carotenoid pigments (B) are expressed as nmol g^{-1} of organic matter; and chironomid assemblages (C) are represented as relative percentages of the total number of head capsules enumerated. Pigments include β-carotene (all algae and plants), alloxanthin (cryptophytes), fucoxanthin (chromophytes), diatoxanthin (diatoms), chlorophyll b (chlorophytes), lutein–zeaxanthin (chlorophytes + cyanobacteria), myxoxanthophyll (filamentous blue-green algae or cyanobacteria), and aphanizophyll (N$_2$-fixing cyanaobacteria). Large blooms of the latter are considered to be especially serious problems in these waters. The dashed lines indicate the approximate times of the onset of European style agriculture in the area in 1890, and the start of tertiary sewage treatment in Regina in 1977. Urbanization increased greatly after 1930. Although there are marked changes in assemblages coinciding with these changes, all these data indicate naturally productive conditions. From Hall et al. (1999b); used with permission of ASLO.

of algae and macrophytes, and suffer from low hypolimnetic oxygen levels. Certainly, given the extensive agriculture, development, and other ongoing anthropogenic activities in the catchments of the QV lakes, a reasonable perception would be that humans were responsible for the deteriorating water quality. However, not all prairie lakes may follow this pattern. The first assessment goal should therefore determine if nutrient levels were high before the period of European settlement.

Detailed paleolimnological analyses from each of the QV lakes showed that they were all naturally eutrophic. Summary diagrams for diatoms, pigments, and chironomids are only presented here for Pasqua Lake (Fig. 11.10), but additional profiles are in Hall et al. (1999b) and A.S. Dixit et al. (2000), and some additional pigment data are given in Fig. 11.11. These analyses clearly showed that nutrient levels were naturally high (from the eutrophic-indicating diatom assemblages), deepwater hypoxia and anoxia were common (from the hypoxia- and anoxia-tolerating chironomid assemblages), and algae and cyanobacteria bloomed (from the fossil pigment data), well before the arrival of European settlers in the late 19th century. These inferences are supported by diary entries of early explorers, who complained that "Thick green slimes impeded the passage of our canoes" (see Allan & Kenney 1978). It appears that prairie lakes such as these, which lie in large, naturally fertile catchments, with possible additional nutrient inputs from groundwater, result in naturally productive systems. This information has important management implications. Although some improvement could be expected (see below), no amount of mitigation will "restore" these systems to clearwater, oligotrophic lakes.

The QV lakes were naturally productive, but the paleolimnological data also showed that water quality further deteriorated with European settlement (Fig. 11.10). For example, diatom shifts indicated even higher nutrient levels in the 20th-century sediments, chironomid communities indicated even more extensive periods of anoxia, and fossil pigment analyses showed that algal abundance, and especially blue-green algae (cyanobacteria), increased further. Water quality is worse now than it was in the early 19th century. What were the triggers for these important changes?

As water quality is affected by multiple and, at times, interrelated and simultaneously co-occurring environmental factors, with multiple stressors occurring simultaneously, it can be a challenging task to disentangle these influences. Variance partitioning analysis (VPA), as described in the Baldeggersee example (see pp. 199–200), can be effectively used in these cases (Borcard et al. 1992). For example, for the QV system, Hall et al. (1999b) collated continuous environmental data from 1890 to 1994 of 83 potentially important variables that might explain the changes recorded in the sedimentary record. Using the example of Pasqua Lake, the first lake in the system to receive sewage from the city of Regina, they found that VPA captured 71–97% of the variance in fossil composition using only 10–14 significant factors, which themselves could be broadly grouped into three major categories: resource use (crop land area, livestock biomass), urbanization (nitrogen in sewage), and climate (temperature, evaporation, river discharge). Resource use and urbanization played the dominant roles in influencing biotic communities over the past century.

Pasqua Lake is just one in a series of lakes in the Qu'Appelle Valley System (Fig. 11.9). Because all the major QV lakes (Hall et al. 1999b; A.S. Dixit et al. 2000), as well as the two QV reservoirs (Hall et al. 1999a), were analyzed using identical paleolimnological techniques, it was possible to undertake some detailed comparative work. A.S. Dixit et al. (2000) compared landscape-scale, limnological changes spanning approximately the past 200 years in four of the so-called "Fishing Lakes" (i.e., Pasqua, Echo, Mission, and Katepwa). They showed that, like

A Inferred Algal Biomass Increase (%)

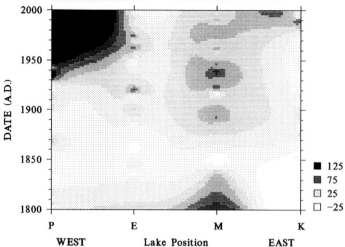

■ 125
■ 75
☐ 25
☐ −25

B Inferred BG Biomass Increase (%)

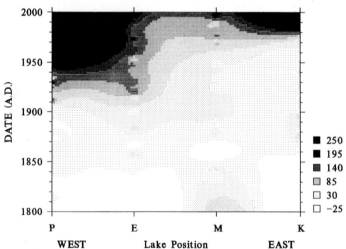

■ 250
■ 195
■ 140
☐ 85
☐ 30
☐ −25

C Inferred N-Fixer Biomass Increase (%)

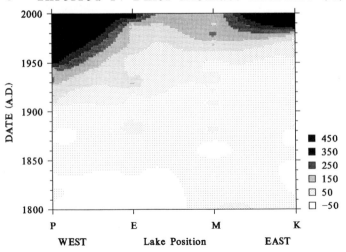

■ 450
■ 350
■ 250
☐ 150
☐ 50
☐ −50

Fig. 11.11 Landscape changes from c. 1780 to 1995 in pigment-inferred total algal biomass (A), as β-carotene; bloom-forming colonial cyanobacteria (B), as myxoxanthophyll; and potentially toxic, N-fixing cyanobacteria (C) as aphanizophyll. The Fishing Lakes of the Qu'Appelle Valley are arranged from west to east: Pasqua (P), Echo (E), Mission (M), and Katepwa (K). All historic changes in pigments are expressed as percentage increases beyond baseline (pre-1880 average) concentrations. The magnitude of algal increase and water-quality change increases with the darkness of shading. Sewage from Regina and Moose Jaw enters the system just upstream of Pasqua Lake, whereas the town of Fort Qu'Appelle contributes effluent between Echo and Mission lakes. Landscape patterns were integrated using distance-weighted, least squares regression. From Dixit, A.S. et al. (2000); used with permission of Blackwell Publishing.

the Pasqua Lake example discussed above, all lakes were naturally productive and that the onset of European agricultural practices (around 1890) had only modest effects on the lakes' biota. However, around 1930, with urbanization and specifically the establishment and expansion of the city of Regina, dramatic changes were recorded in Pasqua Lake, the first lake in the chain that directly received sewage from this urbanization. Even though this lake naturally supported large populations of algae and bloom-forming cyanobacteria, total algal biomass (estimated by the pigment β-carotene) increased more than 350% after 1930 (Fig. 11.11). Similar analyses of the three downstream lakes showed that urban impacts after 1930 were evident in the fossil profiles of carotene and myxoxanthophyll (as a proxy for cyanobacteria), but that large blooms of potentially toxic, N-fixing cyanobacteria (represented by the carotenoid aphanizophyll) only increased over the past 25 years or so (Fig. 11.11). This landscape analysis supported the hypothesis that upstream lakes (such as Pasqua) were effective at reducing impacts of point-source urban nutrients on downstream lakes. Meanwhile, agricultural activities (non-point sources) had relatively limited impacts on eutrophication trajectories of these lakes, and these effects

were ameliorated less effectively by upstream basins. A.S. Dixit et al. (2000) hypothesized that the marked attenuation of urban pollution may result from a high proportion of biologically available nutrients in sewage, as opposed to the less biologically available, organic matter-bound nutrients from terrestrial sources, which may traverse upstream basins prior to mineralization in downstream lakes (Detenbeck et al. 1993).

The combination of paleolimnology, whole-lake mass balances, and long-term ecological research has also provided unique insights on the causes of water-quality degradation in the QV system. For example, catchment-scale analysis of stable isotopes of nitrogen (N) demonstrated that sedimentary $\delta^{15}N$ (which is enriched in urban wastewaters) increased from approximately 6.5‰ during the 19th century to 14‰ by the 1990s, an increase that was linearly correlated with both the mass of N released from the city of Regina and a 300% increase in algal production (as inferred from fossil pigments; Fig. 11.11) since c. 1880 (Leavitt et al. 2006). Further, whole-lake mass balances of isotopes showed that over 70% of ecosystem N was derived from city sources, and confirmed earlier suggestions that N, not P, stimulated blooms of potentially toxic algae

Fig. 11.12 A diagram showing the expected changes in trophic status following river impoundment and reservoir formation. Increased primary production following impoundment and flooding is often referred to as the trophic upsurge. This is typically followed by a period of trophic depression. Subsequent changes are largely driven by other nutrient inputs. In this example, trophic status is increasing during recent times, likely as a result of land-use changes (e.g., agriculture) in the reservoir's catchment. Modified from an unpublished diagram by R. Hall.

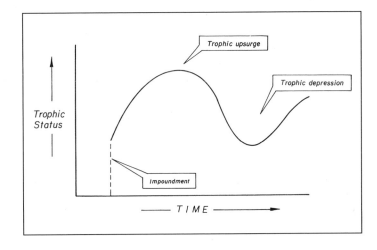

(Hall et al. 1999b; A.S. Dixit et al. 2000). Finally, these data showed that the magnitude of lake response to human disturbance was greatly influenced by lake position within the hydrological landscape – insight that would have been difficult to obtain without application of paleoecological techniques.

11.9 Impoundment and flooding: changes in trophic status due to reservoir formation

There is sometimes a tendency to view reservoirs and natural lakes as similar types of water bodies, as both contain fresh water, phytoplankton, zooplankton, and so forth, and (at some levels) appear to follow similar limnological patterns. As discussed by Thornton et al. (1990), there are in fact many substantive differences. Lakes form through a variety of natural processes, with the origins of most Northern Hemisphere temperate lakes linked to glaciation. Reservoirs, on the other hand, are typically formed by the impoundment of rivers and flooding of the surrounding land. Not surprisingly, these two processes can have important implications for lake trophic status.

As summarized by Marsicano et al. (1995), reservoirs are believed to follow some typical patterns following dam construction (Fig. 11.12). Immediately following reservoir formation and the associated flooding of surrounding land, a period of increased productivity occurs, referred to as the trophic upsurge. This enrichment is largely driven by the influx of nutrients from the inundated reservoir basin. Once conditions stabilize and erosion decreases, the reservoirs enters a period of trophic depression (Ostrofsky & Duthie 1980), when productivity decreases. Trophic status can change thereafter depending on any additional nutrient enrichment from catchment disturbances or any other sources (Fig. 11.12).

A variety of paleolimnological studies have tracked evidence of trophic changes following water-level increases. Marsicano et al. (1995) used a combination of monitoring and paleolimnological data to show that water-quality changes in a Connecticut reservoir followed the expected pattern shown in Fig. 11.12, with evidence of a period of trophic upsurge after dam construction, followed by a period of trophic depression, and then a return to higher trophic levels related to subsequent land-use changes in the reservoir's catchment. Similarly, Donar et al. (1996) recorded a marked increase in eutrophic diatoms following reservoir formation in Michigan, followed by a decline in inferred trophic status.

Flooding as a result of other anthropogenic activities may produce similar eutrophication patterns. Christie and Smol (1996) tracked the limnological repercussions of the 1830s construction of the Rideau Canal (Ontario), a waterway linking Kingston to Ottawa via a series of lakes and the Rideau River, as well as 47 locks. In order to enable ships to use the canal, some lakes had to be deepened by damming and flooding parts of their catchments). One of the lakes flooded in the 1830s was Upper Rideau Lake, where diatom and chrysophyte cyst assemblages indicated marked eutrophication from this time period. A subsequent analysis of four other Rideau Canal lakes similarly recorded eutrophication at the time of canal construction (Forrest et al. 2002).

Paleolimnological data have shown, however, that not all reservoirs follow the generally accepted paradigm shown in Fig. 11.12. By comparing fossil records from two prairie reservoirs in the Qu'Appelle Valley system described earlier (Fig. 11.9), Hall et al. (1999a) suggested that differences in reservoir formation may strongly influence trophic state changes. Lake Diefenbaker is a 500 km^2 reservoir created by the damming of the South Saskatchewan River in 1968. Buffalo Pound Lake, on the other hand, was formed by the flooding of a natural lake in 1952. Lake Diefenbaker generally followed the typical pattern of reservoir ontogeny with an initial short period (~ 4 years) of eutrophication, followed by about a decade of mesotrophy, and then a gradual shift to modern conditions. Meanwhile, in nearby Buffalo Pound Lake, fossil pigment analysis indicated declining algal and cyanobacterial populations following impoundment, whereas chironomids and macrophyte cover expanded. Hall et al. concluded that these differences arose from the extent of impoundment (500 km^2 versus ~ 5 km^2) and the magnitude of subsequent water-level fluctuations (6.3 m versus < 1 m) in the river versus natural lake system.

11.10 Tracking changes in epilimentic versus hypolimnetic conditions: an example of spatial decoupling of limnological processes in different lake strata

The typically observed scenario with eutrophication is that, with the addition of nutrients, primary production increases in the lake's surface waters, eventually followed by the settling out of this excess organic matter to the deep portions of the basin, where decomposition results in lower hypolimnetic oxygen levels. If nutrient inputs decline, algal and macrophyte standing crops typically decrease, and deepwater oxygen concentrations increase. Certainly, in the majority of cases, this coupling of concurrent changes in epilimnetic and hypolimnetic conditions is evident, and many management strategies and models are based on these frequent observations. However, in some cases, epilimnetic processes may become decoupled (at least partially) from processes occurring in hypolimnetic waters, and vice versa. By using indicators that are characteristic of spatially separated habitats (e.g., planktonic diatoms and chrysophytes from epilimnetic waters; profundal chironomids from hypolimnetic waters), paleolimnologists can track changes in these spatially separate lake strata.

Little et al. (2000) combined diatoms, chironomids, and historical records to compare past changes in epilimnetic nutrient concentrations and hypolimnetic anoxia during the eutrophication and remediation efforts of Gravenhurst Bay (Ontario, Canada). Water quality declined markedly with the arrival of European settlers in the mid-1800s, with diatom-inferred phosphorus concentrations reaching their highest values in the first half of the 20th century. Deepwater oxygen levels, quantitatively inferred from chironomid head capsule assemblages, indicated that anoxia became more prevalent coincident with this eutrophication. Anoxia became most severe around the 1950s, when profundal taxa were extirpated. Sewage inputs to the bay greatly decreased after 1972, and the diatom species composition faithfully tracked this recovery with the return of more oligotrophic taxa and declines in diatom-inferred phosphorus. Benthic chironomids, however, showed no parallel recovery. These observations are consistent with recent monitoring data showing declines in epilimnetic phosphorus concentrations, but continued anoxia in the bay's deep waters. The reason for the persistent anoxia is not clear. Regardless of the cause, studies such as this show that surface and deepwater strata may be "decoupled" during eutrophication and subsequent remediation efforts, and further reinforce the value of using multiple indicators to track responses in different lake regions.

11.11 Regional assessments of environmental change in lakes and reservoirs: the Environmental Monitoring and Assessment Program (EMAP)

EMAP is a major initiative of the US Environmental Protection Agency (US EPA) "to take the pulse of the nation's ecological resources" (EPA 2000, p. 1). Its goal is to produce report cards evaluating the conditions of US natural environments, similar in some respects to the reporting done for the economies of the world. The hope is that such data will help resolve environmental problems before they become ecological disasters.

This, of course, would be a major initiative for any country, let alone one as large as the United States. Moreover, as has already been shown by some examples in this book, it is at times difficult to find "representative sites" for environmental change, whose trajectories can then be extrapolated to neighboring lakes. How similar will estimates of environmental degradation and/or recovery be in different lakes, even if they are subjected to similar stressors?

The above task is not a simple one. It must include sampling many sites (and then even these will only be a small subset of the total number of lakes), and selecting representative sites that could then be used to extrapolate to a larger population of lakes. If these data can be gathered, regional trends can be estimated and a statistical summary and assessment of resources can be provided.

The concept of using a statistically chosen set of lakes to make regional estimates was introduced in Chapter 7 on acidification with the PIRLA project, where regional estimates were made of the extent of lake acidification in the Adirondacks (Cumming et al. 1992b). A similar procedure was employed in part of the EMAP project to study the extent of lake and reservoir degradation in the northeastern United States. First, it is important to gather a sample of representative lakes. EMAP applied similar logic and statistical designs to those used in political or other polls that attempt to sample representative populations of randomly selected voters or consumers or whatever. Data gathered from these interviews or questionnaires are then used to make regional estimates and predictions. Similar statistical designs and sampling protocols can, with some modifications, be developed for lakes and other systems. One could also employ a weighted statistical design, where certain chosen lakes are

weighted more heavily in the final analyses, as they represent more common types of lakes, and so forth. The approach is certainly challenging, but feasible.

But how does one take and interpret the "pulse of the nation's lakes and reservoirs" if one does not know what the healthy pulse was? What were the systems like before human impacts? How much change has already taken place? Paleolimnology can provide these data.

One component of the surface waters component of EMAP in the northeastern United States included a paleolimnological assessment using diatoms. Given the large number of sites that are required to make regional assessments, it was not possible (given the financial and personnel resources allotted) to undertake detailed paleolimnological assessments on all the sampling sites. As described earlier, the "top/bottom approach" (Box 4.1) is a reasonable compromise when a large number of sites must be analyzed. This "before-and-after" snapshot approach was followed for the EMAP lakes (Dixit & Smol 1994; Dixit et al. 1999).

EMAP had developed a probability-based sampling design so that data for specific lakes could be used to infer trends for entire populations of lakes in the region. Diatoms were selected as biomonitors for 257 lakes and reservoirs in the northeastern United States, which were divided into three ecoregions: the Adirondacks, the New England Uplands, and the Coastal Lowlands/Plateau. Using diatom species assemblages preserved in the surface sediments, robust transfer functions were developed for lakewater pH, total phosphorus (TP), and chloride (the latter was related to the salting of roads in winter as a de-icing agent). As I have included EMAP in the eutrophication chapter, I will only summarize the TP inferences here, but other variables are discussed in Dixit et al. (1999). Using the top/bottom approach, it was clear that many of the presently eutrophic lakes and reservoirs had greatly increased in TP since before the time of significant cultural impact (Fig. 11.13). However,

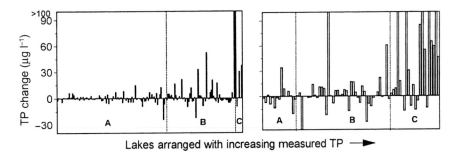

Fig. 11.13 Diatom-inferred total phosphorus (TP) changes in lakes (left) and reservoirs (right) of the northeastern United States. Sites are arranged according to present-day TP, with the most oligotrophic sites to the left (zone A, TP < 10 μg l⁻¹; zone B, TP = 10–30 μg l⁻¹, zone C, TP > 30 μg l⁻¹). The TP changes are based on diatom assemblages preserved in the surface sediments (present-day) and pre-industrial sediments of the lakes. For the reservoirs, the bottom samples were the deepest lacustrine sediments recovered (i.e., sediments deposited near the time of reservoir formation, typically early in the 20th century). The TP inferences are calculated from a diatom transfer function developed for this region. From Dixit et al. (1999); used with permission of NRC Research Press.

the TP inferences showed that presently oligotrophic lakes have not changed markedly over the past approximately 150 years. TP increases were more widespread in the reservoirs.

With the EMAP sampling design, it was also possible to project these inferences onto regional and ecoregional scales. There are many ways in which these data can be presented, but a common method used for other index values, such as economic measures, is to use pie diagrams (Fig. 11.14). One question posed was whether the percentage of oligotrophic (defined as TP < 10 μg l⁻¹), mesotrophic (10–30 μg l⁻¹), and eutrophic (> 30 μg l⁻¹) lakes and reservoirs had changed since pre-industrial or background times. When one combines all the study lakes and reservoirs and divides them into these three eutrophication categories, there are, surprisingly, only small shifts between the above categories. Lakes that would have been classified as naturally mesotrophic, for example, may have increased in TP, but not often over the 30 μg l⁻¹ level.

When one analyses each ecoregion separately, more changes are obvious. Although lakes in Adirondack Park had experienced marked acidification (Dixit et al. 1999), no significant eutrophication has occurred on a regional scale since the 1850s. Lakes in the New England Uplands

show some modest shifts to more eutrophic lake categories, but reservoirs in this ecoregion have shown, on average, a decrease in inferred nutrients since the times of their initial construction. Meanwhile, the proportion of lakes and reservoirs classified as eutrophic has increased in the Coastal Lowlands/Plateau region.

When the analyses illustrated in the previous two figures are compared, it can be seen how different data presentations can emphasize different components in a study. The top/bottom inferences illustrated in Fig. 11.13 show how individual sites can be studied. Clearly, if you are living adjacent to one of the lakes or reservoirs that showed striking changes in TP, you have reasons for concern. However, if you look at a regional assessment (Fig. 11.14), many of the changes average out and are less obvious, especially when the lakes are clumped into the three broad lake classification categories.

11.12 Assessing the ecological status of Irish candidate reference lakes

As noted earlier (p. 19), the European Union (EU) Water Framework Directive (WFD) requires

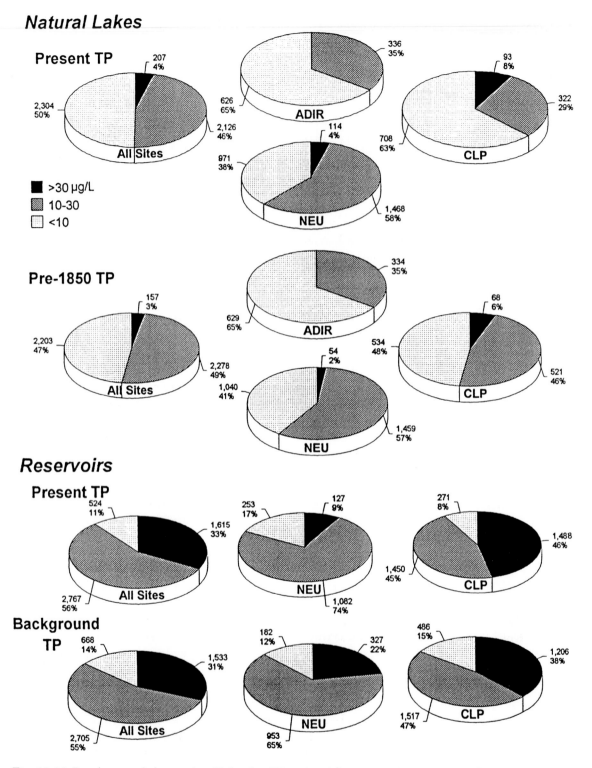

Fig. 11.14 Populations of oligotrophic (defined as TP < 10 μg l^{-1}), mesotrophic (10–30 μg l^{-1}), and eutrophic (> 30 μg l^{-1}) EMAP lakes and reservoirs for present-day (surface-sediment samples) and background conditions, based on diatom inferences. ADIR, Adirondacks; NEU, New England Uplands; CLP, Coastal Lowlands/Plateau ecoregions. From Dixit et al. (1999); used with permission.

member states to establish type-specific reference conditions for all water bodies. This poses a problem for many regions, as most lakes have been significantly impacted by humans, and so it is difficult to find candidate lakes that can be used as reference sites in these space-for-time analyses. The Irish Environmental Protection Agency nominated 76 oligotrophic and meso-oligotrophic water bodies, thought to have had minimal human impacts, as candidate reference lakes (CRL) for the Irish ecoregion. To determine if in fact these CRL were representative of undisturbed limnological conditions, Leira et al. (2006) applied diatom-based and other paleolimnological techniques in a "top/bottom" analysis for 35 CRL. Although 11 of the sampled lakes did not show any significant differences between core top and bottom samples (and thus supporting their inclusion as valid reference lakes), over two-thirds of the CRL sampled recorded biologically important deviations from reference conditions (i.e., the "bottom" sediment samples). Acidification and eutrophication were identified as the main environmental stressors. The study demonstrated how paleolimnology can be used to verify the validity of reference lakes, and also highlighted some of the challenges posed by establishing reference conditions, especially in regions where human activities were already quite intense by the mid-19th century (a period often assumed to represent reference conditions).

11.13 Extending regional assessments into the future: the value of surface sediments to long-term monitoring programs

I have added this section on the use of paleolimnology in long-term monitoring programs in the eutrophication chapter, and after the EMAP and Ireland example, as it tends to follow logically from the previous section on regional assessments. However, the points made below are equally applicable to other limnological problems, such as acidification, other contaminants and so forth, and similar approaches can be used to study these issues.

To continue with the medical analogy introduced earlier in this book, few would argue with the proposition that it is important to continue monitoring the health of a patient (or a lake). Is the patient (or lake) continuing to deteriorate, in steady state, or beginning to recover? With human health, this is typically done by medical check-ups and metrics such as blood pressure, body temperature, and so forth. With environmental issues, this is done primarily by monitoring programs, and metrics such as limnological variables and biological indicators. However, there are very many sites to monitor, and limnological variables can change rapidly even within one lake, and so the number of sites and the required frequency of monitoring becomes a major logistic concern. The social and political will is rarely there to finance these costly undertakings. Consequently, little effective monitoring is ongoing. We do not know "the state of our environment."

It is time to explore new, more cost-effective, approaches to these problems. We have no choice: we need these data, but society seems to be unwilling or unable to support standard monitoring programs on regional scales. Paleolimnological approaches have much to offer.

Two factors often hamper attempts at initiating regional-scale monitoring programs: the large number of sites, and the frequency of conventional sampling that is required to obtain reliable estimates of change. Both are serious concerns. The question of which representative sites to monitor (as not all sites can be included in a sampling program) can be assessed using the probability and statistically based designs used in the EMAP program. Typically, this would still leave a large number of sites that should be studied, even if it is a relatively small subset of the original population of lakes. Given limited resources, the next question is: What should be monitored and at what frequency?

Sampling frequencies can be set on a tiered system (Fig. 11.15). At least some primary reference

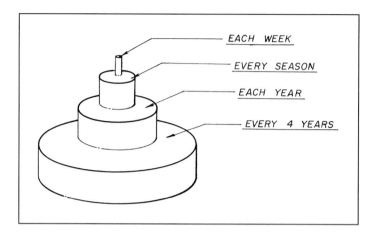

Fig. 11.15 A diagram showing the possible sampling frequencies that a long-term monitoring program might follow. Because of the large number of sites, it is rarely logistically possible to sample all sites on a regular basis. Only a few lakes will be sampled at high frequency, with progressively larger proportions of the population of lakes sampled at coarser time intervals (e.g., only once every 4 years). Surface sediments contain spatially and temporally integrated snapshots of the physical, chemical, and biological changes that had occurred in the system, and can be effectively used in a long-term monitoring program by resampling the surface sediments every 4 years or so. From Smol (1992); used with permission.

sites should be monitored at high frequency (e.g., weekly) so as to better understand processes and fine-scale changes. Resources may allow for a subset of additional sites to be sampled at a moderate frequency, such as every season; and then more sites sampled only annually, perhaps during some "index period," such as spring or midsummer. However, given resource restraints, the vast majority of sites can typically only be sampled once every 4 years or so. The question then must be asked: What variable(s) and/or indicator(s) can be sampled in a large suite of lakes (or reservoirs, or rivers) that would provide a meaningful integration of limnological changes that had occurred in the previous few years? My response would be surface sediments, and the calibration approaches designed by paleolimnologists should be adapted to long-term monitoring programs.

Water chemistry variables are notoriously "noisy," with large variations and fluctuations occurring seasonally (sometimes even daily) and from year to year. Similarly, many biological communities, such as algae and invertebrates,

can often also change abundances seasonally, temporally, and spatially in a lake. Hence, spot water samples that are collected perhaps once every 4 years or so, although still useful to some extent, would have limitations for interpretation and would not integrate the changes that had occurred since the previous sampling interval of 4 years ago.

Surface sediments, on the other hand, can potentially provide temporally and spatially integrated "snapshots" of the biota and chemistry of aquatic systems. Surface sediments are typically accumulating and integrating information from the entire water column, from different regions of the lake, 24 hours a day, every day of the year. In many respects, the surface sediments are archiving a "running average" of what has been happening in the lake. To use the EMAP example just described, there are now over 250 lakes and reservoirs in the northeastern United States for which: (i) robust transfer functions are available to quantitatively infer limnological variables of interest, such as pH, TP, and trends in salinity;

and (ii) background environmental conditions have been inferred, so that we know how much the systems have changed. By returning to these sites, on a 4-year rotation from the time of initial sampling, and resampling the surface 1 cm or so of sediments, re-analyzing them for diatoms (or other biological, chemical, or physical indicators), a continuous record of integrated limnological changes can be established. The amount of environmentally relevant information that one could eventually assemble in this manner is quite impressive, and the cost is a fraction of more standard approaches.

The initial phases of, for example, the EMAP paleolimnological program (Dixit & Smol 1994; Dixit et al. 1999) and Irish programs described earlier, as well as other regional paleolimnological assessments (e.g., Ramstack et al. 2003, 2004; Bennion et al. 2004) have the potential to expand to this type of long-term monitoring program. However, at the time of writing, to my knowledge no such resampling program has yet been implemented.

11.14 The role of macrophytes in lake dynamics: exploring the role of alternative equilibria in shallow lakes

To date, most paleolimnological studies have focused on relatively deep systems. This emphasis has been driven largely by concerns that shallow lakes may be more susceptible to bioturbation and other sediment-mixing processes. To a certain extent, this is true: Sedimentary profiles in shallow lakes often show greater mixing than in deepwater systems due to wind-induced currents and/or more active benthic communities. On the other hand, shallow productive lakes often have much higher sediment accumulation rates, which may offset some of the problems of minor sediment mixing. In any event, the degree of mixing can be estimated using geochronological and other techniques. Despite some challenges, it is now clear from a growing number of studies

that paleolimnology can provide many important insights into shallow-lake dynamics (e.g., Anderson & Odgaard 1994; Sayer et al. 1999; Ogden 2000; Bennion et al. 2001). This is important, as shallow, non-stratifying lakes are often much more common than deepwater systems and in fact dominate the freshwater reserves of many countries.

New theoretical concepts, such as discussions related to alternative equilibria (Scheffer et al. 1993), have heightened interest in shallow lakes. As described in Box 11.2, shallow lakes appear to have two alternative equilibria: a clear state dominated by aquatic macrophytes, and a turbid state characterized by high phytoplankton growth (Scheffer et al. 1993; Scheffer 1998). If this theoretical framework is correct, it has profound implications for lake and reservoir management. If eutrophication proceeds to the turbid state, overall water quality is severely affected. The importance of rooted aquatic macrophytes in maintaining the clearwater state has, until now, been generally under-appreciated. In fact, many lake restoration efforts have included removal of macrophytes as an unsightly symptom of eutrophication. This practice may be unwise, as it may allow a system to switch equilibria to the phytoplankton-dominated, turbid state. Although fewer studies have been completed on shallow lakes, paleolimnology can be used to assess the alternative equilibria hypothesis, as well as track trajectories in shallow-water systems.

To assess the role of macrophytes in influencing trophic dynamics in shallow lakes, paleolimnologists must be able to track, in at least a qualitative sense, past changes in rooted aquatic vegetation. The most direct approach is to use macrofossils, although this has, thus far, rarely been done in a detailed manner (Birks 2001). More often, macrophyte abundance is tracked indirectly using epiphytic indicators, such as diatoms that characterize the epiphyton on macrophytes, or some other proxy. For example, Thoms et al. (1999) found a strong relationship between the percentages of littoral chydorid Cladocera remains in the surface sediments of

Box 11.2 Alternative equilibria in shallow lakes

The pivotal role of aquatic vegetation (often referred to as macrophytes by limnologists) in influencing overall lake trophic dynamics was not fully recognized until recently (e.g., Moss 1990; Scheffer et al. 1993; Scheffer 1998). Shallow lakes can often be differentiated into two equilibrium states: a relatively clearwater, macrophyte-dominated state; and a turbid, phytoplankton-dominated state, which precludes extensive rooted macrophyte growth. Important limnological feedback systems are believed to center around the interaction between submerged vegetation and turbidity (Figs. 11.16 & 11.17). Rooted macrophytes tend to maintain the lake in a clearwater, non-turbid state – conditions that are required for rooted macrophytes to thrive. If the lake is allowed to shift to a turbid, phytoplankton-dominated state, macrophytes are less competitive due to light limitation. An important component of this argument is that the vegetation itself influences the overall limnological characteristics of the lake system. For example, high macrophyte abundances reduce the resuspension of sediment material, provide a day-time refuge from planktivorous fish for zooplankton feeding on phytoplankton, compete with phytoplankon for nutrients, and possibly release allelopathic substances that are toxic to phytoplankton. Therefore, large stands of macrophytes would be expected to help maintain a lake system in one equilibrium state: the clearwater, macrophyte-dominated state.

With increased nutrient input, a macrophyte-dominated lake may have considerable "inertia" to change (i.e., be more stable) than a deepwater system with few rooted macrophytes (Fig. 11.17). With nutrient enrichment, the latter would respond rapidly to eutrophication with increased phytoplankton growth, whilst the macrophyte-dominated system would exhibit a hysteresis to this nutrient addition, as the vegetation would stabilize the clearwater state (for the reasons noted above). Indeed, lakes with dense macrophyte growths tend to have higher transparency than deeper lakes of similar nutrient concentrations. However, with continued nutrient enrichment, a threshold will be reached even in macrophyte-dominated lakes, and the clearwater equilibrium state will eventually shift (often abruptly) to the alternate phytoplankton-dominated, turbid state.

Equilibrium shifts have important implications for lake restoration programs. Just as the macrophyte-dominated systems have a hysteresis to switching equilibrium states toward the phytoplankton-dominated system, attempts at restoring a lake from the turbid state back to the clearwater, macrophyte-dominated system will be hampered by an equally strong hysteresis in the opposite direction. It has long been recognized that restoring eutrophied, turbid, shallow lakes is difficult. As discussed by Scheffer et al. (1993), biomanipulation methods, such as reductions in fish stocks and hence increasing phytoplankton grazing pressure, have been used in some shallow systems to enforce the switch back into a macrophyte-dominated equilibrium. Lowering the water level for a period of time is another option. Many of the ideas related to alternative equilibrium states are still new, and would benefit from long-term perspectives made possible through paleolimnological analyses.

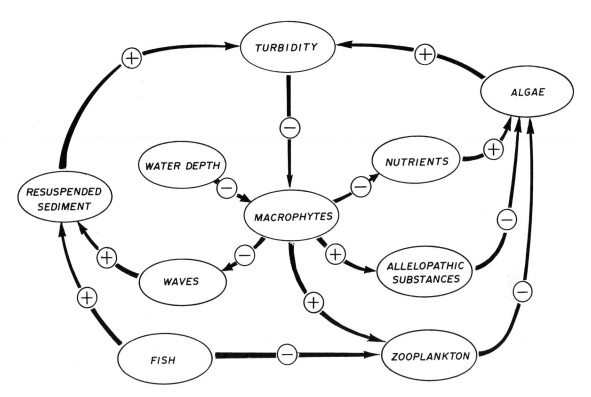

Fig. 11.16 Some of the main feedback loops (both positive and negative) thought be responsible for the existence of alternative equilibria in shallow ecosystems. Rooted aquatic macrophytes are key components of the system. See text for details. Modified from Scheffer et al. (1993).

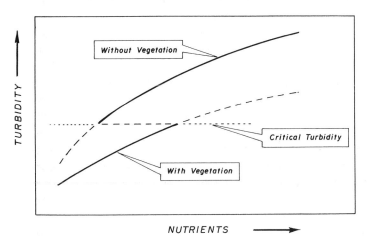

Fig. 11.17 An illustration of the alternative equilibrium turbidities of shallow lakes, with respect to changing nutrient levels. With increasing nutrient inputs, shallow lakes with abundant rooted aquatic vegetation will not easily switch to the turbid, phytoplankton-dominated state, as compared to lakes without vegetation. However, eventually a critical nutrient level will be surpassed and phytoplankton biomass will increase, passing a critical turbidity threshold. Thereafter, rooted macrophytes are less competitive due to light limitation, and the system switches to the turbid equilibrium state. Modified from Scheffer et al. (1993).

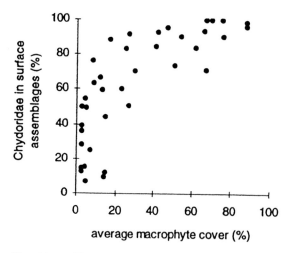

Fig. 11.18 The percentage of littoral chydorid Cladocera remains in the surface sediments of 38 Australian billabongs related to the average percentage of submerged macrophyte cover. From Thoms et al. (1999); used with permission of Blackwell Publishing.

Australian billabongs in relation to aquatic macrophyte cover (Fig. 11.18), and then Ogden (2000) used this relationship to track historical variations in macrophyte cover associated with catchment disturbances. Brodersen et al. (2001) identified distinct chironomid communities associated with different macrophyte classes in Danish lakes, and Odgaard and Rasmussen (2001) proposed that egg-cocoons of certain leeches can be used to track past submerged vegetation. Meanwhile, Schelske et al. (1999) used diatoms and the ratio of total carbon to total nitrogen (TC/TN) in the organic matter of a ^{210}Pb-dated sediment core from shallow Lake Apopka in Florida, to test whether these proxies could be used to track past macrophyte abundance. This large (125 km^2), shallow (mean depth 1.7 m) lake is of special interest to the topic of alternative equilibria because the lake was clear and dominated by rooted macrophytes before 1947. However, anecdotal reports note that the lake shifted rapidly to a phytoplankton-dominated (i.e., turbid) lake in 1947, suggesting a shift in equilibria. The logic behind using diatoms was that, with high macrophyte abundance, benthic

taxa should dominate, but with a shift to the primarily phytoplankton state, planktonic taxa should be more abundant. They calculated the ratio of planktonic to benthic diatoms in the core, and found a significant change around 1947. The reasoning for the second proxy (TC/TN) is based on the observation that algae typically have less carbon and more nitrogen per unit biomass than macrophytes, and so the TC/TN of organic matter in sediments should estimate the relative proportions of past algal to macrophyte abundance. As with the diatoms, the TC/TN ratio of organic matter declined around 1947, concomitant with a shift to a phytoplankton-dominated system.

With tools such as those described above, paleolimnologists can begin to explore some of the problems associated with alternative trophic equilibria. The role of macrophytes in the nutrient dynamics of shallow lakes was also explored by Brenner et al. (1999), who retrieved four sediment cores from macrophyte-dominated Orange Lake (mean depth < 2 m) in north-central Florida (USA). ^{210}Pb-dated cores were used to evaluate spatial and temporal patterns of bulk sediment and accumulation. Using multiple lines of evidence, they showed that the lake was naturally productive. However, even though nutrient loading had increased over the last approximately 70 years, diatom-inferred phosphorus levels declined after c. 1930. This coincided with parallel increases in sediment storage of total phosphorus and non-apatite inorganic phosphorus. Increases in periphytic diatoms documented the expansion of macrophytes over this time period, most notably the introduction of *Hydrilla*. The authors concluded that the much expanded macrophyte beds (which now cover about 80% of the lakebed) were responsible for nutrient removal from the water column (and hence lower diatom-inferred phosphorus levels) and the accelerated nutrient sequestering in the sediments. These changes are consistent with the predictions of the alternative equilibria hypothesis.

Paleolimnological methods can also be used to reconstruct trophic histories of shallow lakes extending back to the time of their inception. For example, Karst and Smol (2000) used diatoms, pollen, and other proxy indicators to track the postglacial (i.e., > 11,000 yr BP) history of Lake Opinicon, a shallow (mean depth < 5 m), macrophyte-dominated lake in southern Ontario (Canada). This lake is part of the Rideau Canal system, a series of lakes, canals, and locks that were built around 1830 as a military and transport shipping route. It is now used exclusively for recreation and boating. Several paleolimnological studies have been completed on deep, stratifying lakes in the region, which showed marked limnological changes in response to past vegetation and climate changes. Such studies have also tracked the more recent effects of cultural eutrophication from settlement in the region beginning in the 19th century, including the land-use changes and flooding from the canal construction in 1830 (e.g., Christie & Smol 1996, Forrest et al. 2002). If the alternative equilibria hypothesis was correct, one would predict that a shallow, macrophyte-dominated lake such as Lake Opinicon would not respond as quickly or to the same extent as similarly affected, deep-water, phytoplankton-dominated lakes that do not have the stabilizing effect of vegetation.

Although changes in fossil diatom assemblages were recorded in response to past natural and anthropogenic changes at Lake Opinicon, these shifts were relatively modest (Karst & Smol 2000). Similarly modest changes were recorded in the chironomid assemblages (Little & Smol 2000). The authors suggested that the apparent inertia to changes in trophic status supports the alternative equilibrium hypothesis, with Lake Opinicon existing in a clearwater, equilibrium state dominated by macrophytes since early in its development. Such results also suggest that the removal of macrophytes for aesthetic reasons may be a poor management decision indeed, as this may result in a switch to a phytoplankton-dominated, turbid state.

Recent paleolimnological studies on other lakes, which have experienced much greater eutrophication than Lake Opinicon, have recorded shifts in equilibrium states. Brodersen et al. (2001) used macrophyte remains, along with chironomid and diatom indicators, to reconstruct the recent history of shallow (mean depth 1 m), hypereutrophic (TP = 310 µg l⁻¹) Lake Søbygaard, Denmark. Species successions indicated that the lake was already eutrophic in the 16th and 17th centuries, but became hypereutrophic with additional nutrient inputs from anthropogenic sources over the past few hundred years. The eutrophication process was recorded by diatom-inferred TP changes, but also by macrophyte shifts from *Chara*, to *Ceratophyllum* and *Potamegeton* dominance, and finally to the present turbid equilibrium state consisting of no submerged macrophytes and dominance by phytoplankton. The concurrent changes in chironomid assemblages were believed to reflect the macrophyte succession.

Although eutrophication and increased phytoplankton production are often considered primary causes of decreased light penetration in lakes, shading can also be related to abiotic turbidity from erosion. Paleolimnological evidence has suggested that the scarcity of aquatic macrophytes in many Murray River (Australia) billabongs is a relatively recent phenomenon, as many of these river floodplain lakes apparently supported extensive macrophyte growth up until about 100–150 years ago (e.g., Ogden 2000; Reid et al. 2002). Nutrient enrichment, leading to increased shading from phytoplankton, is often invoked as the most likely cause of decreased rooted aquatic macrophyte growth. However, using paleolimnological data from Hogan's Billabong, Reid et al. (2007) provided evidence that widespread erosion in the late 19th century, leading to abiotic turbidity, was a critical factor in contributing to decreased water transparency, and so triggering a switch in alternative stable states in this shallow water system.

With heightened interest in shallow lakes from both theoretical limnologists and lake managers, there are many new opportunities for paleolimnological research in this area. Challenges certainly remain, such as determining the optima and tolerances of many shallow-water taxa (Bennion et al. 2001), which may be more problematic than collecting similar data for planktonic species. Nonetheless, as data continue to be gathered, it is becoming clear that important questions can be addressed using the sedimentary records of shallow systems.

11.15 Lake Baikal: tracking recent environmental changes in a unique limnological setting

Although limnologists occasionally (and probably incorrectly) use the word "unique" to describe lakes they are working on, few would argue with the proposition that a truly unique limnological ecosystem is Lake Baikal, situated in the center of the Asian continent in southeast Siberia. This ancient lake also drains a substantial portion of northern Mongolia. As summarized by Mackay et al. (1998), Lake Baikal is by far the oldest extant lake in the world, as it exists in a tectonic valley that began to form c. 30 million years ago. Because of its great depth of over 1600 m, it contains the largest single source of fresh water on this planet, accounting for some 20% of the Earth's total liquid freshwater reserve. Surveys estimate that the lake currently supports over 2500 extant species of plants and animals, of which an astounding 75% are believed to be endemic.

In recent years, considerable international attention has centered on the potential environmental degradation of Lake Baikal, believed to be related to the increased population and industrial growth in its watershed, coupled with other pollution sources from logging and chemicals used in agriculture. Concerns have particularly focused on eutrophication issues, and the effects of pollution from effluents of partly treated sewage and factory discharges on the endemic flora and fauna. Due to the lack of long-term monitoring data, controversy exists concerning the extent of environmental degradation, as well as to the effects of combined stressors on this lake system.

In order to address the above concerns, in a comprehensive study, more than 20 sediment cores collected along the length of the lake were retrieved and analyzed for a large suite of paleolimnological variables. The diatom data are presented in Mackay et al. (1998), and Flower (1998) summarizes most of the other analyses, which were published in a dedicated issue of the *Journal of Paleolimnology* (volume 20, issue 2). In summary, these paleolimnological studies showed that a record of atmospheric pollution since the 1930s, especially in the form of spheroidal carbonaceous particles (SCPs), exists in all the cores taken throughout the length of the lake. Contamination is greatest in the south basin, where industry and coal mining is more prevalent, and is lowest in the more pristine, middle basin region. Metal contamination is slight, with copper and zinc only increasing since the mid-1980s. Lead concentrations, however, showed some low-level contamination beginning in the 19th century. Sediment magnetic measurements indicated a relative increase in the supply of weathered topsoil to the lake.

Diatom species analysis, however, did not record any marked eutrophication signals (Mackay et al. 1998). In fact, in the deepwater sediment cores, no evidence was found that the endemic Baikal diatom flora was being affected. However, in the shallower waters, and especially those near the Selenga Delta and near the Baikalsk pulp and paper mill, there are some localized changes in diatom assemblages indicative of increasingly eutrophic waters.

Collectively, these data provide strong evidence that widespread pollution is not *yet* a major problem in Lake Baikal. Although some environmental degradation from air pollution and other stressors is occurring, eutrophication problems, at this point in time, were only identified in isolated bays. Clearly, though, these early changes are bellwethers of possible future environmental degradation. As Flower (1998, p. 113) warns in the conclusions to their studies, "This is not however a complacent view and future monitoring of the status of the Lake Baikal ecosystem by an integrated quality controlled program employing recent paleolimnological and direct biogeochemical techniques is advocated."

11.16 Atmospheric deposition of nutrients: tracking nitrogen enrichment in seemingly pristine high-alpine lakes

Almost all eutrophication studies have focused on phosphorus, as this element is the limiting nutrient in most temperate freshwater systems. However, nitrogen may be more important in certain limnological settings, such as high-alpine or Arctic lakes, where nitrogen is often in low supply and may be more limiting than phosphorus. Furthermore, almost all eutrophication research has centered on nutrient inputs from point inflows, drainage from land, and/or groundwater. An additional vector for nutrient transport is the atmosphere, although it has generally been ignored or considered to be a minor and insignificant source. Paleolimnology can be used to challenge some of these assumptions.

It is now widely recognized that the natural global nitrogen cycle has been dramatically altered by humans, with enhanced nitrogen deposition (primarily NO_3^- and NH_3) from anthropogenic emissions, which can be related to industrial, urban, and agricultural growth (Vitousek et al. 1997). The Colorado Front Range of the Rocky Mountains (USA) is situated to the west of the rapidly expanding Denver – Fort Collins urban axis. Atmospheric nitrogen deposition on the eastern side of the Range reaches 3–5 kg N ha^{-1} yr^{-1} from urban, agricultural, and industrial sources, compared with 1–2 kg N ha^{-1} yr^{-1} on the western side (Baron et al. 2000). Some recent, short-term monitoring programs have suggested that nitrogen levels have been increasing. Important questions, however, remain unanswered. For example, are the many seemingly pristine lakes contained in this isolated, high-alpine environment changing, even though there are no significant watershed disturbances? If so, when did these changes begin, and how fast are they occurring?

A.P. Wolfe et al. (2001) explored the potential fertilizer effects of atmospheric nitrogen deposition in seemingly pristine high-alpine lakes in the Colorado Front Range using combined diatom, isotope, and geochemical paleolimnological approaches. They reasoned that diatoms would be amongst the first group of algae to respond to nitrogen fertilization, given their acknowledged sensitivity to changes in resource allocation (Interlandi & Kilham 1998). Indeed, their reconstructed diatom changes were striking. In their two study lakes (Sky Pond and Lake Louise, elevations 3322 and 3360 m a.s.l., respectively), a typical, oligotrophic, alpine diatom assemblage dominated the flora in the pre c. 1900 sediments (Fig. 11.19). However, the two well known planktonic indicators of elevated nutrient levels, *Asterionella formosa* and *Fragilaria crotonensis*, became much more common between c. 1900 and c. 1950. Then, between c. 1950 and c. 1970, *A. formosa* became the dominant taxon.

Based purely on a qualitative assessment of the above species changes, a diatomist would conclude that the lakes began to respond to fertilization in the early half of the 20th century, and this trend was particularly pronounced after c. 1950. But what was the source of these nutrients? As the lakes are isolated, with almost no human contacts in their catchments, atmospheric deposition was suspected. Nitrogen isotopes were used in the study to help identify the sources of the enrichment.

In both cores, sediment $\delta^{15}N$ (i.e., the normalized ratio of $^{15}N/^{14}N$) values are relatively stable between 4 and 5‰ prior to c. 1900 but, coincident with the diatom changes, begin to decrease shortly thereafter, by about 1‰ in the first half of the 20th century, with further decreases after c. 1950 of about 1.0 and 2.5‰ in Sky Pond and Lake Louise, respectively (Fig. 11.19). Wolfe et al. concluded that the increasingly lighter sedimentary nitrogen isotope signals

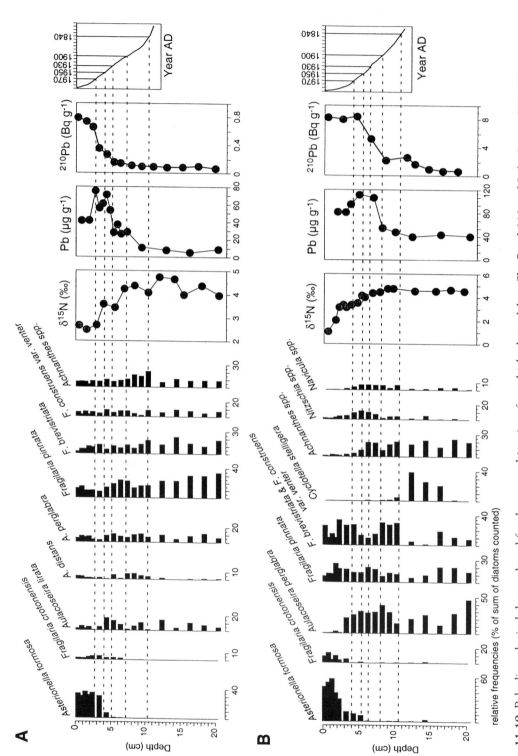

Fig. 11.19 Paleolimnological data gathered for the recent histories of two high-alpine lakes: Sky Pond (A) and Lake Louise (B), Colorado Front Range. The dominant diatom taxa are expressed as relative frequencies, the sediment $\delta^{15}N$ as the normalized ratio of $^{15}N/^{14}N$, total lead [Pb] as concentrations, and the ^{210}Pb profile. Depth–age relationships for the two cores are shown to the left, calculated using ^{210}Pb data. From A.P. Wolfe et al. (2001); used with permission.

were tracking the additions of atmospherically transported, isotopically depleted nitrogen compounds from the industrialized regions east of the study sites. They confirmed that these stratigraphic changes were not simply within the boundaries of natural variability by analyzing diatoms and nitrogen isotopes in a full, post-glacial sediment core from Sky Pond, and showed that the changes occurring in the 20th century were unprecedented in the lake's history.

Several anthropogenic sources were possible for this nitrogen enrichment, and so other geochemical techniques were used to tease apart some of the potential sources. For example, elevated stable lead concentrations [Pb] in the profiles documented evidence of the widespread mining and smelting activities in the Denver and Rocky Mountain region from the early 1900s until c. 1930 (Fig. 11.19). Thereafter, the [Pb] signal largely reflected emissions from leaded gas use, which was further confirmed by the post-1970 declines in sedimentary [Pb], coinciding with the removal of lead additives in North American gasoline. However, the decoupling of the [Pb] and the $\delta^{15}N$ isotope signal in the post c. 1970 sediments indicated that other sources, besides nitrogen compounds emitted from automobile exhausts, are reaching these high-alpine environments.

Agricultural activities were suspected to be the most likely sources of nitrogen enrichment over the lakes' recent histories. The post-1950 period of marked diatom and isotope changes corresponds with the intensification of agricultural practices, animal husbandry, and population growth in the Denver – Fort Collins region. Furthermore, the Haber–Bosch process was widely implemented in the manufacture of commercial N fertilizers after 1950, leading to massive applications of nitrogen to soils. Although fossil fuel combustion and other industrial sources may emit a wide range of $\delta^{15}N$ isotope signals, emissions from commercial fertilizers always carry a light $\delta^{15}N$ isotope signature (e.g., the Haber–Bosch process uses N from the air, and so has a $\delta^{15}N$ of 0 ± 3; Kendall 1998). It is likely that growing agricultural activities were primarily responsible for the accelerated nitrogen enrichment recorded in the study sites, even though they are at elevations of > 3300 m a.s.l., an altitude that is above previously suggested altitudinal limits for anthropogenic deposition by upslope air-mass trajectories.

As noted in Chapter 7, nitrogen deposition could also eventually cause lake acidification. However, paleolimnological work in the region could not, as yet, detect evidence of acidification (Baron et al. 1986), suggesting that critical loads had not yet been exceeded in this environment.

The Wolfe et al. study showed that remote and seemingly pristine and undisturbed sites are not escaping anthropogenic disturbances, and indeed may be quite vulnerable to airborne nutrient enrichment. Agricultural sources were the primary suspects thus far, but urban and industrial emissions are also important and increasing. In some respects, this work is analogous to the examples in Chapter 9 regarding long-distance transport of other contaminants. Paleolimnological approaches can be used to assess these changes, which may be so incrementally small from year to year that it would be difficult to document these trends employing standard monitoring programs that typically operate over only a few years. Sedimentary approaches can also be used to identify sensitive sites before further damage occurs. Since there is little evidence that nitrogen emissions will be declining, this problem will only likely become more severe in the future.

11.17 From the Tropics to the poles: eutrophication is a global problem

Paleolimnological approaches are now being applied to a wide array of eutrophication-related problems. However, keeping in mind that eutrophication is often considered to be the most serious water-quality problem on this planet, one cannot help noticing the geographical bias of many of my examples. Almost all the paleolim-nological applications for eutrophication, and indeed for many of the problems discussed in this book, are from North America and Europe. Certainly paleolimnology has much to offer other regions, and especially developing nations where water-quality problems are often severe, monitoring data are typically lacking, and there is a need for cost-effective approaches to be implemented. Although much work remains to be done, some projects have been initiated in these geographical regions. I include examples of two studies here – one from tropical Lake Victoria (East Africa) and another from Meretta Lake (Cornwallis Island, Nunavut) in the Canadian High Arctic – showing how paleolim-nology can be used to assess eutrophication trends in these (as yet) poorly studied environments.

Lake Victoria, whose shoreline borders five countries in East Africa, is the largest fresh-water lake in the Tropics by area. The lake is of tremendous economic (e.g., fisheries) and ecological importance, but it has also gone through a series of profound changes over recent decades, including the typical symptoms of eutro-phication, such as increased algal biomass and the replacement of diatoms by cyanobacteria as the dominant phytoplankton group (Kling et al. 2001). Further ecological changes have been linked to the explosive population increase of Nile perch (*Lates niloticus*) in the 1980s, a piscivore originally introduced to the lake in the 1950s and 1960s, and extirpation of several hundred species of endemic cichlid fishes. Clearly, the lake is subjected to multiple stressors, but many changes appear to be linked to the deforestation (Fig. 11.20) and intensified agriculture through-out Lake Victoria's catchment. The lack of long-term monitoring data has prompted several paleolimnological investigations regarding the eutrophication (e.g., Hecky 1993; Verschuren et al. 1998, 2002) and contaminant (Lipiatou et al. 1996; Campbell et al. 2003) history of this important water body.

Fig. 11.20 Trade in charcoal for domestic use is a principal cause of forest clearance on islands within Lake Victoria. Photo by Dirk Verschuren.

Hecky (1993) and Verschuren et al. (2002) used biogenic silica (BiSi) and diatom assemblages preserved in an offshore ^{210}Pb-dated sediment profile to elucidate the ecological changes in Lake Victoria linked to cultural eutrophication and restructured algal and fish communities. They showed that the increase in phytoplankton diatom production (estimated with BiSi) began in the 1930s, coincident with marked growth of the human population and accelerated agricultural activities in the lake's catchment. However, silica depletion since the 1980s may have stalled additional diatom growth, and to some extent diatoms were replaced by the proliferation of cyanobacteria, another indicator of eutrophication. The diatom species assemblages were relatively stable from the base of the core (c. AD 1820) until c. 1940, but then changed in composition, providing further evidence for recent nutrient enrichment. Verschuren et al. (2002) then used the abundance submicrofossil ratios of two important chironomid species (i.e., *Procladius brevipetiolatus*, which is intolerant of anoxia, and *Chironomus imicola*, which is anoxia-tolerant) to indicate that deepwater oxygen concentrations began to deteriorate in the early 1960s. The nature and timing of these ecological changes, and their links to the known historical record of population and agricultural growth in the catchment, showed that bottom-up effects of excess nutrient loading (as opposed to the top-down effects of the proliferation of the introduced Nile perch in the 1980s) were primarily responsible for the current poor water quality of Lake Victoria. Seehausen et al. (1997) also concluded that eutrophication directly contributed to the loss of cichlid species through its effects on water clarity and the impairment of sexual selection by the cichlids. Moreover, the progressive loss of an oxygenated deepwater habitat may have facilitated the demise of demersal endemic fish species and other taxa (Fig. 11.21), as this deepwater

Fig. 11.21 Nile tilapia (*Oreochromis niloticus*) and lungless snails (*Bellamya unicolor*) that were washed ashore after being killed due to a deep-mixing event that brought anoxic bottom water to the surface of Lake Victoria. Photo by Dirk Verschuren.

refuge from predation by the Nile perch would have been severely reduced or eliminated due to eutrophication. The environmental trends recorded in the sedimentary record are especially troubling, as the United Nations (1995) projects a doubling of the regional population by the year 2020. Clearly, new land-use strategies and other mitigation efforts will be required to save this important ecosystem from further deterioration.

In contrast to temperate and tropical regions, cultural eutrophication is not generally considered to be a major water-quality issue in polar regions, as human settlements are spaced widely apart and have low populations. There are, however, some exceptions.

Much of what we knew about High Arctic eutrophication was based on work initiated in the late 1960s and early 1970s as part of the International Biological Programme (IBP). During IBP, two Canadian High Arctic lakes in the hamlet of Resolute Bay (Cornwallis Island, Nunavut) were chosen for detailed study of freshwater production: ultra-oligotrophic Char Lake and eutrophic Meretta Lake (Rigler 1974). Culturally eutrophied lakes are still quite rare in Arctic regions, but Meretta Lake was an exception (see also pp. 171–2), as it received sewage from a small number of buildings from 1949 to 1998. The Department of Transport base (locally referred to as the "North Base"), associated with the local airport, discharged its gray water and sewage through a central collecting pipe and then via a utilidor onto the land (Fig. 11.22). Thereafter, the discharge flowed through a series of streams that ultimately drained into Meretta Lake. Schindler et al. (1974) summarized Meretta Lake's water-quality variables during the short monitoring time window of the IBP program but, as pre-eutrophication conditions were not known, it was not possible to place current limnological conditions into a historical context. As Rigler (1974) noted in the final report on this IBP

Fig. 11.22 This utilidor system, which was recently dismantled, is releasing sewage and gray water from the Resolute Bay (Cornwallis Island), North Base into a series of streams and ponds. The effluent ultimately reaches Meretta Lake, about 1.6 km downstream.

project, "The conclusions we can draw are limited because the original condition of Meretta Lake is unknown."

The absence of long-term monitoring data prompted Douglas and Smol (2000) to undertake a diatom-based paleolimnological study of Meretta Lake. Using a high-resolution sediment core collected in 1993, they showed that eutrophication from the "North Base" had significantly affected diatom assemblages; however the species changes were markedly different from those recorded in nutrient-enriched temperate lakes. Despite the increase in nutrients, periphytic diatoms continued to overwhelmingly dominate the assemblage, further confirming that extended ice covers (discussed further on pp. 265–7), and not nutrients, were precluding the development of large populations of planktonic diatoms.

In 1993, when the Douglas and Smol (2000) sediment core was collected, the North Base still had a moderate number of users. However, during the 1990s, the number of people using the North Base declined steadily, and the entire utilidor system was dismantled in 1999 and sewage was thereafter trucked to a designated area of the hamlet's dump (Fig. 11.23). Douglas and Smol (2000) measured summer nutrient levels from 1991, and tracked the decline in eutrophication throughout the 1990s. Comparing the diatom assemblages preserved in the surface sediments of a core collected in 2000 (Michelutti et al. 2002a) with those at the surface of the 1993 core (Douglas & Smol 2000) revealed recent diatom species shifts that were consistent with nutrient reductions. These recent assemblage shifts also showed that there were no significant lags between lakewater nutrient concentrations and the deposition of diatoms in the profundal zone in these largely ice-covered lakes. In addition, changes in periphytic assemblages collected

Fig. 11.23 The sewage and water truck that services the hamlet of Resolute Bay (Nunvaut). The truck delivers water from Char Lake to the water tanks of local residents, but also pumps out their sewage and gray water holding tanks and transports to the local dump for land disposal. Hopefully the two tanks are never confused!

each summer since 1991 from Meretta Lake's littoral zone also tracked the recovery from eutrophication, as taxa characteristic of more oligotrophic environments increased relative to more eutrophic diatoms (Michelutti et al. 2002b).

Paleolimnological approaches will likely become more important in water-quality assessments in high-latitude regions, as they have been in many other parts of the world. For example, Canadian Inuit birth rates are the highest of any population in the developed world, and so concerns over issues of waste disposal and other environmental problems will only escalate in the future.

11.18 Does eutrophication result in increased ecosystem variability and reduced predictability?

Ecologists have suggested that perturbations should increase the variability of ecosystems and reduce predictability. However, these concepts are based primarily on theoretical grounds, and on only limited and short-term observations. Cottingham et al. (2000) used the annually resolved fossil pigment record from the sediments of artificially fertilized Lake 227 in the Experimental Lakes Area of Ontario (Leavitt et al. 1994) to test the hypothesis that enrichment increases the variance and reduces the predictability of algal communities. Similar to the Lake 223 example discussed earlier (Box 11.1), Lake 227 was the subject of a well-documented, whole-lake fertilization experiment that began in 1969.

Cottingham et al. compared fossil pigment composition before and after fertilization of Lake 227 using a median-log Levene's test and indeed showed that variability increased significantly with cyanobacteria and many algal groups. Predictability, estimated using dynamic linear models, simultaneously decreased.

Similar types of paleolimnological analyses should be done on other systems, as it is difficult to make large-scale generalizations from one lake, but these preliminary data suggest that human activities such as eutrophication may increase inherent community variability. This potentially has several consequences, such as destabilizing lakes and possibly obscuring the impacts of other simultaneous stressors, such as climatic change.

11.19 Nutrient management using a combination of modeling and sedimentary approaches

Just as modelers and paleolimnologists have combined their efforts to provide more defensible management plans when dealing with acidification (see pp. 115–16) and other environmental problems (Fig. 2.5), there is considerable potential for blending these two approaches with eutrophication work. For example, O'Sullivan (1992) used diatoms, geochemistry, and historical records from two coastal lakes in southwestern England to show that the lakes had become progressively more eutrophic since 1945 due to increased agricultural development and municipal sewage loading. O'Sullivan then used nutrient export coefficient models (Jørgensen 1980), coupled with long-term records (1905–85) of land-use, fertilizer application, and livestock census data from the catchment to estimate historical changes in phosphorus loading. Using Vollenweider's lake model (1975), he then concluded that, before 1905, P loads were at or near permissible levels for a shallow lake of that kind, but since 1945 loads had increased to about 10 times the acceptable limit for lakes of that size and volume. O'Sullivan concluded his assessment by using the same nutrient model (Vollenweider 1975) to evaluate the effects of various restoration strategies on annual N and P loads. He estimated that total removal of P from sewage and detergents would still result in P loads that were about twice the OECD permissible limits. Diffuse sources from agriculture also had to be reduced. He concluded that, in addition to treating the point sources of nutrient pollution, buffer strips of woodland at least 15 m wide along the rivers draining into the lake would be needed to decrease diffuse nutrient inputs.

Meanwhile, Rippey et al. (1997) used diatom-inferred P concentrations, geochemistry, and mass-balance models to demonstrate that flushing rate and internal nutrient loading can play a strong role in regulating eutrophication. Jordan et al. (2001) used similar approaches in a small lake in Northern Ireland to show that diffuse phosphorus loading began to increase after the 1940s, when the catchment was converted exclusively to grassland agriculture.

Paleolimnological data can also be used to evaluate process-based models, although only a few examples exist thus far. For example, several types of empirical and process-based models are increasingly being used by lake and watershed managers to quantify lake responses to changing land-use practices. No one doubts that these modeling projections are important for setting policy guidelines for shoreline development (e.g., the number of permissible cottages or other dwellings, types of septic systems, etc.). Nonetheless, the predictive abilities of most of these models have rarely been rigorously assessed, as long-term monitoring data are typically lacking. Paleolimnological inference data can be used to evaluate these models. For example, diatoms have begun to be used in paleo-model comparisons (Hall & Smol 1999; Reavie et al. 2002) to evaluate the predictive abilities of the Ontario Ministry of Environment & Energy Lakeshore Capacity Trophic State Model (TSM: Dillon et al. 1986; Hutchinson et al. 1991). This mass-balance model attempts to predict TP concentrations in a lake's current and "natural" (i.e., prior to human development) states. Poor agreements between the diatom-inferred and TSM hindcasted TP values would provide reasons to reconsider some of the assumptions and parameters used in the model. On the other hand, strong agreements between paleolimnological and TSM inferences would certainly increase the credibility of any model-based management strategy.

There is considerable potential to expand these modeling/paleolimnology comparison exercises to many other regions and applications. For example, efforts are currently under way to model hypolimnetic oxygen deficits and other symptoms of eutrophication. Paleolimnological approaches can be used to provide independent data to evaluate these models (e.g., chironomid-inferred hypolimnetic oxygen levels), and also provide insights into how scientists and managers can fine-tune their models when model-paleo comparisons disagree.

11.20 Linking evolutionary changes to shifts in trophic status

With new advances, there are some exciting possibilities for combining genetic techniques with those used by paleolimnologists. For example, Weider et al. (1997) showed that diapausing egg banks of zooplankton, which can be recovered from dated lake sediment cores, have the potential to remain viable for decades or even centuries. They recovered diapausing ephippial eggs from the *Daphnia galeata-hyalina* group from a sediment core from Lake Constance (Germany), and used electrophoresis to examine long-term changes in the genetic composition of the hatching pool. Their data showed that significant shifts had occurred in the genetic structure of the population, which were correlated with trophic state changes in the lake over the past 25–35 years. These types of studies could potentially be used to study micro-evolutionary processes, and relate them to past environmental changes in aquatic systems.

11.21 Summary

Excessive nutrient enrichment is the root cause of cultural eutrophication, which has many negative repercussions on aquatic systems. Although lakes and rivers naturally receive nutrient inputs from their catchments and the atmosphere, many human activities have greatly accelerated the eutrophication process by, for example, sewage inflows, runoff from agricultural fields, and industrial effluents. Phosphorus is the element most limiting to algal and cyanobacterial growth in most temperate lakes. Some of the obvious symptoms of eutrophication include excessive growths of algae, cyanobacteria, and aquatic macrophytes, but a myriad of other related problems may also occur (e.g., declines in deepwater oxygen levels). Paleolimnological approaches have been developed to track many of these limnological changes using a variety of physical, chemical, and especially biological techniques. For example, the abundances and distributions of diatom species have been used to construct transfer functions that can infer past epilimnetic

nutrient levels, and chironomid head capsules can be used to reconstruct past deepwater oxygen levels. Paleolimnological research has shown that many lakes have been adversely affected by nutrient inputs, but that some water bodies naturally had productive waters, even before marked human interventions. This has important implications for setting realistic mitigation targets. Equally important, paleolimnological approaches can be used to study lake ecosystem responses to nutrient reductions (Battarbee et al. 2005). New developing research areas include studies on shallow, macrophyte-dominated lakes; assessing the role of atmospheric deposition of nutrients in oligotrophic lakes; the application of paleogenetic techniques to resting stages preserved in sediments; and studying eutrophication patterns in new geographical regions, such as the Tropics.

12

Erosion: tracking the accelerated movement of material from land to water

As water wears away stones
and torrents wash away the soil,
so you destroy man's hope.
Job 14: 19

12.1 Erosion and human activities

If you look out the window of an airplane flying over any part of this planet that is hospitable to humans, you will likely see how effective our species has been at carving and rearranging the landscape. Our "footprints" are almost everywhere, from plowed fields, to cut forests, to paved landscapes. Once native vegetation cover is disrupted and the soil profiles are disturbed, increased rates of wind and water erosion are almost inevitable. Dearing and Jones (2003) estimate that sediment delivery from land to receiving water bodies is typically five to ten times higher (and sometimes much greater) in catchments affected by major human activities. Although new soil is constantly being produced by physical, chemical, and biological processes, it is being renewed at an extremely slow rate, estimated at about 1–2 mm every century. In many regions, erosion greatly exceeds the rate of soil production. For example, erosion rates in the United States average about 6 mm a year (Hidore 1996); alarmingly, about one-third of

the world's arable land has been lost by erosion since the middle of the 20th century (Pimentel et al. 1995). As Dearing (1994) notes in his excellent review on reconstructing the history of soil erosion, we live on "an eroding planet."

Accelerated soil erosion has serious consequences for terrestrial ecosystems. Farmland productivity decreases, natural habitats are destroyed, and our infrastructure (such as houses and buildings, especially near coastlines) becomes less stable. Lakes and other water bodies are downhill of their catchments, and therefore are also seriously affected, acting as vectors or sinks for eroded soil material and other matter. For example, the Mississippi River now carries about 2 million tons of topsoil every 24 hours past the city of New Orleans. This is equivalent to over 20 cm (8 inches) of topsoil from over 550 ha (1376 acres) of farmland being eroded and lost every day of the year.

Increased sediment loads have serious direct and indirect limnological consequences. For example, a higher sediment load in flowing waters increases the scouring effect of the water. Accelerated sedimentation can affect aquatic habitats, such as fish spawning grounds, as well as many other services we take for granted from our lakes and rivers. Furthermore, increased erosion not only delivers higher sediment loads to lakes and rivers directly affecting water clarity and quality, it also increases the export of contaminants and nutrients. As such, the topic of erosion overlaps many other environmental problems discussed in this book, such as eutrophication, forestry, and contaminants. Nonetheless, given that the approaches used to study erosion are often somewhat different from those used by paleolimnologists dealing with other problems, I have included a separate chapter on this topic. The tools that are often used to interpret erosional sequences are primarily those of the physical geographer, such as analyses of sedimentation rates, the composition and texture of the sediments, magnetic susceptibility, and so forth. My focus here will be on the physical movements of material into lakes, and not the chemical signatures they contain. The latter very important environmental issues are covered in other chapters.

Contemporary, process-based studies can answer many important questions related to erosion at broad spatial scales, but many questions remain unanswered due to the short time scales (days to sometimes a few years) that are characteristic of these studies. Furthermore, as summarized by Dearing et al. (1990), while such short-term studies may address problems associated with the frequency and magnitude of changes, there are no assurances that short-term records can identify threshold conditions. Nor can they examine long-term reactions and recoveries to past changes, whether they be natural (e.g., climatic) or human related (e.g., agriculture and other land-use disturbances). Sediments offer the additional advantage that they can be analyzed and quantified on a basin-wide scale, as sediment cores can be sampled at different spatial and temporal scales set by the investigator.

Paleolimnologists can reconstruct long-term trends in erosion data from lake and reservoir sediments using three major approaches (Dearing & Foster 1993). First, assuming that reservoir surveys were conducted at the time of impoundment (or from maps of the original valley) and at other times in the reservoir's development, one can compare survey data collected at different time periods, and calculate the average sediment volumes between different dates, and so estimate the rate of infilling from erosional and other processes. Second, by sampling, dating, and analyzing individual lake sediment cores, sediment accumulation rates (at least at that one point in the lake where the core was taken) can be estimated over time. These data can be augmented by proxy erosion data, such as mineral magnetic measurements (see below), and other physical, chemical, and biological data. Third, by taking multiple cores in a lake, and then correlating them to a master, dated sequence, total sediment masses or yields can be estimated through averaging or three-dimensional mapping of the accumulation rates. To some extent, this third approach can be facilitated by using new developments in seismic sequencing of sedimentary profiles (Scholz 2001).

Using the above approaches, paleolimnological assessments can help answer questions such as: What activities have resulted in large erosional events? Were these sustained processes or episodic events? Have different types of land-use practices altered the frequency and intensity of erosion? Have similar activities in different environmental settings (e.g., under different climatic regimes) resulted in similar responses? How well do sediment models perform over long time scales at inferring erosion processes in different landscapes? Sedimentary records have been instrumental in answering these questions in lake, river, and reservoir environments. Below are a few applications illustrating these techniques.

12.2 Land clearance and the increased rate of sediment accumulation in lakes

A pioneering study on using lake sediments to track soil erosion was by Davis (1976), who used

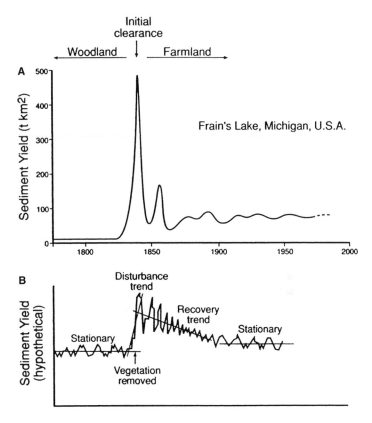

Fig. 12.1 (A) Estimates of sediment yield at Frains Lake, Michigan, over the past two centuries, showing the rapid increase in the erosion at the time of land clearance for agriculture. (B) The Frains Lake sediment yield data, showing the different states and responses in the sediment system to disturbances and then recovery. Based on data from Davis (1976), with figures from Dearing (1994); used with permission of Blackwell Publishing.

pollen analysis to reconstruct past vegetation changes and related this to early land clearance and settlement in the catchment of Frains Lake (Michigan). Based on sediment accumulation rates and estimates of sediment composition, she calculated that mineral matter was lost from the Frains Lake catchment at a rate of about $9\,t\,km^{-2}\,yr^{-1}$ prior to the arrival of European settlers. However, erosion rates increased 30–80 fold in the mid-19th century following forest clearance and the shift to cultivation (Fig. 12.1). After an approximately 30–40 year period, erosion rates leveled off to a new steady state, but were still about one order of magnitude higher than natural, pre-disturbance rates. Studies such as these are a useful guide to what one might expect with land-use changes in similar catchments from other regions.

12.3 Magnetic proxies of soil erosion

The Frains Lake example focused primarily on the amount of sediment accumulating over time. But what were the sources of these sediments? Techniques are now available to help decipher the provenance of material accumulating in a lake basin. Magnetic proxies often lead these initiatives. Oldfield (1991), who was instrumental in developing many of these approaches, provides an interesting perspective on the development and use of environmental magnetism. Meanwhile, Maher and Thompson (1999) have compiled a series of chapters dealing with the paleoenvironmental applications of magnetism, including a useful primer on terminology and approaches (Maher et al. 1999). Sandgren and Snowball

(2001) summarize methodological aspects. The reader is referred to these references for more details on procedures, definitions, and applications.

Although interpretations of magnetic data may be complex and at times difficult, the measurement techniques are relatively simple, inexpensive, non-destructive, and can be carried out in the field or the laboratory. As Dearing (1999) recounts in his excellent summary of magnetic proxies in lake sediments, the first attempt to use this approach by Thompson et al. (1975) in Lough Neah (Northern Ireland) was certainly a classic study, but also a lucky choice of sampling sites for a pioneering investigation. They found that changes in magnetic susceptibility correlated well with pollen changes indicating land clearance and presumably erosion. They concluded that the changes in magnetic susceptibility were measuring the content of titanomagnetite grains derived from the surrounding basalts during periods of increased erosion when the catchment was deforested. With high concentrations of ferrimagentic minerals in its catchment, Lough Neah was certainly a fortuitous starting point for the baptism of a new approach! We now know that many lake basins would not allow such straightforward interpretations of the magnetic data.

Magnetic minerals can be brought into a basin from four major sources (Dearing 1999). Two are allogenic, derived from outside the lake basin: (i) atmospheric-transported particles (e.g., pollutants, wind-eroded material, tephra); and (ii) river-transported detrital materials. Two authigenic sources are (i) bacterial magnetosomes and (ii) authigenic iron sulfides. Meanwhile, reductive diagenesis is an authigenic destructive process. Dearing (1999) summarizes the different physical, chemical, and observational methods used to distinguish these sources and processes. Therefore, like most techniques used by paleolimnologists, interpretations are rarely straightforward, but many insightful studies have been completed. As with other paleolimnological studies, the strongest inferences come from paleolimnological studies that use a spectrum of techniques. A few examples, illustrating the range of approaches, are summarized below.

12.4 Quantifying the erosional response to land-use changes in southern Sweden

One of the earliest multidisciplinary paleolimnological studies attempting to quantify the effects of land-use changes on erosional processes was conducted in southern Sweden, where Dearing et al. (1987) used a combination of magnetic susceptibility, radiometric and paleomagnetic dating, and core correlation, coupled with the analysis of pollen and diatoms preserved in the sediments of Havgårdssjön. Explanatory variables for interpreting the sediment influx records were quantitative data of land-use change gleaned from inventories and census documents for farms in the lake's catchment.

In order to obtain meaningful sedimentation patterns in a lake basin, multiple cores should to be used. This requires a considerable amount of resources dedicated to radiometric dating of the sediment cores (e.g., ^{210}Pb), or the use of paleomagnetic measurements or some other easily measured parameter to correlate several cores to some well-dated master core(s). In Havgårdssjön, Dearing et al. (1987) used a grid system to determine their sampling strategy to collect 47 cores, which were then correlated using magnetic susceptibility to map sediment influx on a basin-wide scale. They found that anthropogenic activities have greatly increased the transport of material into the lake. For example, estimated sediment yields for the period approximately 5000 to 2000 years ago was 0.25 t ha^{-1} yr^{-1}, which is typical of stable, deciduous forest catchments. Some minor susceptibility peaks around 2950 and 650 BC may have been related to minor watershed disturbances from late Neolithic and late Bronze Age activity, but the changes were modest and equivocal. However, from c. 50 BC onwards, sediment yields increased markedly, reaching approximately 0.86–2.50 t ha^{-1} yr^{-1} by AD 950–1300. This 3.5- to 10-fold increase coincided with more intensive agriculture with the introduction of a new technology: the heavy-wheeled plow. Peak magnetic susceptibility values at c. AD 1300 are matched with the historical record of the construction of the Turestorpsön castle, and presumably its associated farmland.

The Havgårdssjön data thereafter indicate declines in erosion, with a drop in sediment yield to about 0.5 t ha^{-1} yr^{-1} (which was still about double natural levels) and a decrease in magnetic susceptibility for the period AD 1300–1550. Agrarian depression during this early part of the so-called "Little Ice Age" has been documented in many northwest European locations. However, determining whether erosion decreased as a result of less intensive land-use activities or due to colder temperatures is difficult, as the latter can directly affect sediment mobilization and transport, and both factors probably occurred simultaneously. Most likely, erosion declined due to both cooling and the related agricultural depression.

The establishment of two farms and a village by AD 1600 coincided with increased susceptibility measurements, peaking around AD 1712, coupled with increased sediment yields of about 1.75 t yr^{-1}. Pollen data reveal that cereal crops increased at this time, and so agriculture was once again a major activity in the catchment. From AD 1682, the land-use records document increasing proportions (about six-fold) of plowed land, which is mirrored by paleolimnological inferences of higher erosion rates. For example, sediment yield increased 1.5-fold from AD 1682 to 1979.

Dearing et al. (1990) found similar human-induced patterns of erosion in a related study from nearby (~ 20 km away) Lake Bussjösjön. This record, however, extended back approximately 10,000 years, and could broadly be split into four major periods of sediment responses to different sets of environmental conditions. The earliest part of the core records high levels of sediment movement from the catchment, which began during periglacial times, but continued until c. 9250 yr BP (Fig. 12.2A). The high magnetic susceptibility of the sediments (Fig. 12.2B) confirms that this material was not topsoil, but most likely the movement of previously deposited unweathered material. This is followed by a 6500-year period of relative catchment stability, with the establishment of a well-established forest cover, as indicated by the pollen data. Sediment load had greatly decreased from the early postglacial period, and had shifted to primarily topsoil erosion, as evidenced by the low susceptibility (Fig. 12.2B).

Similar to the Havgårdssjön example described earlier, human influences on catchment erosion became evident around 2500 years ago; this third major sedimentary zone extended to about 250 years ago. Initially, the sediment load increased due to major deforestation, although this sediment loss was still much lower than modern agriculture. The sediment source was interpreted to be subsoil during this zone, based on the magnetic data, indicating that early human impact caused changes to the hydromechanics of the catchment channels. However, from about 50 to 300 yr BP, there is a shift to greater losses of topsoil and finer particles. Since the sediment load is still high (Fig. 12.2A), it would appear that the absolute amount of eroded topsoil increased at this time, and it was not simply a relative decrease in the amount of eroded subsoil.

Over the past approximately 300 years, the channels in the Lake Bussjösjön catchment have only eroded significantly during times of disturbance during drainage operations, as shown by the load (Fig. 12.3) and source (Dearing et al. 1990) data. Sediment yields dropped in the middle part of the 20th century, with a reduction in the overall surface drainage network, but increased again, largely due to subsoil erosion as a result of undersoil drainage, in the latter part of the century, with more intensive agriculture. Since 1970, erosion has decreased.

Since the time of these Swedish studies, a few new approaches and refinements have been introduced to the field of paleolimnology (e.g.,

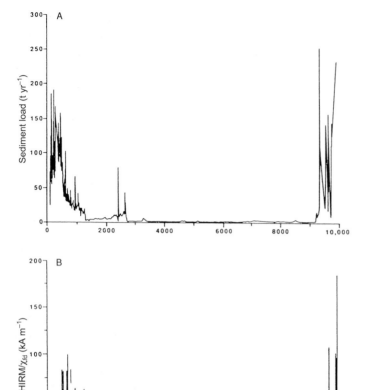

Fig. 12.2 (A) Loading of minerogenic sediment (t yr^{-1}) to Lake Bussjösjön, Sweden, from 10,000 to 250 yr BP. (B) Record of the magnetic properties of the sediments, shown as a ratio of secondary ferromagnetic minerals (ultrafine magnetite and maghemite), which may be detected using the measurement of frequency-dependent magnetic susceptibility (χ_{fd}) and magnetic remanence after forward saturation (HIRM). Values of χ_{fd} are marginally higher in bulk topsoils than in subsoils, with a high concentration in clays. Meanwhile, unweathered parent material shows magnetic properties typical of antiferromagnetic minerals (goethite and hematite), with high levels of magnetic remanence after forward saturation (HIRM). The ratio of these magnetic properties (HIRM/χ_{fd}) most clearly distinguishes between sediments from topsoil or deeper origins. From Dearing et al. (1990); used with permission of John Wiley & Sons Limited.

chapters in Maher & Thompson 1999; Last & Smol 2001a,b). Nonetheless, these pioneering studies showed how useful sedimentary records are at extending our time scales for a better understanding of erosional and related processes. Similar approaches have been used to explore a variety of other issues. For example, Sandgren and Fredskild (1991) used pollen analysis and magnetic measurements to track erosion rates in southern Greenland. Following a period of rapid erosion following deglaciation (similar to the Swedish example cited above), the soils stabilized with the establishment of a dwarf shrub tundra vegetation cover. However, this equilibrium was interrupted in about AD 1000 with the arrival of the Norsemen, who introduced cattle, following which the vegetation cover was disrupted.

With the disappearance of the Norse in about the 15th century, the soils restabilized, but then were once again disrupted by human recoloniza-tion and the introduction of sheep, primarily early in the 20th century. Using an eight-century paleoenvironmental perspective, Dearing (1992) showed that mining activities between AD 1765–1830 and 1903–85 had greatly increased the rate of sediment load from a Welsh catchment. Riverine and reservoir systems can also be studied. For example, Foster and Walling (1994) tracked sediment transport in a South Devon (UK) river by using the record preserved in a downstream reservoir. They showed that sediment yield had increased four-fold since World War II, from about 20 to about 90 t km^{-2} yr^{-1}; this was attributed to increased livestock numbers and

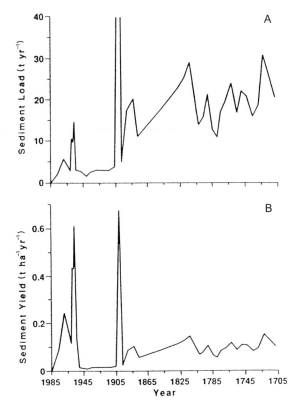

Fig. 12.3 (A) Loading of minerogenic sediment ($t\ yr^{-1}$) to Lake Bussjösjön, southern Sweden, from AD 1700 to 1985. (B) The record of minerogenic sediment yield from the catchment ($t\ ha^{-1}\ yr^{-1}$). From Dearing et al. (1990); used with permission.

grazing intensity. Data can be used to evaluate process-based models, assuming the temporal and spatial resolution is sufficient for paleo-model comparisons.

12.5 Contrasting erosion rates in Mexico before and after the Spanish conquest: a 4000-year perspective

The example from Frains Lake used to introduce this chapter documented increased erosion as a result of European style land-use changes occurring in North America. For most regions of the United States and Canada, the past 200 years or so covers the period of significant changes in land-use activities that may result in accelerated erosion, and periods younger than the 19th century can often be considered as "controls" representing "pre-impact" conditions. This is not the case in some regions of the world, where time frames restricted to the past two centuries will not capture the full range of repercussions from land-use changes, as human impacts may have occurred much earlier. One such region is Mexico, which supported a large indigenous population with a highly developed system of agriculture long before the arrival of the Spanish in AD 1521. Pollen data show that maize agriculture had rapidly spread across Middle America after about 4000 yr BP.

Parts of the central Mexico highlands are considered to be degraded ecosystems as a result of intense land erosion. The general thinking has been that the Spanish conquerors encountered a near pristine landscape at the time of their arrival in the early 16th century, and it was the implementation of European-style agriculture (i.e., plow farming with the use of draft animals) that was the root cause of the severe erosion that this area has experienced. Movements were being initiated to return agricultural practices in this region back to the traditional methods used by the pre-contact communities. But were these pre-16th century, traditional agricultural systems any less susceptible to the ravages of erosion in this highly susceptible, mountainous region? O'Hara et al. (1993) addressed these questions by using a combination of data sources from geological, archeological, and paleolimnological records. They took 21 sediment cores from Lake Pátzcuaro, a relatively large (126 km²), intermontane (2036 m a.s.l.) closed lake basin in the Michoacán highlands. Using sediment stratigraphy, bulk density, magnetic susceptibility, loss-on-ignition, carbonate content, and sediment chemistry, six main lithological units were identified.

The paleolimnological data pinpointed three major periods of forest clearance and accelerated soil erosion that occurred *before* the period of European contact. These erosion episodes were identified by discrete layers of red clay in the sediments, which were also delineated by high levels of magnetic susceptibility, iron, aluminum, and silica, as well as pollen from maize, grasses, and weeds. An early, short-lived erosion period was identified between 3640 and 2890 yr BP. A second, more intense erosion period occurred in the late Preclassic/early Classic periods, between 2530 and 1190 yr BP. O'Hara et al. estimated that about 13.4 Mt of sediment was deposited in the lake at an average rate of about 10,300 t yr^{-1} between 2500 and 1200 yr BP (about double the pre-agricultural rate). Sediment accumulation rates were higher in the cores collected from the northern basin of the lake, suggesting that the main late Preclassic/early Classic agricultural sites were concentrated in the northern part of the lake's catchment.

Paleoenvironmental data document a prolonged drought in the region between c. 1200 and 850 yr BP. This period coincides with a period of catchment stability, where erosion rates dropped sharply to ~ 4500 t yr^{-1}. Presumably, agriculture was much reduced or totally abandoned in the area at this time. However, shortly after the arrival of the Purépecha, in the later Postclassic period (850–350 yr BP), accelerated erosion was renewed, with an estimated 24.5 Mt of sediment deposited in the lake at an average rate of 29,000 t yr^{-1}. Land clearance for agriculture, as well as timber harvesting for ritual, construction, and military purposes, greatly increased erosion, even during this pre-Columbian period.

This study documented that marked erosion in the lake's catchment predated the time of European contact in Mexico, and showed the long-term impacts of indigenous farming on a tropical lake. What is perhaps most surprising is that it is hard to distinguish the impact of European-style plow agriculture from the effects of indigenous farming. In fact, estimated sediment influx rates suggest that, if any change occurred, erosion rates apparently decreased following the arrival of the Spanish in the area in AD 1521. Depopulation, coupled with forest regeneration, may have resulted in the reduced soil loss.

A major conclusion was that a return to traditional farming techniques would not solve the problem of environmental degradation from accelerated erosion in regions such as this. Similar approaches can be applied to other regions and types of land use to gain a better understanding of human influences on erosion. Other examples of the effects of pre-Columbian agricultural and urbanization (e.g., Mayan city building) on tropical and subtropical America are summarized in Binford et al. (1987).

12.6 Contrasting the effects of human impacts from natural factors on sediment deposition in the Nile River Delta, using a 35,000-year paleoenvironmental history

In many fluvial systems, a detrimental impact of human activities is accelerated soil erosion and hence increased sediment loads to rivers. However, some human activities, such as dam construction, can decrease downstream sediment transport, and therefore negatively impact delta ecosystems. One such region is Egypt, where the construction of the High Aswan Dam has resulted in a myriad of ecological problems. One repercussion of the 1964 impoundment of the dam was that fluvial sediments were no longer transported northward to the delta, and the balance between marine and river processes was totally modified. Specifically, the loss of fluvial sediments has resulted in the loss of source material for delta formation and maintenance, and so now coastal processes dominate, such as Mediterranean waves and currents, which are gradually eroding and destroying the delta. Many other stressors have also affected the river and its delta, such as salinization and pollution from human activities, and natural processes such as climatic shifts and sea-level changes.

Few countries are as dependent on a single freshwater source as Egypt is on the River Nile. The population is concentrated along the Nile

Valley and Delta, and all Egyptian agriculture must be irrigated. The Nile Delta represents two thirds of Egypt's habitable land, and so any loss of the northern delta plain, whether from coastal erosion, salinity increases, or pollution, is of critical economic, social, ecological, and political importance. Given the strategic nature of this region, Stanley and Warne (1993) undertook an ambitious project to reconstruct the past 35,000-year history of the river and delta so as to assess the relative influences of natural and anthropogenic factors on delta development. Such baseline analyses are important, as effective coastal management strategies can only be developed and implemented once the impacts of humans can be distinguished from natural environmental changes.

Stanley and Warne (1993) undertook sedimentological analyses of almost 4000 samples from nearly 100 radiocarbon-dated sediment cores from across the northern Nile Delta. The region has seen tremendous changes over the millennia, long before human impacts were apparent. For example, during portions of the late Pleistocene (< 30,000–12,000 years ago), when sea levels were low, the delta was as much as 50 km north of the current coastline. Climatic oscillations were recorded in the sedimentary profiles by accumulations of sediments from different source areas. The most worrisome changes, however, were those of recent times resulting from human-related disruptions of sediment transport processes due to dam construction. Before impoundment of the High Aswan Dam, the River Nile transported approximately 124×10^6 metric tons of sediment to the delta each year. Sediment input is now minimal, with a negative balance in sediment budget along the delta margin. The delta is eroding.

The paleoenvironmental data can be used to help plan for the future of the region. The geological data suggest that some of the natural trends reconstructed for the past few millennia will continue into the next century. For example, sea level is expected to rise at least 1 mm a year, and subsistence rates along the coast are

expected to continue at 1–5 mm a year. These two factors, when combined together, led to the prediction that a relative sea-level rise of from 12.5 cm to 30 cm will occur by AD 2050, and so a significant portion of the delta will be submerged. These natural factors, coupled with the previously discussed loss of source sediments for the delta due to dam construction, do not bode well for the future of the region, especially in light of Egypt's large rate of population growth.

12.7 Determining the effects of hydroelectric dams and river impoundments on flooding histories from oxbow lake sediments

As discussed in various sections of this book, diverting or damming river systems may result in a suite of negative environmental impacts, ranging from changes in sediment discharges to deltas (see the previous section), to hampering the movement of species (e.g., p. 294), to changes in the quality and quantity of downstream water (e.g., pp. 304–7). However, in some cases, the ecological effects are not easily discernable on downstream aquatic habitats, especially when ecosystems are subjected to multiple stressors. One controversial area is the Peace–Athabasca Delta (PAD) in northern Alberta (Canada), a highly productive and internationally recognized northern boreal ecosystem. The PAD is characterized by a myriad of oxbow lakes, wetlands, and other deltaic features that depend on periodic flooding from the Peace and Athabasca rivers (often as a result of ice jams) to replenish and recharge this floodplain ecosystem (Fig. 12.4). In 1968, a large hydroelectric dam was constructed on the British Columbia portion of the Peace River, and since that time there have been claims that the impoundment had so reduced the river flow that flooding frequency has been greatly reduced, with concomitant ecosystem degradation (reviewed in Wolfe et al. 2006). However, as long-term hydrological records predating the

Fig. 12.4 An aerial view of a portion of the Peace–Athabasca Delta in northern Alberta. The periodic flooding of the river recharges the oxbow lakes in the floodplain, and contributes to the productivity and diversity of the ecosystem. Photograph taken by Brent Wolfe.

construction of the dam were not available, it was difficult to resolve this growing environmental and political controversy.

In an attempt to assess the impacts of both natural (e.g., climatic change) and anthropogenic (e.g., river regulation) influences on the PAD, Wolfe et al. (2005, 2006) undertook comprehensive paleolimnological studies on a variety of floodplain lakes that would be subjected to different flood regimes based on local geographical factors, such as elevations. For example, floodplain lakes that are almost at the same elevation as the river bank would be expected to be replenished more frequently, as even lowstand flooding events would inundate these water bodies. Lakes at higher elevations would be subjected to less flooding frequency, as only with larger floods would these lakes be recharged. By analyzing the sediments from a wide spectrum

of floodplain lakes, and tracking proxy indicators that could be linked to river flooding, a paleo-flood frequency map could be developed.

Using laminated lake sediments spanning approximately the past 300 years from oxbow lakes, Wolfe et al. (2006) reconstructed past flood events for portions of the PAD. Although a variety of complementary physical, chemical and biological paleolimnological approaches were employed, I will focus on only one aspect of their study: the use of sediment core magnetic susceptibility measurements to reconstruct past flooding (i.e., high-energy) events. The oxbow lake sediments cores were characterized by visually distinct banding (Fig. 12.5). The dark bands corresponded to a variety of other proxies indicating high-energy deposits, which would most likely be deposited in the oxbow lakes during flood events. These dark bands are also clearly

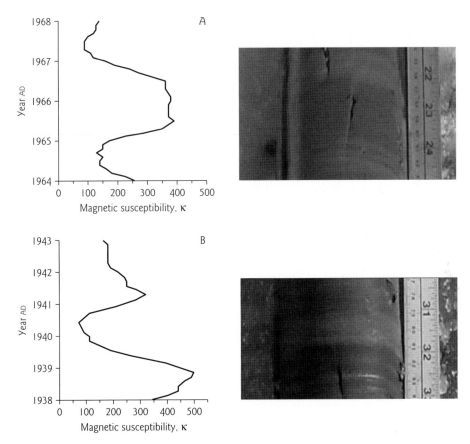

Fig. 12.5 (A) Detail of a sediment core from Peace–Athabasca Delta (PAD) oxbow lake PAD 54, showing that the peaks in magnetic susceptibility are associated with the dark-colored laminations, which were interpreted to represent high-energy (i.e., flooding) events. (B) Similar detail of a sediment core from PAD 15, showing the troughs in magnetic susceptibility associated with light-colored (i.e., non-flooded) laminations. From Wolfe et al. (2006); used with permission of John Wiley & Sons Limited.

demarked using magnetic susceptibility measurements (Fig. 12.5A). Similarly, light-colored bands corresponded to paleolimnological proxies indicating low-energy environments, and corresponded to troughs in the magnetic susceptibility measurements (Fig. 12.5B). Using these paleolimnological data, Wolfe et al. (2006) concluded that, contrary to expectations, they could find no compelling evidence that post-1968 river regulation had any discernable effect on overall flooding frequency of the PAD. Instead, their data indicated that multi-decadal periods of lower (e.g., mid-1700s) and higher (e.g., late 1800s and early 1900s) flood frequency, and that

climatic variability, and specifically an ongoing warming and drying trend over the past century, could best explain overall declines in flooding events.

12.8 Historical experiments from the African Rift Valley: linking sediment discharge to deforestation, the slave trade, ivory caravans, and elephants

Lake Tanganyika (Burundi, Tanzania, Zambia, Republic of Congo), the largest of the African rift lakes, supports an extraordinarily diverse biota. The lake is the habitat for more than 1500 species, over one-third of which are believed to be endemic. Environmental concerns have been raised, however, as the lake has been and continues to be affected by a variety of human impacts. In a series of nine papers summarized by Cohen et al. (2005a,b), a diverse array of paleolimnological techniques were employed to track the limnological consequences of past human disturbances on this large, complex tropical lake ecosystem. Given the size and complexity of the Lake Tanganyika basin, a main focus of the study was to describe the nearshore effects of increased sediment loads from erosion, and to link different human interventions with the sedimentary records preserved in delta sediments draining affected watersheds in the northeastern portion of the lake. To achieve these goals, this multidisciplinary team used a comparative, "historical experimental" approach by mapping and coring six river deltas of comparable size, but draining different watersheds of the Lake Tanganyika ecosystem. The deltas were strategically chosen to represent a gradient of human-related impacts, from very little influenced to heavily developed catchments.

Rates of sediment accumulation were much higher in the heavily populated watersheds, many of which followed different historical developments. The paleolimnological data indicated that extensive deforestation occurred in the late 18th and early 19th centuries in the northern part of the Lake Tanganyika catchment, in the late 19th and early 20th centuries in the northern part of current-day Tanzania, and in the mid-20th century in central Tanzania. With the arrival of the slave trade and ivory caravans in the central portion of Lake Tanganyika in the early 19th century, both human and elephant populations declined precipitously. Human densities decreased due to the associated rapid transmission of commutable diseases, the forced removal of inhabitants by the slavers, as well as famines. Meanwhile, elephants (which are very effective vectors for deforestation) were culled for their

ivory. Thus, forest covers in these regions were largely maintained. Meanwhile, deforestation was far more extensive in the northeastern part of the lake system, as these areas were part of a highly regulated and densely populated kingdom that kept out the slave and ivory caravans.

Past ecosystem disturbances also left a biological legacy. Records from relatively undisturbed deltas continued to support diverse benthic invertebrate and fish communities, whereas a lower diversity (and even extirpation) of some benthic taxa was recorded in the heavily disturbed delta systems. Impacts of sedimentation also appeared to be less severe in deltas draining small watersheds than those with very steep sub-lacustrine slopes. Certain activities, such as the large-scale agriculture of cassava – a crop that can be grown on very steep and easily erodable slopes – in the early 20th century, greatly accelerated the inwash of sediments. Climate was also shown to influence sediment loads. For example, past climate-related fluctuations in lake levels altered sediment delivery patterns to the deltas, but in general these changes were dwarfed by the human-related effects described above. However, periods of extreme weather events, such as very high rainfalls in the early 1960s, led to marked increases in sediment delivery from the deforested catchments, with biological consequences in the affected deltas.

Based on their wealth of paleolimnological data, Cohen et al. (2005a) argued that there were predictable relationships between watershed impacts and limnological effects. These findings have important implications for future management plans for the Lake Tanganyika ecosystem.

12.9 Roman invasions, Napoleonic Wars, and accelerated erosion from the catchment of Llangorse Lake, Wales

Llangorse Lake, the largest lake in the southern half of Wales, has been the subject of considerable paleolimnological research on the effects of road construction by the Romans almost 2000 years ago (Jones et al. 1978, 1985, 1991; Jones 1984). During the Roman campaign against the Silurians (AD 74–8), General Frontinus built permanent roads linking forts across Wales. One of these roads was constructed in this lake's catchment. Agricultural activities also increased after this time. The considerable body of pale-olimnological data assembled for Llangorse Lake recorded marked increases in erosion with con-struction, as evidenced by accelerated sediment accumulation, elevated percentages of sediment-ary clastic materials, as well as a suite of changes in the chemical and biological proxies (e.g., inferred higher turbidity affecting the diatom assemblages), concurrent with the period of road construction and other activities in the lake's catchment. Similar paleoenvironmental assess-ments showing how road construction can trigger erosion range from the multidisciplinary paleo-limnological studies summarized by Hutchinson and Cowgill (1970) on the effects of building a Roman road, the Via Cassia, in 171 BC, to the repercussions of 1930s highway construction on the limnology of small Ontario lakes in a Canadian park (Smol & Dickman 1981).

A subsequent study of Llangorse Lake's more recent sediments offered additional information on the limnological repercussions of increased erosion – in this case, accelerated transport of met-als to the lake as a result of wheat farming. Jones et al. (1991) recorded elevated concentrations of copper and zinc in the lake's sediments beginning in the 18th century. There are no known ore deposits or industries near the lake, and so aerial transport of these metals to the catchments was implicated, with the most likely source being the non-ferrous smelting industry in the Swansea area, about 80 km upwind of the lake. Smelting was under way in the 17th century, and grew to such an extent that it was the world centre for copper smelting during the late 18th and early 19th centuries. Some of these airborne metals would be deposited in the lake and incorporated into the sediments, and indeed the lake's sediments track the increase in metals coincident with the

known smelting activities at this time. However, as noted below, subsequent land-use changes may have further accelerated the transport of terrestrial-bound metals into the lake.

Urban populations increased rapidly during the Industrial Revolution, which led to increased pressures on food production. Grass and grain pollen, indicating local farming, increased syn-chronously in the Llangorse sediments, with elevated metal concentrations in the late 18th century. The Napoleonic Wars added further pressure for grain production, with poor climatic conditions and Britain's isolation as a result of naval blockades. To meet the demand for wheat, farmers extended cultivation to marginal lands, which included the catchment of Lllangorse Lake. Cu and Zn concentrations, along with Ti, an indicator of erosion (Engstrom & Wright 1984), suggest that the increased erosion from plowing also accelerated the delivery of catch-ment metals into the lake. Not only did erosion increase the amount of material being deposited into the lake, but also the metal pollution load – an early example of multiple stressors.

12.10 Determining the influence of human-induced changes in vegetation cover on landslide activities

As shown in the previous examples, removing native terrestrial vegetation can accelerate the transfer of material from land into water. Most of this erosion is occurring slowly. However, there are instances where massive amounts of soil, rocks, and other debris are transported downhill at a very fast rate, such as during a land or mudslide in mountainous regions. These erosion events can have devastating effects on the people and property in their paths, as well as in any receiving waters that are inundated by massive amounts of terrestrial-derived materials. Community groups, local authorities, and scien-tists are keenly interested in knowing what types of human activities will decrease slope stability, which can then lead to these catastrophic events.

Dapples et al. (2002) addressed some of these issues at Schwarzee, a small lake in the western Swiss Alps, which has been prone to large landslides. Historical accounts record major erosion events, such as a 1994 landslide that destroyed 41 buildings, but the question remained: Has this area always experienced landslides, or have human activities accelerated the frequency of these events? What are the relative roles of natural, climate-induced changes versus human-related activities, such as removal of vegetation? Using a combined paleolimnological approach, Dapples et al. reconstructed vegetation changes over the past approximately 7000 years using pollen analyses, and tracked past landslide events using radiocarbon dating methods and dendrochronology on fossil wood samples, coupled with grain-size analysis, X-ray diffraction analysis, and geochemical techniques. They found that the progressive replacement of forested areas with pastures and meadows, beginning about 4300 years ago in this area, but accelerating after about 3650 years ago, has been steadily lowering slope stability. For example, 36 turbidity events were recorded in the sediments over the past approximately 2000 years, compared to only 16 in the previous 4300 years. These data provide strong evidence for human-induced changes in landslide activity, which must be considered in addition to climate-controlled processes.

12.11 Tracking erosion resulting from massive terrestrial degradation related to acute point source acidification

Disturbances such as Roman roads, Mayan buildings, and pre-Columbian Mexican and European-style agriculture have modified the terrestrial landscape, increased soil erosion, and affected the catchments of lakes and streams. Few examples, however, can compare to the stunning ecological destruction linked to massive acidic and metal emissions from the Sudbury (Ontario, Canada) area smelters that fumigated the region in the early to middle part of the 20th century. The limnological effects of the massive acidic and metal deposition were highlighted in Chapters 7 and 8. However, the local terrestrial environment was also greatly altered and denuded of most vegetation, especially in the area near the smelters. The region was often referred to as a "moonscape," and in fact NASA astronauts came to Sudbury to train for their space missions, as the local shield geology and the lack of vegetation was believed to be a close analogue to the landscape they would encounter on our moon (Fig. 7.10)!

Not surprisingly, the de-vegetation resulted in greatly accelerated erosion, which was estimated to be about two orders of magnitude higher in Sudbury during the period of massive acidification than in other regions of North America. A striking series of aerial photographs taken since 1928 of the Kelly Lake delta attests to the amount of soil and other material eroded from the landscape (Fig. 12.6). The location of this lake is ideal for tracking erosion, as it lies immediately downstream of the commercial and residential core of the city, as well as the Inco mining, milling, and smelting complex. Junction Creek, which flows into Kelly Lake, drains most of the watershed and forms the delta. As summarized by Pearson et al. (2002), the lake has acted as a settling pond for sediment (much of it with a very high metal content) and other materials eroding from the Sudbury area for over a century. In fact, Kelly Lake's strategic location has likely mitigated more widespread ecosystem pollution, as the delta is the last major settling area before the contaminant-laden water would flow into the Great Lakes.

Using aerial photographs such as those in Fig. 12.4, Pearson et al. (2002) mapped changes in the delta from 1928 to 1999, during which time the delta top expanded by approximately 161,000 m^2, or an average of 2270 m^2 annually. The most rapid period of erosion occurred between 1928 and 1956, when about 110,000 m^2

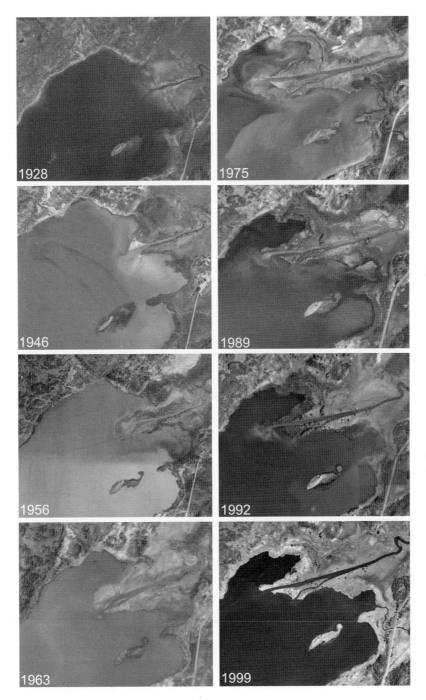

Fig. 12.6 Growth of the Junction Creek Delta into Kelly Lake from 1928 to 1999. Junction Creek drains the industrial and municipal core of Sudbury, including the Inco mining, milling, and smelting complexes. Aerial photographs taken in 1928, 1946, 1956, 1963, 1975, 1989, 1992, and 1999 show the rapid progradation of the delta due to severe erosion following de-vegetation in the area. From Pearson et al. (2002); used with permission.

were added to the delta top, at about 8000 m^2 per year. Deforestation in the catchment coincided with this massive erosion period. The delta has stabilized since 1992.

The Kelly Lake delta has acted as a relatively effective storehouse for some of Sudbury's contaminants. However, with the predicted increases in climatic warming expected in boreal regions as a result of increased emissions of greenhouse gases, places such as Sudbury should soon expect lower lake levels. One consequence of this will be that deltas such as the one at Kelly Lake, which have not only been accumulating sediments but also massive amounts of pollutants such as metals for over a century, will be more exposed to redistribution into the receiving bodies of water with dropping lake levels.

12.12 Trajectories of vulnerability: from nature-dominated to human-affected to human-dominated ecosystems

Human activities can result in a broad gradient of ecosystem impacts from negligible effects to profound and often very negative influences. Messerli et al. (2000) explored this "trajectory of vulnerability" in nature–human interactions, and proposed the hypothesis that "for any human society there is a historical beginning with highly vulnerable societies of hunters and gatherers, passing through periods with less vulnerable, well-buffered and highly productive agrarian-urban societies to a world with regions of extreme overpopulation and overuse of life-support systems, so that vulnerability to environmental changes and extreme events is again increasing." To assess the sequence and magnitude of human–environmental interactions, long-term data are required.

Dearing et al. (2007) used the rich paleoecological and historical data of a lake-catchment ecosystem in Erhai, a large lake from a densely populated region of southwest China, which is vulnerable to flooding, erosion, and tectonic

activity, to document the effects of over 9000 years of human impacts. One question they posed was: At what point in the Erhai catchment history did it switch from a "human-affected" to a "human-dominated" ecosystem? The paleoecological data identified several key periods of impacts, which also provided important insights on the resilience and sustainability of this ecosystem. From a terrestrial perspective, major deforestation and grazing 4800 years ago led to fundamental changes in species composition and the functional ecology of the forests, as well as increased sediment delivery and fluvial discharge from the mountains. From a hydrological perspective, the introduction of paddy field irrigated farming by about 2000 years ago markedly changed the aquatic ecosystem. From a geomorphic perspective, surface erosion and gullying from the exploitation of the surrounding mountain slopes were registered as major human interventions by about 1600 years ago. Sadly, each of these impacts has caused essentially irreversible changes in this now clearly "human-dominated" ecosystem.

12.13 The potential role of mammals, other than humans, in influencing erosion patterns

Humans are undoubtedly the dominant biological agents responsible for accelerated erosion, but they are not necessarily the only agents. Rapidly increasing populations of large ungulates, such as elk and bison, can also increase erosion by overgrazing vegetation, trampling grounds, and other processes. Herds are dramatically increasing in some protected areas, such as reserves and parks, due to hunting restrictions and declines in other predators. Selected culling has been suggested as a management practice to help preserve the overall integrity of some of these protected lands. However, once again, documentation of the potential negative effects of large ungulates on the landscape is often sparse, circumstantial, and controversial (Engstrom et al. 1994; Hamilton 1994).

Yellowstone National Park, whose boundaries overlap three US states (Idaho, Montana, and Wyoming), was established in 1872, making it the first and oldest national park in the world. The northern range is so named because, in winter, many of the animals that make the park famous (i.e., elk, bison, antelope, deer) move from the higher-elevation forested regions to the valleys, where snow cover is less deep and forage is more accessible. Following the creation of the park, herds were encouraged to expand through winter feeding programs, bans on hunting, and controls of native predators. However, populations were increasing rapidly, and so a hunting and trapping program was implemented between 1935 and 1968. Public pressure resulted in a ban on these activities in 1969, and populations began to soar again, possibly resulting in problems associated with overgrazing and erosion.

Many important management questions have been raised. For example: Are such large numbers of ungulates a natural component of the northern range? Are these populations within the natural "envelope" of variability, and so are not adversely affecting the integrity of the park? Have past population fluctuations, as a result of the management attempts noted above, affected the park in a discernible way? A landscape history of the region was required to address these issues.

Engstrom et al. (1991) attempted to answer these questions by applying multidisciplinary, paleolimnological approaches to eight small lakes located in the northern range. Of particular interest was to reconstruct environmental changes that occurred before the Park was established, and then track the effects of the different management strategies that had been implemented (such as the expansion of herds from 1872 to 1934, followed by culling between 1935 to 1968, and then renewed expansion in the subsequent three decades). If the effects of past changes in herd size can be discerned in the paleolimnological record, then expanding ungulate populations

may indeed have to be considered more closely in management and conservation strategies.

The researchers argued that if expanded herd populations were affecting the park, then repercussions should be manifested in the sedimentary record by: an increased load of silt from eroded hillslopes and trampled lake margins; declines in the percentage of pollen grains from taxa-sensitive to ungulate grazing, such as aspen, willow, alder, and birch; compensatory increases in pollen from weedy species, indicative of soil disturbance, such as ragweed; and changes in autochthonous limnological indicators, such as diatoms, reflecting nutrient enrichment from manuring near the lake shore, as well as the indirect effects of increased erosion. Although some small changes were noted in the sediment profiles of the eight study lakes, Engstrom et al. concluded that current ungulate grazing in the northern range did not have a strong direct or indirect effect on the vegetation and soil stability in the catchments, or on the resulting water quality, relative to natural variability driven by climate and other processes.

12.14 Determining the factors responsible for peat erosion

Related to the topic of accelerated transport of material into lakes is blanket peat erosion – a widely recognized environmental problem in several regions, and perhaps most severe in the United Kingdom and Ireland. An understanding of the causes and trajectories of this problem is required for a coherent and sustainable management strategy for these important ecosystems. However, as summarized by Stevenson et al. (1990), many hypotheses have been put forward to account for blanket peat erosion, including climatic change, the development of inherently unstable peat masses, human-related activities such as burning, grazing, trampling, and pollution (e.g., acidic deposition), or some combination of these factors.

One promising approach to determine the causes and timing of peat erosion, at least in catchments that also contain lakes, is to track the progression of peat erosion by using the sedimentary records of lakes situated in these peatland drainage basins. Stevenson et al. (1990) used this approach to track past peat erosion in two catchments in Scotland. By using the percentage or organic matter in the sediments as a proxy for peat erosion, coupled with palynological (tracking changes in the obligate aquatic macrophyte *Isoetes*) and diatom techniques to track parallel changes occurring in the catchment lakes, they showed that the commencement of peat erosion began between AD 1500 and AD 1700, well before acidic deposition was a problem. Other sources of atmospheric pollution were also ruled out. They concluded that the most likely causes of peat erosion were climatic cooling associated with the Little Ice Age and/or an increase in the intensity of burning.

12.15 Summary

As John Dearing (1994) reminds us, "the way of the Earth's surface is to erode." This natural process, however, can be greatly accelerated by human activities such as agriculture and other land-use practices. As lakes, rivers, and reservoirs are downhill from these disturbances, their sedimentary profiles can often be used to reconstruct erosion trends. Such data have clarified how different land-use practices have influenced erosion rates and the limnological characteristics of receiving waters. Like most paleolimnological studies, these investigations are most powerful when applied using multi-proxy indicators. Such data can also be effectively applied to the development and evaluation of models, which can assist managers to determine which land-use activities would pose the least environmental damage to different ecosystems.

13

Species invasions, biomanipulations, and extirpations

Nature . . . She pardons no mistakes.
Ralph Waldo Emerson (1803–82)

13.1 Biological pollution: tracking accidental species invasions

I suspect most scientists would not expect to find a section on species introductions in a book dealing with pollution. Nonetheless, introductions of non-native organisms (so-called "exotic species") can have devastating effects on ecosystems (Elton 2000) and the economy (Pimentel et al. 2000), and at times can dwarf the negative repercussions of other sources of pollution, such as chemical inflows. For example, once a chemical is dumped into a lake, it will slowly (granted, perhaps over hundreds of years) diminish in concentration by sedimentation or some other process. One gram of chemical pollutant will not reproduce itself to become two grams or ten grams of pollutant. On the other hand, if one introduces a new organism into an ecosystem that is capable of reproduction and it increases its numbers (e.g., it has few or no natural predators in its new habitat), it can greatly alter an ecosystem by changing food web structures and other ecosystem functions. Simberloff and Van Holle (1999) warn that the situation may be even more serious and complex, as additional species

invasions may occur once invaders are established, since the latter may facilitate the establishment of the subsequent entry and establishment of new exotic species (i.e., an "invasional meltdown"). In short, introduction of exotic species can be viewed as a form of biological pollution.

The Great Lakes are a classic example of an "underwater zoo," with about 70% of the recent invaders originating from sites in southeastern Europe, transported to North America in ballast water (Ricciardi & MacIsaac 2000). Although most attention has focused on zebra mussels (Fig. 13.1) and fish such as alewife or lamprey, an entire menagerie of microscopic organisms have also invaded these waterways (Mills et al. 1993, 1994). These species introductions are often poorly documented and their ecological repercussions are, to a large extent, still a "black box." Paleolimnological techniques can be used to study these developing (and seemingly constantly expanding) problems.

For example, the first report of the diatom *Stephanodiscus binderanus* in the Great Lakes system, a species likely introduced from Europe, was by Brunel (1956), who determined that blooms of this taxon caused problems at the Montreal municipal filtration plant (St Lawrence River) in November

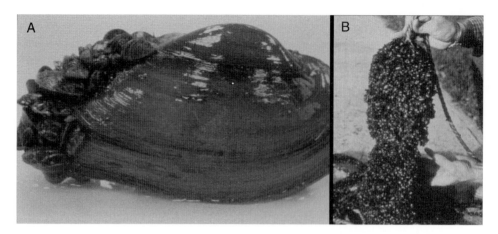

Fig. 13.1 One of the recent exotic invaders that has garnered much attention in North America has been the zebra mussel (*Dreissena polymorpha*), which was first noted in Lake St Clair (Ontario) in 1988. It has since spread profusely throughout the Great Lakes and other systems. (A) This photograph shows how zebra mussels colonize almost any surface, even native mollusks. The density of colonization will increase to the point that the native clam will die. (B) This Great Lakes' coastguard buoy shows how successful these invaders have been at colonizing substrates. Photographs from the Ontario Ministry of Natural Resources; used with permission.

1955, with similar problems reported from this invader by Vaughn (1961) in the Chicago area (Lake Michigan). Stoermer et al. (1985a) used a dated sediment core from Lake Ontario to track its introduction and subsequent proliferation. Reavie et al. (1998) similarly recorded a peak in this diatom around the 1950s in sediment cores from the St Lawrence River. Other diatom invaders to the Great Lakes, such as *Actinocyclus normanii* fo. *subsalsa*, have also been documented (Stoermer et al. 1985a). These invasions are continuing – such as the documentation of *Thalassiosira baltica* in Lake Ontario (Edlund et al. 2000). Paleolimnological analyses indicated that this diatom entered the Great Lakes in about 1988, even though it was not reported until 2000 (Edlund et al. 2000). This timing differential is important, since mandatory ballast water exchange, a procedure thought to greatly reduce successful transfer of exotic species such as *T. baltica*, was not implemented on the Great Lakes until 1993 (i.e., after it had invaded).

Diatom introductions have been identified in other lake systems as well. For example, Harper (1994) used paleolimnological techniques to suggest that the commonly occurring planktonic diatom, *Asterionella formosa*, may have been introduced to New Zealand lakes by Europeans in the 1860s with their unsuccessful attempts at fish introductions.

Paleolimnology can also be used to track microzoological invaders. The spiny water flea, *Bythotrephes longimanus*, a large, predaceous cladoceran zooplankter, is believed to have invaded the Lake Ontario in 1982, when it first appeared in fish diet even though its density was too low to be detectable in plankton samples (Johannson et al. 1991). If present in high enough numbers, it can markedly change the behavior, standing crop, and size composition of its prey, and possibly result in marked alterations in food web dynamics and other changes (Yan & Pawson 1997). A characteristic feature of this exotic taxon is its long, chitinized caudal process, which is typically well preserved in sediments. For example, as this taxon was introduced into Lake Erie in 1984, the presence of these tail spines in lake sediments can be used as a chronological marker for this time period (Keilty 1988). The spiny water flea, with the unintentional help of boaters, anglers, and other vectors, has since been introduced into other lakes in North America, including 35 inland lakes in Ontario (Borbely 2001), and so the spatial and temporal distributions of these introductions can be tracked using paleolimnological techniques. For example, Hall & Yan (1997) reconstructed the spatial distribution and rate of accumulation of caudal processes in sediment cores from Harp Lake (Ontario). They further suggested that the percentage of processes found broken in sediments may be a proxy on past rates of fish predation. Meanwhile, Forman and Whiteside (2000) tracked the dispersal of *B. cederstroemi* in 12 Minnesota (USA) lakes, using the presence or absence of remains in surface sediments. Other introduced species, such as non-native ostracods (Fox 1965), can be studied using similar paleolimnological approaches.

Sediment analyses can also be employed to reconstruct patterns of plant invaders. For example, pollen and/or macrofossil analysis can be used to track invasions of both exotic land and aquatic plants (e.g., Mathewes & D'Auria 1982). Many other applications can likely be implemented as we become more aware of these important issues. Sala et al. (2000) proposed that species invasions will be the single most important factor precipitating biodiversity changes in lakes during this century; the use of paleolimnological approaches will assist scientists reconstruct invasion time-lines and their consequences on native taxa.

13.2 Planned species introductions: determining fisheries status prior to fish-stocking programs

Although more attention is currently focused on the impacts of inadvertent species invasions, humans have been purposely introducing species into different environments for millennia. In freshwater systems, this has often taken the form of fish-stocking programs. The general consensus seemed to be that any fish were good for any lake, especially a fishless lake. This, of course, is a very anthropocentric view.

Since many introductions occurred without the collection of any baseline (i.e., pre-introduction) data, and since many of these manipulations did not include any follow-up monitoring programs, paleolimnological techniques have to be used to reconstruct these missing data sets.

Often one of the first questions asked is: What was the system like before the introductions? This is of utmost importance, for without baseline data, how can a system be "restored"? What is the "target" for your mitigation efforts? Frequently, the question can be as simple as: Were fish in this lake before the introductions?

This was the question posed by Lamontagne and Schindler (1994), in their paleolimnological study of the historical status of fish populations in three Canadian Rocky Mountain lakes. Fish stocking had been prevalent in the lakes and rivers of the Rocky Mountain national parks for almost a century, and so it is now difficult to determine whether fish communities are native, recently introduced, or a combination of the two. However, park managers need to know the original fisheries status of lakes under their jurisdictions to fulfill the Parks Canada mandate of "maintaining ecosystems unimpaired for future generations." Three lakes (i.e., Cabin, Caledonia, and Celestine lakes) in Jasper National Park (Alberta) were identified as important for study, as they were stocked with salmonids in the 1920s and 1930s, but it was unclear what their status was prior to the stocking program. Park managers required reliable "target conditions."

In order to determine if these lakes were naturally fishless, Lamontagne and Schindler (1994) identified *Chaoborus* mandibles preserved in ^{210}Pb-dated sediment cores, using similar approaches to those developed for lake acidification work (see Chapter 7). *C. americanus* is considered to be an indicator of fishless conditions, because the older instars are large and do not migrate vertically, and so are vulnerable to fish predation and are easily extirpated (Uutala 1990). Work in eastern North America had shown that the presence of mandibles of this taxon in sedimentary profiles is a reliable indicator of fishless conditions (e.g., Uutala 1990; Uutala et al. 1994). *C. trivittatus* is also vulnerable to fish predation, and is perhaps best described as "fish sensitive."

The above approaches and relationships were developed in eastern North America. Lamontagne and Schindler (1994) first had to demonstrate that similar species relationships occur in western mountain lakes. They surveyed 43 lakes in Jasper and Banff national parks, and found that *C. americanus* was the only species inhabiting fishless lakes and was only found in three lakes with low fish density. These data confirmed that similar methods could be used to monitor fish populations in these lakes. Paleolimnological analyses of the three sediment cores showed that *C. americanus* was the predominant taxon in Cabin and Celestine lakes in the 19th century, suggesting that these two lakes were originally fishless. Contemporaneously with the fish-stocking program, *C. americanus* (as well as *C. trivittatus* in Cabin Lake) was extirpated from these lakes. Meanwhile, in Caledonia Lake, *C. flavicans* was present in the 19th-century sediments, suggesting that this lake had fish populations prior to stocking.

The above is a good example of how managers (albeit in the 1920s) manipulated lakes without a sufficient understanding of the systems they were dealing with. Most ecologists would agree that adding fish to a naturally fishless lake is not ecologically sound. The food webs and communities present in these naturally fishless lakes have developed over millennia. There is nothing "unhealthy" about a fishless lake, provided that it is naturally fishless. Human prejudices and economic priorities (i.e., fish = anglers = tourism = money) can easily override sound ecological practices. Perhaps these criticisms are too harsh and anachronistic for decisions made in the 1920s, when a "fishless" lake was seen to be "unhealthy" requiring remediation (i.e., stocking). But this problem persists to the present time. With paleo-acidification work, Uutala (1990) found a

lake that was naturally fishless for millennia, yet it had been subjected to (unsuccessful) fish-stocking attempts. Adding fish to lakes has many ecological repercussions. The following example, again looking at mountain lakes in the Rockies, will demonstrate how dramatically many aspects of a lake system can be altered following a fish-stocking program.

13.3 Indigenous or introduced populations? Determining the pedigree of cutthroat trout

A similar question to the one posed in the previous example is whether a lake population represents a native or an introduced strain or subspecies of fish.[1] This has important implications for restocking programs, which hope to reintroduce native strains of fish into lakes. One such example is the case of westslope cutthroat trout (*Oncorhynchus clarki lewisi*) in Glacier National Park in Montana (USA). This trout subspecies has declined throughout much of its historic range in Montana, Idaho, and the western Canadian provinces due to land-use practices, habitat destruction, and competition from the introduction of non-native species. There are now concerted attempts to reintroduce these trout into what are believed to be their native habitats.

However, although a fair number of genetically pure populations have been identified, the choice of fish for the restocking program is problematic because of uncertainty about whether these apparently natural populations are truly indigenous to the lakes in which they now occur.

A healthy and genetically pure population of westslope cutthroat trout was identified in Avalanche Lake, a headwater lake in the park. This population may represent the native strain of cutthroat trout in the valley, and so a potential

---

[1] Carlton (1996) coined the phrase "cryptogenic species" for taxa of unknown origin.

source of broodstock for recovery of the native fisheries in the region. However, the pedigree of this lake's fish stock has been questioned. Although trout were identified in Avalanche Lake during the 1890s, which is several decades before organized hatchery introductions in the region, there are anecdotal accounts of non-authorized fish introductions in the region. The important management question posed was: Is the population of trout in Avalanche Lake indigenous, or did it become established as a result of some undocumented introduction? Verschuren and Marnell (1997) used paleolimnological techniques to answer this question.

As noted previously, fish fossils are notoriously poorly represented in the paleolimnological record, and so indirect approaches must typically be used to infer past fish stocks. Verschuren and Marnell (1997) suggested that fossil diapause eggs (ephippia) of large *Daphnia* species, which are visible and sensitive to fish predation, could be used as an approximate indicator of trout abundance in these lakes: large numbers of zoo-planktivorous trout resulted in very low *Daphnia*

abundance, and hence low concentrations of ephippia in the sedimentary record. Conversely, *Daphnia* thrived in lakes lacking zooplantivorous fish. Using a ^{210}Pb-dated sediment core, Verschuren and Marnell showed that sedimentary ephippia were very rare even before historical records were available for the lake (Fig. 13.2), indicating a healthy population of trout before any potential introductions by humans. These data led to the conclusion that native westslope cutthroat trout had existed in Avalanche Lake since at least the early 18th century, and so the population is indigenous and would make good broodstock for the reintroduction programs in the region.

The paleolimnological record also documented a period from the 1930s to the early 1940s that contained large populations of *Daphnia* fossils, suggesting a disruption of the food web at that time, with zooplanktivory significantly reduced (Fig. 13.2). These changes coincided with a Park Service fish-stocking program that introduced close to half a million fingerlings of another subspecies of cutthroat trout to the lake. Verschuren and Marnell concluded that this

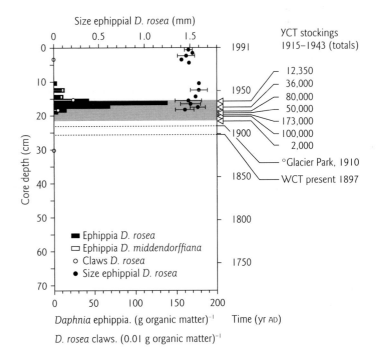

Fig. 13.2 Concentrations (expressed as fossils per gram organic matter) of *Daphnia* ephippia and postabdominal claws, as well as ephippial sizes, in a ^{210}Pb-dated sediment core from Avalanche Lake, Montana. The first documented observation of westslope cutthroat trout (WCT) was in 1897, Glacier Park was established in 1910, and the Park Service introduced non-native Yellowstone cutthroat trout (YCT) to the lake from 1915 to 1943. From Verschuren and Marnell (1997); used with permission of the American Fisheries Society.

severe overstocking resulted in crowding, depletion of available food resources, and increased density-dependent mortality, resulting in a partial collapse of the native fish population, which was ultimately reflected in the fossil zooplankton record. With cessation of the stocking program, the *Daphnia* indicated re-establishment of intense grazing by about 1950.

13.4 The population biology of failed invasions: blending morphological and genetic techniques

In a novel study, Hairston et al. (1999) combined paleolimnological techniques with an analysis of the viability of *Daphnia exilis* resting eggs to record the past invasions of this species in Onondaga Lake (New York), and then used protein electrophoresis to characterize the genetic structure of the population. As was previously shown by Weider et al. (1997), lake sediments preserve many zooplankton eggs that contain genetic material from past populations. Hairston et al. had recovered a large number of *D. exilis* ephippia in the sediments of Lake Onondaga; this species was no longer present in the lake, and in fact was believed to be restricted to shallow, fishless, temporary ponds of the southwestern United States and Mexico. Hairston et al. showed, using paleolimnological techniques, that this taxon was present in Onondaga Lake for several decades in the 20th century (from the mid-1920s to the early 1980s) and although some of the ephippia found in the sediments contained viable embryos, it has not been reported in the water column for the past 15 years or so.

Given the known biogeography of *D. exilis* and the types of habitats in which it typically occurs, it was surprising to find its remains in the recent past of a relatively deep and permanent lake in the northeastern United States. So why would this zooplankter be found in Lake Onondaga? The answer to this puzzle appears to lie in the industrial development and pollution history of the lake

and its drainage basin. Deposits of halite (NaCl) and limestone in the area led to the establishment of a soda ash (Na_2CO_3) industry at the south end of the lake in 1881. Starting in 1926, large waste beds containing high concentrations of $CaCO_3$ and $CaCl_2$ were established along the central western shore of the lake. Paleolimnological studies of the lake document the rise in lakewater salinity at this time by diatom species changes (Rowell 1996). The industrial production of soda ash was terminated in 1986, water column salinity declined, and the taxon was extirpated.

In addition to the morphological study of eggs, Hairston et al. hatched eggs and found many were viable. Genetic analyses using protein electrophoresis showed strikingly low genetic variation in comparison to variations documented for populations in the southwestern United States. These data suggest that a single genotype, likely a single clone, established the *D. exilis* population in Onondaga Lake.

In a similar study, Duffy et al. (2000) used paleogenetics using mitochondrial deoxyribonucleic acid (mtDNA) from cladoceran resting eggs to track the invasion and subsequent failure of another exotic *Daphnia* species into Onondaga Lake. Morphological studies of fossil ephippia from the dated sediment core were somewhat ambiguous, as a seemingly new egg case was found in sediments deposited between c. 1952 and 1983, which could not be linked to any known species from the eastern United States – nor could the resting eggs be hatched to provide positive identifications. Instead, the authors used paleo-genetic techniques by sequencing the mtDNA from the mysterious resting eggs and comparing these DNA sequences to those available from recognized taxa. This led to the surprising conclusion that the unidentified ephippia were from *D. curvirostris*, a Eurasian species that had hitherto only once been recorded in North America, and this was from a lake over 4500 km away in Arctic Canada!

Similar to the *Daphnia exilis* example noted above (Hairston et al. 1999), industrial activity at Onondaga Lake was likely responsible for this invasion. Duffy et al. (2000) showed that *D. curvirostris* was only found in sediments from the mid- to late 1900s, the period of peak industrial activities at this lake. For a long period in the 1900s, the chemical operations around this lake were run by a company that was co-owned by industrialists who had European factories operating throughout the native range of *D. curvirostris*. Duffy et al. suspected that this taxon was introduced into the lake from mud transported on some of the equipment, which was frequently exchanged between operations in Europe and at Onondaga Lake.

Interestingly, both the invasion of *Daphnia exilis* and *D. curvirostris* occurred when Onondaga Lake was most polluted and when native *Daphnia* species were eliminated. As proposed by Lodge (1993), disturbed habitats may make ecosystems more susceptible to invasion by exotic species,[2] and this paleolimnological study would support this contention. Once the lake began to recover and return to more natural conditions, the native taxa were again more competitive, and the two exotics disappeared. Alternatively, the initial invasion may have proceeded only after sufficient propagules, estimated by Hairston et al. to be as few as one, reached the lake during periods when the physico-chemical conditions were suitable for survival.

13.5 Biomanipulations and the trophic cascade: food web alterations due to fish-stocking and removal programs

Ecologists have repeatedly shown that population size and community structure are not only influenced by physical and chemical environments (such as nutrients), but also by food web interactions (e.g., feeding pressure from herbivores and predators). It therefore stands to reason that

if managers manipulated these trophic interactions, then aquatic production could potentially be regulated. As a leading symptom of eutrophication is increased phytoplankton abundance, lake managers have come to realize that biomanipulating food web structure may be an effective way of decreasing algal standing crop, especially in lakes and reservoirs where reductions in nutrient inputs are not feasible.

Carpenter et al. (1985) and Shapiro (1995) provide introductions to the overall principles of cascading trophic interactions in fresh waters. Biomanipulations are usually initiated by changes in fish populations, either by stocking or removal programs, with the overall goal of reducing fish predation on herbivorous zooplankton, which results in a shift to larger-bodied zooplankton herbivores that are more effective grazers of algal stocks. Put more simply, if you increase the biomass of piscivores (e.g., bass), then planktivores (e.g., minnows) will decrease, resulting in higher herbivore biomass (e.g., *Daphnia* zooplankton), and so decreased phytoplankton biomass. The trophic interactions *cascade* through the food web. The process is also sometimes referred to as "top-down" controls of algal biomass, as opposed to "bottom-up" controls (e.g., nutrients). Although the two main types of controls are sometimes presented as opposing schools of thought, cascading trophic interactions and nutrient loading models are complementary, not contradictory (Carpenter et al. 1985).

The effects of biomanipulations and the long-term responses of lake systems to changes in fish communities can be explored using paleolimnological techniques. Leavitt et al. (1989) used fossil pigment composition as a proxy of past algal and bacterial communities and zooplankton remains from the annually laminated sediments of two northern Michigan (USA) lakes (Peter and Paul) that had undergone three major changes in fish community structure (i.e., trout to minnows to bass) since the middle of the 20th century (Fig. 13.3). In 1951, rotenone was used to kill the bass and perch from both lakes, and Paul Lake was stocked with rainbow trout (an omnivore)

---

[2] The hypothesis that biological invasions may be related to disturbance has been challenged by Lozon and MacIsaac (1997).

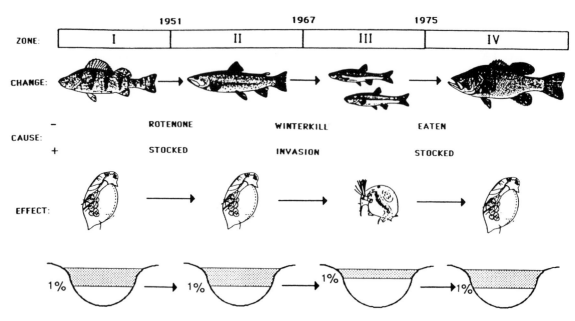

Fig. 13.3 A summary of the effects of trophic manipulations in Paul Lake, Michigan, showing the fish changes from a bass and perch assemblage (zone I), to rainbow trout (zone II), to minnows (zone III), and then to largemouth bass (zone IV). This resulted in marked changes in zooplankton populations, with a shift from larger (e.g., *Daphnia*) to smaller, less efficient grazers (e.g., *Bosmina*) when the planktivorous minnows were most abundant (zone III). During this time, phytoplankton populations were at their highest (and so light penetration was at its lowest), as predicted by the trophic cascade hypothesis. From Leavitt et al. (1989); used with permission of ASLO.

from 1951 to 1961. However, the trout did not reproduce and their numbers were reduced further due to periodic winterkills. By the end of the 1960s, minnows had supplanted the ailing trout population. These efficient planktivores dominated the fish assemblage until 1976, when largemouth bass, effective predators of minnows, invaded the lake via a culvert from Peter Lake and caused the minnow population to crash. By the 1990s, Paul Lake was a virtual monoculture of largemouth bass, with a zooplankton assemblage dominated by large *Daphnia* and *Diaptomus*.

Analysis of annually laminated sediments showed that the change in food web structure and limnological conditions were well recorded by paleolimnological proxies. Due to space limitations, only the Paul Lake data are summarized here (Fig. 13.4). The pre-manipulation zone I (1940–51) is characterized by slowly decreasing

carotenoid concentrations, likely reflecting reductions in algal productivity due to decreased erosion, leaching, and nutrient input during recovery from previous watershed disturbances. The changes in fish community structure in 1951, when the original bass–perch assemblages was removed and replaced by rainbow trout (zone II), had relatively little impact on algal and zooplankton communities, reflecting that both bass and trout are ineffective zooplanktivores. Much more striking changes, however, were observed when the minnows supplanted the trout populations (zone III) with the nearly total replacement of large-bodied *Daphnia* by small-bodied *Bosmina* zooplankton (Fig. 13.4). Based on ecological theory, this shift in herbivores (i.e., large, efficient grazers replaced by simpler, inefficient grazers) should result in increased algal standing crops and decreased light penetration (Fig. 13.4).

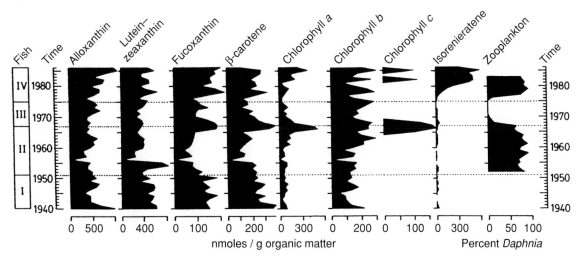

Fig. 13.4 The sedimentary record of algal and bacterial pigments and zooplankton remains from the annually laminated sediments of Paul Lake, Michigan, from 1940 to 1986. The zones represent those described in Fig. 13.3. The "% *Daphnia*" is a percentage based on the number of *Daphnia* fossils divided by the sum of *Daphnia* and *Bosmina* remains enumerated from the sediments. From Leavitt et al. (1989); used with permission.

Consistent with these expectations, pigments characteristic of deepwater chrysophyte blooms (e.g., chlorophyll *c*) decreased dramatically at this transition. The invasion of largemouth bass in 1975 and their subsequent removal of the minnows by 1980 (zone IV) reduced the intensity of size-selective planktivory, and therefore allowed large-bodied, efficient zooplankton grazers such as *Daphnia* to thrive as they had in zones I and II. This shift in herbivores resulted in a lower algal standing crop and a resurgence of deepwater algal taxa, such as chrysophytes (chlorophyll *c*). In addition, a striking increase in the pigment isorenieratene, which is characteristic of brown varieties of green sulfur bacteria in the group Chlorobiaceae (Fig. 13.4), occurred at this time. As these deepwater bacteria are photosynthetic, their abundance provides further documentation that light penetration had increased during zone IV.

Paleolimnology provides the framework by which the long-term effects of food web manipulations can be evaluated. The Leavitt et al. (1989) study showed that changes in fish communities can have long-lasting effects that cascade down to the microbial level of the food web, and that these ecosystem shifts can be captured in the sedimentary record. Although still rarely used in food web studies, similar paleolimnological approaches can be applied to many other settings where the ecological effects of past manipulations have not been well documented.

13.6 Linking fish introductions with eutrophication

Fish introductions can affect predator–prey relationships in lakes, but can also alter fundamental limnological processes, such as nutrient cycling. St Jacques et al. (2005) examined the limnological effects of the 1948 introduction of cisco (also known as lake herring) in Lake Opeongo, a large wilderness lake in Ontario (Canada). Using paleolimnological techniques, St Jacques et al. tracked marked changes in diatom assemblages indicating increased nutrient levels following the fish introductions, with concurrent shifts in cladoceran species composition.

Similar changes were not recorded in near-by reference lakes. The authors concluded that the increased nutrient recycling was driven by the human manipulation of the fish community. From a management perspective, these data also highlight the potential effects of stocking programs on water quality and other aspects of ecological integrity. Similar paleolimnological approaches can be used to study, for example, the environmental effects of fish cage facilities (e.g., Clerk 2001; Clerk et al. 2004), which are an important component of within-lake fish aquaculture operations in some regions.

13.7 Studying species extirpations using paleolimnological techniques

Just as species invasions are important from an ecological and environmental perspective, the losses of species from a habitat or ecosystem are also critical areas of study. Given the media and scientific attention to terrestrial regions such as the Amazonian rain forest, with an additional focus on animals such as birds and mammals, one might think that freshwater communities are not under threat from biodiversity loss. However, there appears to be similar depletion of biodiversity in North American freshwater fauna at least (Ricciardi & Rasmussen 1999), and likely in other freshwater systems.

In other parts of this book, I have discussed examples of animal species extirpations as a consequence of different environmental stressors, such as the loss of fish due to lake acidification using *Chaoborus* mandibles (see Chapter 7) or extirpations due to other forms of contamination (see Chapter 8). Plant extirpations can also be studied with paleolimnological methods. Using plant macrofossils in dated sediment cores from three large lagoons in the Nile Delta (Egypt), H.H. Birks (2002) documented the relatively recent extirpation of the floating water-fern *Azolla nilotica* from Egypt following changes in salinity resulting from increasing controls on the annual flow of the River Nile and increasing eutrophication (Birks et al. 2001).

Paleolimnological data can also be used in other contexts that have important implications for studies of biodiversity. For example, these studies can draw attention to microbial species losses, as opposed to the megafaunal diversity losses that tend to receive much more attention. The ecological repercussions of microbial diversity loss are less well understood, but are certainly worthy of further study.

Stoermer et al. (1989) and Julius et al. (1998), for example, tracked the local extirpation of the planktonic diatom *Stephanodiscus niagarae* from Lake Ontario. This taxon was first formally described by C.G. Ehrenberg in 1845 from the Great Lakes system, at Niagara Falls, as one might guess from its name. Since its description, it was one of the most studied diatoms from the Great Lakes; that is, up until recently, when it steadily began to decrease and eventually seemed to disappear from Lake Ontario plankton records. Julius et al. (1998) used sediment cores, as well as material from many previous paleolimnological studies (e.g., Stoermer et al. 1989), to track its long-term population changes. They found that this diatom was already in serious decline by the 1980s, and diatom valves preserved in sediments dated to about 1988 were the last records of this taxon in Lake Ontario, indicating its final extirpation from the system.

Certainly the continued and varied environmental disturbances occurring in Lake Ontario would be the first reason one should consider as being responsible for this species loss. Interestingly, this taxon withstood the first onslaught of environmental changes in the Great Lakes during the 19th century, which extirpated some very sensitive species (e.g., *Cyclotella americana*) in the 1830s–1850s in the lower lakes, such as Lake Erie (Stoermer et al. 1996).

The paleolimnological data, however, also suggest a second, more biological explanation for this extirpation. By measuring the valves preserved in the sedimentary profiles, Stoermer et al. (1989) and Julius et al. (1998) showed that

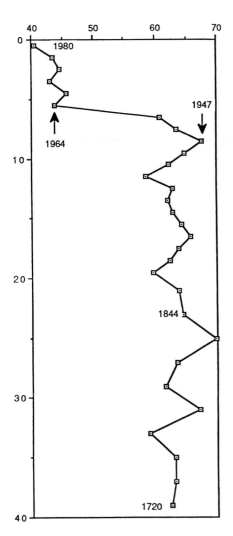

Fig. 13.5 A plot of the mean diameter of *Stephanodiscus niagarae* valves versus sediment depth (measured in cm; the dates are estimated from ^{210}Pb chronology) in a core from Lake Ontario. Stoermer et al. (1989) describe other morphological changes in the valves, in addition to size reduction. From Stoermer et al. (1989); used with permission.

the mean diameter of *Stephanodiscus niagarae* valves began to decrease in the late 1940s and early 1950s (Fig. 13.5). This was a period of intense silica depletion, as a result of phosphorus enrichment (i.e., with increased fertilization from detergents and other effluents, intense algal blooms were stimulated, which resulted in a marked decrease in lakewater silica). This reduction in valve size, due to the lack of sexual reproduction, could be linked to silica depletion. Continued asexual reproduction in the post-1980 population, which would result in further size reduction, may have decreased mean valve diameters below the minimum required for cells to develop. One may have suspected that this diatom would continue to survive in Lake Ontario through continuous re-seeding from Lake Erie, but that turned out not to be the case.

Hence two explanations, not mutually exclusive, were proposed for this extirpation. These types of data are very useful for stimulating future research and also to provide a warning of the possible repercussions, sometimes indirect, of human disturbances, even at the level of an individual diatom species. Sediments contain a wealth of information on such processes, in addition to records of speciation and endemism (Edlund & Stoermer 2000), as well as preserved DNA (Weider et al. 1997). In many respects, this line of work has been underutilized, but as we become more sensitized to biodiversity issues in freshwater systems, it is my view that these approaches should be used more intensely.

13.8 Extirpations on island ecosystems: exceeding the limits for sustainable development

Isolated island communities may be especially vulnerable to extirpations. Easter Island, located approximately 3700 km east of South America in the southeastern Pacific Ocean, certainly warrants the title of "isolated." The island is very small (about 22 km by 11 km) and is composed mostly of volcanic rock. The earliest known human colonizers were Polynesians, who arrived about 1500 years ago. The island is famous for its massive statues (or moai), all carved by the original inhabitants of the island, which at one point numbered over 600 and stood as sentinels along the coastline (Fig. 13.6).

Fig. 13.6 An example of the moai statues built on Easter Island. Photograph by J.A. Keast.

The first European known to have visited the region was Jacob Roggevenn, a Dutch explorer who landed on Easter Sunday (April 5) in 1722. His impressions of the island were not one of an organized society, capable of building these large ceremonial statues. Instead, he saw an impoverished and disorganized community (estimated at only about 2000 inhabitants), living in a devastated ecosystem, with no sources of wood except two small tree species and two types of shrub, and no native animals larger than insects. What happened to this island community? What happened to its resources, such as the wood and fiber required for the machinery to help build and erect these statues? Obviously this society could not have arisen in such a barren landscape.

The history of Easter Island has been pieced together by archeologists and paleoecologists.[3] Archeologists believe that the earliest human activities started in about AD 400–700, with the period of peak statue construction from about AD 1200–1500, with few statues erected after that point. Population estimates range from about 7000 to 20,000 people at this time (Diamond 1995). But then what happened?

Pollen analyses, as well as other paleolimnological approaches, have helped clarify the island's ecological and social collapse (Flenley et al. 1991). For at least 30,000 years before human arrival, and during the early period of Polynesian settlement, Easter Island was a very different ecosystem than it is today. Pollen and macrofossil evidence taken from lake sediment cores reveals the presence of a subtropical forest, with palm trees very abundant (but now extirpated). Such timber would be ideal for transporting and erecting statues and constructing seaworthy canoes, as well as providing food in the form of nuts and edible sap. Other common species included the rope-yielding hauhau tree. Alarmingly, the pollen records from lake sediments also track the destruction of the island's forests, which was well under way by about AD 800. Pollen from palms and other trees declined, and charcoal particles increased. The regeneration of palm trees was further hampered by rats, which the Polynesians introduced to the island, as these rodents also chewed the few remaining palm nuts. Shortly after AD 1400, the Easter Island palm went extinct.

The forests were gone, and now the social and political repercussions of this ecosystem disaster began to ripple through all aspects of Easter Island's society. Archeological evidence showed that porpoise meat was a major dietary staple

---

[3] Dumont et al. (1998) provide paleolimnological evidence of some alternate views of the island's history, focusing on species introductions from South America and Europe.

during the island's affluent period. With no wood available to make harpoons or sturdy boats, this marine food resource was no longer accessible. Shortly thereafter, land birds, another food staple, disappeared from the island, as did some seabirds. With the trees gone, soil erosion increased, and production from land-based crops declined. Cannibalism soon followed. Around AD 1700, rival clans began toppling each other's statues. Their organized society began to collapse.

As Diamond (1995) comments, the trees did not just suddenly disappear, but vanished over decades of over-exploitation. One wonders what the islanders were thinking as they cut down the last of the palm trees. The Easter Island story is an example of exceeding the limits of sustainable development and the disasters that may follow. When reading these accounts, I am reminded of what the Queen said in Lewis Carroll's novel *Alice's Adventures in Wonderland*: "It's a poor sort of memory that only works backwards." We should learn from our past mistakes. But how much have we really learned?

13.9 Summary

Water-quality managers and scientists have only recently become sensitized to the vast ecosystem changes that can occur following exotic species invasions. Thus far, paleolimnological approaches have only rarely been applied to these developing problems, but many potential applications exist. Sedimentary studies have several advantages. For example, as some invading taxa may only be present in the water column for a short period of time during the year, they may be missed using standard water-sampling programs, whilst lake sediments effectively preserve records of even short-lived populations. Paleolimnological studies can also be used to track limnological changes resulting from intentional introductions, such as those used as part of biomanipulation programs. However, an equally pressing problem is the loss of biodiversity (i.e., the extirpation or extinction of taxa), some of which can be related to competition with or predation by exotic species, but also from habitat destruction and other anthropogenic activities. Paleolimnological analyses can often assist in determining the timing and causes of these species losses. As shown by E. Stoermer and his colleagues, working on the Great Lakes, paleolimnology has been useful in identifying emerging issues, such as the decline and eventual extirpation of *Stephanodiscus niagarae* in Lake Ontario, or the introduction of *Thalassiosira baltica* and other taxa.

14

Greenhouse gas emissions and a changing atmosphere: tracking the effects of climatic change on water resources

The fault, dear Brutus, is not in our stars,
But in ourselves . . .
William Shakespeare (1564–1616), *Julius Caesar*, I. ii. 140–1

14.1 Climatic change, pollution, and paleolimnology

The inclusion of a chapter on climatic change may seem out of place in a book on water pollution. Nonetheless, it is clear that climatic changes have myriad effects on water quality and other pollution issues (as discussed more fully below), and so a short chapter describing some of the ways that paleolimnologists track climatic changes is warranted. However, in a chapter of this size, it is impossible to cover this subject in a comprehensive way. I therefore refer the reader to a variety of sources dealing with paleoclimatology. Foremost amongst these would be Bradley's (1999) textbook (with a new edition nearing completion), which summarizes the ways in which climatic trends can be reconstructed from terrestrial, marine, and aquatic archives. Alverson et al. (1999) edited a series of review papers dealing with climate proxies, many of which are relevant to the discussion presented here. Several sections of

Cohen's (2003) paleolimnology textbook focus on climate reconstructions. In addition, many chapters in DeDeckker et al. (1988), Chivas and Holmes (2003), Stoermer and Smol (1999), Smol et al. (2001a,b), Last and Smol (2001a,b), other volumes in the *Developments in Paleoenvironmental Research* book series, as well as a number of review articles (e.g., Smol et al. 1991; Battarbee 2000; Smol & Cumming 2000) summarize other paleoclimatic approaches used by paleolimnologists.

14.2 The need for paleoclimatic data

A central theme of this book has been the need for long-term environmental studies. Compared to limnological monitoring, meteorological data are often of higher quality, with continuous temperature records available for many major urban centers covering the past century or more. In fact, temperature measurements have been collected for a few cities since the 18th century, dating back to the early usage of the temperature scales developed by Daniel G. Fahrenheit (1686–1735) and Anders Celsius (1701–44). Nonetheless, even these data are insufficient to address many questions posed by environmental scientists and policy makers, as the available meteorological records are still too short and sparse, and have been collected using variable observational standards. A spatial representation of the available pre-20th century records identifies another shortcoming – most are from a few major cities in Western Europe, with some additional records from the eastern United States. The fact that all the pre-19th century (and in fact almost all of the pre-20th century) data are from urban centers raises another potential concern, namely that these urban meteorological records have been influenced by heat generation from the cities themselves, or the so-called "urban heat island effect." However, researchers have shown that its effect on global mean temperature estimates is small (Parker 2004), and in fact these data have been used to correct meteorological data to produce homogeneous records. Long-term climate records, on appropriate spatial and temporal scales, are therefore required to answer questions such as: What are the types and ranges of natural variability at different time-scales? How have human influences, such as the accelerated releases of greenhouse gases (Box 14.1), affected climate? Has the frequency and intensity of extreme climate events increased over recent years? Furthermore, as scientists and policy makers are using complex climate change models to develop scenarios of future climate change (e.g., Intergovernmental Panel on Climate Change 2000), long-term historical meteorological data are valuable for evaluating and modifying existing models (e.g., Goosse et al. 2005, 2006; Fig. 2.5). Clearly, without a long baseline of meteorological data, many questions remain.

Because lakes, ponds, and rivers are important sentinels of climatic change, and because they accumulate records of past environmental conditions in their sediments, it is not surprising that paleolimnologists have developed methods to tease apart climatic inferences from the information preserved in sedimentary profiles.

Below, I briefly review some of the key limnological indicators and approaches used to track climatic trends. But first, to emphasize the urgency associated with collecting reliable long-term climate data, I outline the key environmental problem currently facing humanity: greenhouse gas induced warming.

14.3 Human-influenced climatic change: our greatest environmental problem

There is no doubt that striking climatic changes have occurred in the past which were not influenced by humans. For example, only approximately 15,000 years ago there would have been more than a kilometer of ice over where my office now is, as at that time most of Canada was in the grip of one of the great Ice Ages. The global climate can and does change due to a variety of natural forcing mechanisms, such as secular variations in the Earth's orbit, fluctuations in the Sun's solar output, changes in ocean circulation patterns, as well as other factors such as volcanic activity (and the ejection of tephra ash and sulfur aerosols into the atmosphere), and natural changes in the composition of our atmosphere. The legacies of past climatic changes are recorded in a variety of landscape features and natural archives, such as moraines left by past glacial ice sheets, bones of dinosaurs, giant ferns and other biota, large underground coal and oil deposits, and sediments.

Although natural climatic changes have occurred on our planet for billions of years, human activities are now major drivers of climate. Ever since humans learned how to harness fire, our species has begun to influence the carbon balance on this planet (although, at that early stage in our technological development, our effect was still very minor). However, with the onset of the Industrial Revolution, coupled with the burning of coal and oil on a large scale, as well as other industrial and agricultural activities, *Homo sapiens* began changing (unintentionally) the physical and chemical properties of our atmosphere (and thereby the weather patterns on our planet) by increasing the concentration of greenhouse gases, such as carbon dioxide (CO_2). Greenhouse gas concentrations have certainly changed in the past due to natural causes, and paleoenvironmental work has been critical in assessing and evaluating these natural changes (Bradley 1999). However, the human footprint on recent climate changes is now clearly evident, due to the elevated concentrations of greenhouse gases, resulting in an anthropogenically enhanced greenhouse effect (Box 14.1). As T. Flannery (2005) cogently noted: "We are now the weather makers."

Box 14.1 The greenhouse effect

As early as 1824, the French mathematician Jean Baptiste Joseph Fourier (1768–1830) proposed a link between climate and the concentrations of certain gases in the Earth's atmosphere. Later that century, the prescient Swedish scientist, Svante Arrhenius (1859–1927), calculated that fossil fuel combustion may eventually lead to warming of the atmosphere, and in 1896 proposed a relationship between the concentrations of carbon dioxide (CO_2) and temperature. Arrhenius concluded that the reason why the average surface temperature of our planet was about 15°C was due to the infrared absorption capacity of gases such as water vapor and carbon dioxide, or the "natural greenhouse effect" (Fig. 14.1). These greenhouse gases, which include methane, nitrous oxides, and ozone, absorb infrared radiation emitted by the Earth's surface. Without this atmosphere of greenhouse gases and absorption of long-wave radiation, the average temperature on our planet would be about −17°C, or too cold for life as we know it to have evolved.

The concentrations of greenhouse gases have changed significantly over time due to a variety of natural mechanisms. However, it has only been over the past approximately 150 years that the burning of fuels by humans has been of sufficient intensity to shift the balance between the natural sources and sinks of greenhouse gases. Although direct atmospheric monitoring data of greenhouse gas concentrations have only been available since the 1950s, scientists have reconstructed past concentrations of atmospheric gases by analyzing gas bubbles preserved in ice cores from polar regions (e.g., Siegenthaler et al. 2005; Spahni et al. 2005). With widespread burning of coal and other fossil fuels, carbon dioxide emissions have been steadily increasing since the time of the Industrial Revolution, so much so that current atmospheric CO_2 concentrations have increased by more than 30% since the middle of the 19th century. Schiermeier (2006) estimated that humanity's burning of coal since 1750 has released about 150 billion tonnes (gigatonnes) of carbon into the atmosphere. In addition, other greenhouse gases, including methane and nitrogen dioxide, have sharply increased in the atmosphere due to human activities.

We now live in an "enhanced greenhouse effect" world; the Earth is getting warmer, and the projected warming and associated problems forecast for the coming decades are tremendously alarming from environmental, economic, and social perspectives (McCarthy et al. 2001; Stern 2007).

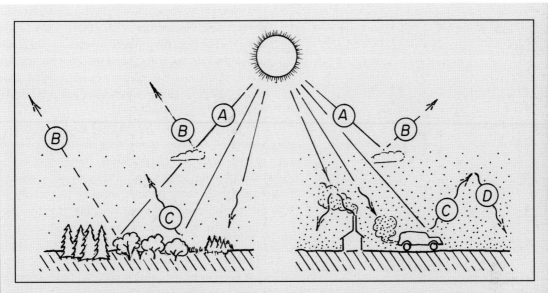

Fig. 14.1 A schematic diagram illustrating the greenhouse effect. The ultimate source of energy is solar radiation (A), some of which can be reflected (B) by, for example, cloud cover or high-albedo surfaces, such as ice and snow cover. A portion of this incoming radiation, however, is absorbed by the planet. Some of this energy is then re-emitted into the atmosphere as infrared or long-wave radiation (C). With increased concentrations of greenhouse gases (right side of figure), such as carbon dioxide (CO_2), the atmosphere becomes less "transparent" to outgoing long-wave radiation, as greenhouse gases effectively absorb infrared radiation. As a result of this imbalance, more heat is trapped in the atmosphere (D) as well as in the oceans.

14.4 Paleolimnological methods to track climatic change

14.4.1 Using biology to infer climate

The distributions of all biota are at least partly related to climate. With the pressing needs for long-term climatic inferences, paleolimnologists have used the environmental optima and tolerances of species to reconstruct climate trends. Two general approaches have been used: direct and indirect inferences.

14.4.1.1 Direct inferences of temperature from biological indicators

As climate is often an overarching driver influencing the distribution of vegetation and terrestrial insects, it is not surprising that some of the earliest paleoclimate research was done by palynologists and then entomologists, as reviewed by Bennett and Willis (2001) and Elias (2001), respectively. However, especially since the 1990s, a large number of paleolimnological approaches have been developed that use aquatic organisms to reconstruct temperature and related climatic variables.

Direct inferences of climate include the construction of transfer functions from surface-sediment calibration sets (see Chapter 6) that attempt to infer variables such as lake water or air temperatures directly from the species distributions of indicators. The main assumption with these types of analyses is that species composition is either directly related to temperature, or that assemblages are related to an environmental

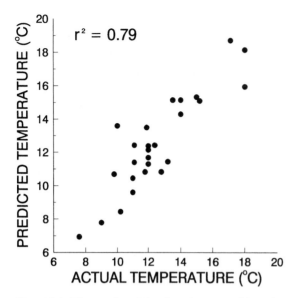

Fig. 14.2 The results of the first chironomid-based transfer function developed from a surface-sediment calibration set of Labrador (Canadian) lakes. Following this initial study, many new transfer functions have been constructed (e.g., Brooks 2006; Walker & Cwynar 2006). From Walker et al. (1991); used with permission.

variable that is linearly related to temperature. Paleolimnological studies using chironomids have been the most often used, following early work by Walker et al. (1991), who developed the first transfer function to directly infer lakewater temperature from head capsules preserved in the surface sediments of Labrador (Canada) lakes (Fig. 14.2). Although initially a controversial approach, similar studies were soon completed for many other lake regions, and chironomid-based temperature inferences are now widely used in paleolimnological studies. Recent applications from Eurasia and North America have been summarized by Brooks (2006) and Walker and Cwynar (2006), respectively.

Other paleolimnological indicators have also been used to reconstruct temperature. For example, Pienitz et al. (1995) and Vyverman and Sabbe (1995) proposed that diatoms could potentially be used to infer water temperature. These approaches have been applied in a number of studies (e.g., Smol & Cumming 2000; Kumke et al. 2004; Weckström et al. 2006). Other indicators, such as chrysophycean cysts, Cladocera, and ostracods, have also been used in temperature reconstructions (e.g., Brown et al. 1997; Lotter et al. 1997a, 1999; Curry & Delorme 2003; Kamenik & Schmidt 2005).

There has been some debate in the paleolimnological literature concerning the relative merits of, for example, chironomids versus diatoms as paleoclimate indicators. Although all scientific hypotheses must be interrogated repeatedly to assure the most accurate and precise interpretations, in my view some of these discussions have been naïve and not overly productive. There are far more pressing and constructive questions that need to be debated, as opposed to which indicators are "better" at reconstructing climate variables. There are no "winners"; the distributions of *all biota* are at least *partly influenced by* climate. Instead, more fruitful discussions should focus on determining *which taxa* are most closely related to *which meteorological variables*, under *which limnological conditions*, as these relationships will clearly change under different environmental settings. As summarized by Smol and Cumming (2000), temperature–biology relationships are rarely straightforward, as environmental variables such as "July epilimnetic water temperature" or "average July air temperature" are most likely just the measured limnological variables capturing most of the statistical variance in the species assemblages enumerated in the specific training set under study. In fact, the taxa may be more closely tracking other (unmeasured) environmental variables that are only indirectly related to temperature (as described in the next section), such as stratification patterns, ice cover, chemical changes, and so on. This distinction, however, is not critical from a paleoclimatic perspective, as long as temperature is related to these changes, and as long as temperature can be reconstructed in a statistically robust and ecologically sound manner. Such quantitative inferences should be interpreted as overall *trends* in past climate, rather than absolute temperature changes, primarily because the "real" temperature optima are difficult to estimate in studies that span incomplete temperature gradients and relatively few calibration sites. Of course, these biotic inferences are more defendable when combined with results obtained from other proxy data (Birks & Birks 2006).

14.4.1.2 Indirect inferences of climate from biological indicators

Although direct inferences of meteorological variables provide quantitative data that would be most useful for other scientists and the public at large, many paleolimnological inferences of climate are more indirect. Common approaches include inferring shifts in limnological variables (e.g., water chemistry, lake ice cover, etc.) that are related to climate. Similar paleolimnological approaches across climatic regions can be used to track very different environmental gradients. For example, inferences in polar regions have focused on past lake ice conditions, whereas near high-latitude or -altitude treeline ecotones, researchers have developed methods to infer past lakewater dissolved organic carbon (DOC), as this variable can be linked to the density of coniferous trees (such as spruce) in a drainage basin (which itself is related to climate). In closed-basin lakes from arid and semi-arid regions, past lakewater salinity and related water chemistry variables, which can be reconstructed from fossil algal and other assemblages, are often closely tied to the balance of evaporation and precipitation (i.e., drought frequency). Below, some of these approaches are discussed in more detail.

Tracking past lake and pond ice cover using biological indicators

Polar and high mountain regions are believed to be especially sensitive to climate changes due to a variety of positive feedback mechanisms, many of which are related to changes in albedo or the reflectivity of the ocean and landscape. With warming, more and more of the highly reflective snow and ice cover melts, exposing greater proportions of the darker-colored ground or water

(which has a lower albedo, and hence absorbs more solar radiation). This leads to more warming and so more melting of the ice and snow cover, and so forth. A positive feedback system is in operation. For these reasons, considerable attention on climate change research has focused on high-latitude and high-altitude ecosystems. However, long-term meteorological observations are often especially sparse from these regions, and so indirect proxy methods must be used in lieu of these missing data sets. Fortunately, many paleolimnological approaches are available for cold environments (e.g., Pienitz et al. 2004).

The duration of ice and snow cover, which is closely related to temperature and other climate variables, exerts a powerful influence on the physical, chemical, and biological properties of high-latitude and alpine lakes and ponds (Smol 1988; Fig. 14.3). Although most notably affecting the amount and quality of light available for photosynthetic growth of phytoplankton and periphyton, shifts in ice cover also affect nutrient chemistry, mixing regimes, gas exchange, as well as a suite of other limnological processes (Douglas & Smol 1999; Spaulding & McKnight 1999; Smol & Cumming 2000).

At its simplest level, the changing environmental conditions can be summarized as follows. Under colder conditions, a thicker and longer-lasting raft of ice and snow characterizes lakes (Fig. 14.3A), and only a small moat of open water may thaw in the littoral zone. With increasing temperatures and reduced ice conditions, progressively more and more of lake habitats become more amenable to photosynthetic growth (Fig. 14.3B). Eventually, with continued warming, the entire raft of ice and snow may melt (Fig. 14.3C), thus allowing phytoplankton and related biota to be more competitive (Sorvari et al. 2002; Rühland et al. 2003). Due to changing albedo and positive feedback mechanisms, the relationships between ice melt and temperature are not linear: As more ice thaws, the water warms faster, thus accelerating additional ice melt. By tracking the known habitat requirements of different taxa (e.g., diatoms), past inferences of ice cover (and hence

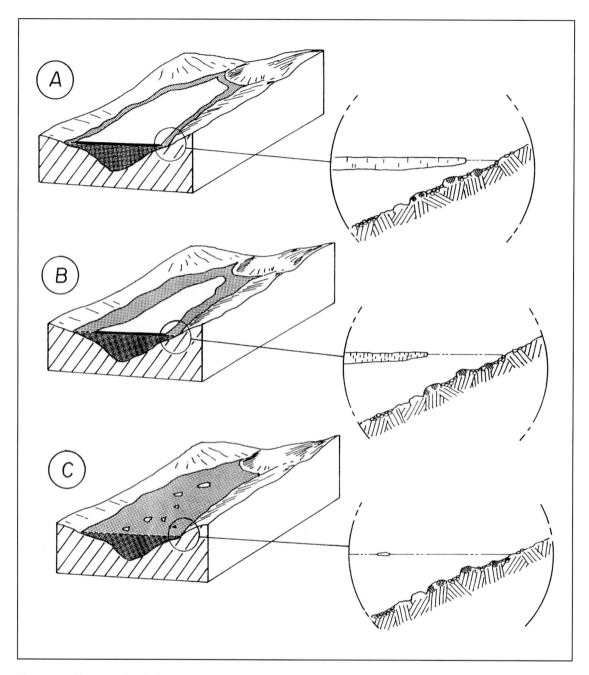

Fig. 14.3 Climate-related changes in ice cover exert important influences on the limnological conditions present in high latitude and high altitude lakes. This drawing shows changing ice and snow conditions on an Arctic lake during relatively cold (A), moderate (B), and warm (C) conditions. During colder years, a permanent raft of ice may persist throughout the short summer (A), precluding the development of large populations of phytoplankton, and restricting much of the primary production to the shallow, open-water moat. With progressively warmer conditions, more of the lake becomes available for autotrophic growth. Many other physical, chemical, and biological changes occur in lakes that are either directly or indirectly affected by snow and ice cover (Douglas & Smol 1999). From Smol (1988); used with permission.

climate) can be made. Of course, other limnological variables are also changing with warming, such as thermal stratification patterns, which may also potentially be tracked using shifts in planktonic diatom assemblages (Sorvari et al. 2002).

Similar mechanisms relating to the extent of ice cover and its effects on biota operate in shallow ponds, although planktonic diatom communities rarely develop due to the shallow nature of these water bodies. However, even a slight temperature change can greatly affect the open-water season. As described by Douglas and Smol (1999), a warmer summer, and hence a longer growing season, provides more opportunity for overall algal production to be higher, and for a more complex and diverse algal assemblage to develop (e.g., moss epiphytes; tube- and stalk-dwelling diatoms, as opposed to simply adnate forms). With warmer summers, changes in hydrology and water chemistry also take place (Douglas & Smol 1999). With increased snow melt and a deeper thaw in the permafrost layer, the pH, alkalinity, specific conductivity, and nutrient content of the water entering lakes and ponds from the catchment also changes. All the above climate-related variables can be tracked using biological assemblages (Smol & Cumming 2000).

Douglas et al. (1994) were the first to record striking diatom changes that could most easily be explained by warming, beginning in the 19th century, from several ponds on Ellesmere Island (Nunavut, Canadian High Arctic). Subsequent chironomid analyses from the same sediment cores also tracked food web changes indicative of warming (Quinlan et al. 2005). Although initially controversial (Smol & Douglas 2007), many studies applying similar approaches to those used by Douglas et al. (1994) were subsequently completed. This eventually allowed Smol et al. (2005) to undertake a meta-analysis of 55 sedimentary profiles from across the circumpolar Arctic. They showed that widespread post-1850 changes in diatoms and other bioindicators reflected recent climatic warming. The "human footprint" on climate could be discerned from Arctic lake and pond sediments.

The examples cited thus far were all from high-latitude lakes and ponds. However, similar mechanisms have been used to decipher past climate trends in other ecosystems characterized by extended ice-cover conditions, such as high mountain lakes (e.g., Lotter & Bigler 2000; Thompson et al. 2005). Although less pronounced in temperate regions, changing climate will also alter the periods of ice cover, thermal stratification, and related variables, which can be tracked using diatoms and other indicators (e.g., Harris et al. 2006; Hausmann & Pienitz 2007).

Tracking past river discharge using biota

River discharge is another environmental variable that may be linked to climate (e.g., spring snow melt) and related pollution issues, as flowing waters may both transport and dilute contaminants. Past river flows can be reconstructed, at least in an indirect and qualitative way, using biological proxies.

Realizing that flow regimes are important determinants of river and stream diatom (e.g., Potapova 1996) and invertebrate assemblages (e.g., Gandouin et al. 2006; Greenwood et al. 2006), a variety of paleolimnological approaches have been developed to use this ecological information. For example, working on a High Arctic river/lake system on Northern Ellesmere Island (Nunavut, Canada), Ludlam et al. (1996) recognized that certain diatom taxa characterized inflowing river habitats (i.e., lotic taxa), whilst other diatoms dominated the littoral zone of the downstream lake (i.e., lentic taxa). They proposed a measure called the *Lotic Index*, which was simply calculated as the relative numbers of river diatoms from the genera *Hannaea* and *Meridion*, divided by the total number of pennate diatoms. They then applied this index to the assemblages preserved in the annually laminated sediments from the downstream lake to infer past trends in river discharge, which they correlated with other paleolimnological indicators of past discharge. Antoniades and Douglas (2002) used similar approaches on Cornwallis Island (Nunavut, Canada).

Reconstructing the discharge of larger rivers is more challenging. In certain cases, fluvial lakes in river systems provide adequate sedimentation basins (e.g., Reavie et al. 1998). It is also possible to extract past environmental and climatic information of past river flow and hydrology by analyzing diatoms and other proxies in the sediments of delta lakes. The physical, chemical, and biological characteristics of delta floodplain lakes, such as those of the Mackenzie Delta (Northwest Territories, Canada) or the Peace–Athabasca Delta (see pp. 237–9), are tightly linked to their associated river systems, which can be used by paleolimnologists to gauge past river flooding. For example, Hay et al. (1997) studied the diatoms collected from the surface sediments of 77 Mackenzie Delta calibration lakes, representing three categories of river influence: (i) lakes having continuous connection with the river (i.e., no closure), (ii) lakes that flood every spring but lose connection with the river during the summer (i.e., low closure), and (iii) lakes that flood only during extreme river flooding events (i.e., high closure). The degree of flooding/connectiveness with the parent river greatly influences lakewater chemistry and other variables (Lesack et al. 1998). Hay et al. (1997) used these limnological data to develop a diatom-based transfer function that used delta lake diatoms to provide a tool for estimating long-term changes in river influence. Michelutti et al. (2001b) subsequently applied this transfer function to interpret the diatom assemblages preserved in sediment cores from eight delta lakes with differing connections to the river. They concluded that overall trends of past river influences on the floodplain lakes could be inferred but, not surprisingly, these deltaic systems are dynamic and challenging environments. Similar approaches are being used to track past river discharge in the Peace–Athabasca Delta (Wolfe et al. 2005, 2006).

Trees, water chemistry, and shifting ecosystem boundaries

Ecotones are sensitive locations to study climatic change, since even slight changes in temperature and precipitation may have pronounced impacts on these ecosystem boundaries and their associated aquatic systems. Several studies have focused on examining limnological changes across the Arctic and alpine treeline and tundra ecotone (e.g., Lotter et al. 1999) and, to a lesser extent, near forest–grassland transitions (e.g., Bradbury & Dean 1993).

Arctic or alpine treeline to tundra ecotones have received considerable attention from paleolimnologists, as shifts in these boundaries can result in striking limnological changes (Lotter et al. 1999). For example, as the position of the Arctic or alpine tree line is controlled to a large extent by climatic factors, and as the abundance of coniferous trees in a watershed can significantly alter the limnological characteristics of the lakes draining these catchments, paleolimnologists have developed methods to reconstruct these terrestrial–aquatic linkages. Diatom indicators have been used most often, where surface-sediment calibration sets have been developed to reconstruct, at least in an approximate way, past concentrations of dissolved organic carbon (DOC). As described more fully in the following chapter, the concentration of colored DOC in lakes is determined by many factors, but a major allochthonous source is humus from trees, and especially conifers such as spruce (*Picea*). As transfer functions can be developed to link lakewater DOC concentrations with diatom assemblages (e.g., Lotter et al. 1999), paleolimnologists can then reconstruct past DOC levels (and hence past changes in coniferous vegetation) from dated sediment cores, and thereby also infer climatic trends. These types of reconstructions, however, have other applications aside from paleoclimatic research, such as the study of paleo-optics and the reconstruction of past ultraviolet light penetration (see Chapter 15).

Links between lakewater pH and climate

Paleolimnological studies from soft-water high-altitude lakes suggest that past climatic conditions may also be recorded in diatom-inferred pH changes (Psenner & Schmidt 1992). Sommaruga-Wögrath et al. (1997) studied the acid-base status of 57 alpine lakes by comparing samples taken in 1985 and 1995. They identified increases in sulfate and decreases in nitrogen concentrations in many lakes, despite atmospheric trends in the opposite direction. Additionally, most lakes had increased in pH, base cations, and silica. These observations, in conjunction with the > 1°C increase in temperature since 1985, led Sommaruga-Wögrath et al. (1997) to suggest that temperature-related processes, such as increased weathering and biological activity, may have been responsible for the observed changes. A comparison of diatom-inferred pH from an alpine lake and air temperature records from 1778 to 1991 provided further evidence for a coupling between climate and lakewater pH (Sommaruga-Wögrath et al. 1997; Koinig et al. 1998a). The proposed mechanisms for the increase in pH may be related to longer ice-free seasons, and/or increased light, temperature and nutrients, all of which would enhance biological production (Sommaruga-Wögrath et al. 1997). Psenner and Schmidt (1992) further suggested that during warm and dry years, there would be an increased potential for in-lake alkalinity generation, since the increased production would help promote anoxic conditions at the sediment/water interface. Finally, pH could be depressed during times of extended ice cover because of CO_2 oversaturation, decreased in-lake production of alkalinity, and reduced deposition of base cations directly to the lake (Koinig et al. 1998a,b). Similar approaches have been applied to other paleoclimatic studies (e.g., Wolfe 2002; Schmidt et al. 2006). Regardless of the exact mechanism that results in pH increases during warmer periods, the correlational evidence suggests there is potential for interesting paleoclimatic records using these methods. Of course, these relationships increase the complexity of tracking pH changes related to anthropogenic acidification (Chapter 7).

Shifts in precipitation–evaporation (P/E) ratios reflected in biological proxies

Long-term records of changes in precipitation and evaporation (e.g., drought frequencies) are amongst the most important climate variables required by environmental scientists and policy makers. Closed-basin lakes in semi-arid to semi-humid regions are amongst the most sensitive recorders of past precipitation–evaporation (P/E) ratios due to changes in water levels and chemistry, and hence the lakes' biota. At a superficial level, the types of limnological changes occurring with shifts in P/E appear quite simple (Fig. 14.4); however, as is often the case, there can be many complicating factors in nature that can obscure the relationship between P/E and the indicators preserved in lake sediments.

In its simplest form, a closed-basin lake experiencing drought condition in, for example, the North American prairies can be compared to a pot of soup left on a stove set at low heat, with the lid removed. If you check on the soup every 15 minutes or so, you will observe two progressive changes. First, as water continues to evaporate, the volume of the soup will decrease over time. The same happens in nature with a closed-basin lake during drought conditions – the water level lowers. Second, if you taste the soup at intervals, you will soon realize that it is getting saltier and saltier the longer you leave it on the stove top, for the same reason that the volume is decreasing (i.e., a distillation is occurring as water is evaporated). Similar processes occur in closed-basin lakes over the course of a hot, dry summer – the water chemistry changes markedly, with evaporative concentration of the solutes (Fig. 14.4).

A variety of paleolimnological approaches are available to track the environmental changes noted above, including isotopic (described below), geomorphological, geochemical, and paleobiological approaches. For example, old beaches that may indicate past high water stands, similar to bathtub rings, can be traced and dated using geomorphological techniques (e.g., Hendy 2000; Schustera et al. 2005). Meanwhile, dendroecology can be used to, for example, reconstruct lower water-level stands using submerged tree stumps (e.g., Stine 1994). A rich legacy of chemical precipitates and mineralogical data are stored in

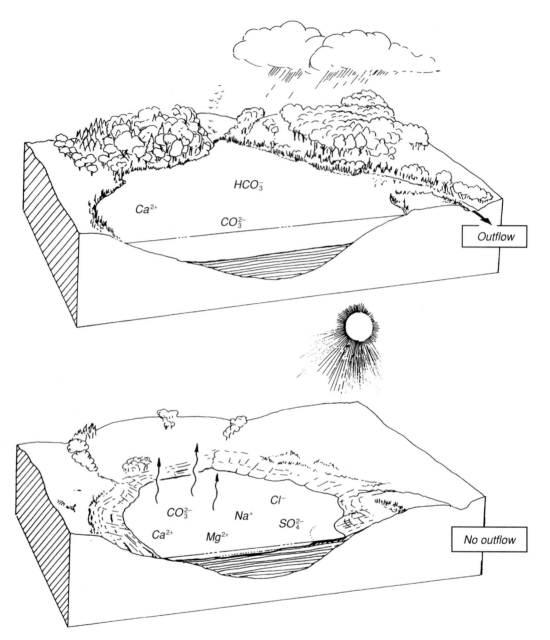

Fig. 14.4 A schematic diagram showing some of the limnological responses of a closed-basin lake to changes in precipitation and evaporation. The upper diagram summarizes lake conditions during wet periods, when lake levels are high and, in some cases, the lake may have an outflow. Because the lake is receiving more precipitation and less evaporation is occurring, the lake is relatively fresh. Most freshwater lakes are dominated by $CaCO_3$. The lower diagram summarizes conditions during periods of drought. Lake levels are lower and salinity increases, as evaporation exceeds precipitation. Typically, the longer the duration and severity of drought, the higher are the salinity levels. Inland saline (athalassic) lakes can vary widely in their ionic composition. From Smol and Cumming (2000); used with permission of PSA.

lake sediments that provide additional information on past in-lake processes resulting from decreased P/E (Boyle 2001; Last 2001b; Last & Ginn 2005). Important insights are also made possible from stable isotope analyses (Leng & Marshall 2004).

Not surprisingly, limnological changes associated with fluctuating water levels and altered chemistry can also be tracked using species assemblage changes. For example, algae (e.g., Wolin & Duthie 1999; Moser et al. 2000) and invertebrate indicators (e.g., Korhola et al. 2005) have been used to infer water depth. However, in closed-basin lakes that are sensitive to changes in P/E, some of the most striking limnological changes occur in water chemistry; most notably changes in ionic concentrations following prolonged evaporation. In fact, following prolonged droughts, closed-basin lakes may become quite saline (referred to as athalassic lakes; see the lower part of Fig. 14.4), with salinities far exceeding those present in the oceans. Of course, the relationships between P/E and lakewater variables are never as simple as those shown in Fig. 14.4. Water chemistry will be influenced by a spectrum of factors, including the balance between water inputs from precipitation, stream inflow, surface runoff, and groundwater inflow, and between outputs from evaporation and from stream and groundwater outflow (e.g., Fritz et al. 1999). Some lakes will be more responsive to climatic change than others, and factors such as lake hydrology, as well as basin and watershed morphometry, have to be considered. Nonetheless, many robust transfer functions have been developed and applied to reconstruct past salinity levels and, to some extent, the ionic composition of past salinity. Most of this work has been done using diatoms (e.g., Fritz et al. 1999), but other indicators have been used, such as chrysophytes (Zeeb & Smol 1995), cladocerans (Bos et al. 1996), ostracods (Mischke et al. 2007), chironomids (Eggermont et al. 2006; Heinrichs & Walker 2006), pigments (Vinebrooke et al. 1998), as well as other biomarkers.

14.4.2 Isotopes and climatic reconstructions

Stable isotopes have many uses in paleoenvironmental studies, including some of the earliest work on reconstructing climatic trends (McCrea 1950). Stable isotope data often provide important complementary information to the proxies described thus far (e.g., Ito 2001; Talbot 2001; B.B. Wolfe et al. 2001; Leng & Marshall 2004; Leng et al. 2006). Many approaches are used, but oxygen isotopes remain one of the mainstays in paleoclimatic research.

In its simplest form, stable isotope analyses in paleolimnology use environmentally mediated fractionation of isotopes. For example, oxygen is present as primarily two stable isotopes: ^{16}O and ^{18}O (^{17}O is also present, but is uncommon). The water molecules containing the lighter isotope (i.e., ^{16}O) are more easily evaporated, and so, after a long period of warm and dry weather, the water remaining in a lake will be enriched with the heavier isotope (i.e., ^{18}O), as proportionally more of the lighter isotope (^{16}O) has evaporated (Fig. 14.5). Similar to the P/E studies cited previously, some of the strongest isotope signals are preserved in the sediments of closed lake basins, where lakes primarily lose water via evaporation, and therefore contain sediments with variable and generally high $\delta^{18}O$ values. However, my description of isotope fractionation is highly simplistic, as many complications can obscure signals (Leng & Marshall 2004). Nonetheless, like other proxy indicators, when interpreted carefully, stable isotopes can provide important information on past climatic trends as well as other ecosystem processes.

Isotope ratios from specific organic compound classes, such as lipids (e.g., Huang et al. 2004), may further refine paleoclimatic inferences. Moreover, the use of isotope signatures preserved in morphological fossils, such as ostracod shells (e.g., Donovan et al. 2002; Schwalb 2003) or diatom frustules (e.g., Jones et al. 2004; Rosqvist et al. 2004), represent other important avenues for research.

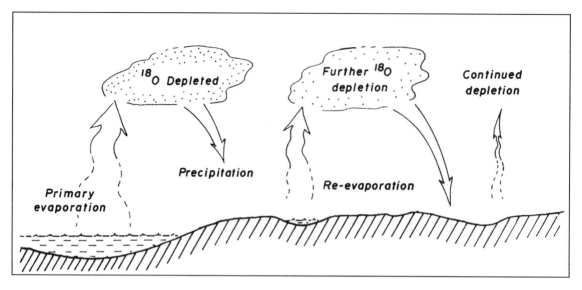

Fig. 14.5 Oxygen isotope fractionation occurring due to evaporation from a lake. Water molecules containing the lighter oxygen isotope (i.e., ^{16}O) are more easily evaporated, and so the water remaining in a lake will be enriched with the heavier isotope (i.e., ^{18}O), as proportionally more of the lighter isotope (^{16}O) has evaporated. The process can continue with further evaporation. This is a highly simplified schematic diagram of the possible processes, as other factors can of course alter these isotopic measures.

14.4.3 Other geological techniques to track climatic change

As noted earlier, the physical, geochemical, and mineralogical information stored in lake sediments provides a wealth of paleoenvironmental data that can be linked to past climatic changes (e.g., Boyle 2001; Last 2001a,b; Lamoureux and Gilbert 2004; Last & Ginn 2005). In fact, all the paleoclimatic approaches described earlier in this chapter, where biological proxies were emphasized, can also be investigated using geological methods. For example, athalassic lakes archive important geochemical and mineralogical data reflecting past evaporation and precipitation ratios (Last & Ginn 2005). Using varved sediments, high-resolution reconstructions of past river inflows and sediment accumulation rates can be linked to climate (e.g., Francus et al. 2002; Smith et al. 2004; Lamoureux et al. 2006). New approaches, such as the use of alkenones (e.g., Liu et al. 2006),

are also being developed. Furthermore, as with stable isotopes, a variety of methods have been employed to study the geochemistry of individual morphological remains, such as ostracod shells, to gain further paleoclimatic information (e.g., Chivas & Holmes 2002). As with all paleoenvironmental studies, multidisciplinary investigations will provide more confident interpretations of climatic trends (e.g., Battarbee et al. 2004).

14.5 Climatic change and pollution

The preceding sections have summarized some of the ways in which paleolimnologists reconstruct past trends in climatic variables. However, I conclude this chapter by returning to the main theme of this textbook, and emphasizing some of the ways in which climatic changes affect water quantity and quality.

Climate influences, directly or indirectly, every water-quality problem discussed in this book. The combined effects are often cumulative, worsening the environmental impacts (Schindler 2001). For example, there are many links between eutrophication and climatic change (Blenckner et al. 2006). These include some direct effects of higher temperatures, such as the many physical and chemical reactions that are influenced by warming. Phosphorus recycling is closely related to temperature, as are processes of nutrient mineralization and release from sediments. Warmer temperatures and reduced ice cover can lead to stronger and more prolonged thermal stratification, which can enhance periods of deepwater anoxia, which in turn can greatly affect internal loading of nutrients as well as the distribution of biota in a lake's deep waters. With lower water levels, which may occur with higher rates of evaporation, nutrients and other pollutants are further concentrated (Schindler & Donahue 2006). The problem is exacerbated if flushing rates are also reduced (Rippey et al. 1997).

As acidic precipitation (Chapter 7) and many other contaminants (e.g., Chapters 8–10) can be transported via the atmosphere, climatic changes will influence contaminant pathways and processes (Macdonald et al. 2005). For example, the cycling of temperature between summer and winter is the main driver of the so-called "grasshopper effect" that is responsible for the manner in which some POPs are transported outward from their points of release (Box 9.2). Increased melting of ice on mountain glaciers is a locally important process that releases pollutants (Blais et al. 2001) that have been archived during cooler, more contaminated periods. The biological consequences will, in most cases, be additive (cumulative) and multiplicative (synergistic), with higher contaminant loads and biomagnification in aquatic ecosystems (Wrona et al. 2006). Climatic changes will also affect biovectors that may influence contaminant transport (Box 9.3). Furthermore, if lake levels and river flows are lower due to increased evaporation, there is less water available to dilute pollutants, and so concentrations are higher. Meanwhile, as the two main agents of erosion (Chapter 12) are wind and precipitation, changes in climate will affect soil loss (e.g., Kelley et al. 2006). Paleolimnological perspectives may also provide insights on biodiversity loss (Chapter 13) due to warming, such as the vulnerability of slow-growing, cold-water endemic diatoms in Lake Baikal (Mackay et al. 2006).

Drought can also result in some surprising side-effects. For example, Yan et al. (1996) showed that declines in lakewater pH and DOC, and resultant increased UV penetration (Chapter 15), were attributed to the re-oxidation of sulfur compounds (previously deposited from local smelting activity) from a lake's exposed littoral zone following a period of reduced precipitation. A reduced winter ice cover will also allow far more ultraviolet radiation to penetrate into the water column, which will lead to additional repercussions (Wrona et al. 2006). Some of the synergistic effects of climatic change and other multiple stressors are discussed more fully in the subsequent chapter.

14.6 Summary

No scientist – at least, no reputable scientist – ever said that reconstructing climate was simple. Nonetheless, paleolimnologists have developed a number of original and robust methods to track overall trends in climatic change. These transdisciplinary approaches are providing information on past modes of climatic change, as well as demonstrating the influence of humans on the Earth's climate. Greenhouse gas induced climatic change represents the most serious problem facing humanity today. As W.S. Broecker warned us, "The climate system is an angry beast, and we are poking it with a stick."[1]

---

[1] Wallace S. Broecker; quoted in *The New York Times*; "If climate changes, it may change quickly," January 27, 1998, pp. C1–2.

Moreover, as climate exerts an over-arching influence on a broad spectrum of physical, chemical, and biological processes, paleoclimate data also provide important information pertaining to water quantity and quality, contaminant transport and processes, and the synergistic and cumulative negative effects of climatic change on other environmental problems.

15

Ozone depletion, acid rain, and climatic warming: the problems of multiple stressors

Nature is a collective idea . . .
Henry Fuseli (1741–1825)

15.1 The challenges of multiple stressors

There is now ample evidence that our planet is undergoing accelerated warming. The root cause of this unplanned global experiment has been the dramatic increase in anthropogenic burning of fossil fuels, such as coal and oil, and the subsequent release of greenhouse gases (e.g., carbon dioxide) into our atmosphere. As discussed in the previous chapter, paleolimnology is playing a very important role in the study of long-term climatic change. Clearly, a warming planet has tremendous and often negative environmental repercussions that are both direct and indirect. This chapter deals with some of the indirect effects and is used to illustrate the problem of "multiple stressors" (i.e., several environmental problems culminating into a larger, potentially more serious calamity).

Our planet has entered a period of accelerated warming because of increased concentrations of greenhouse gases; I do not believe there is any real controversy remaining here (at least not much *scientific* controversy), despite the concerted efforts of industrial-sponsored obfuscators. However, greenhouse warming is only one of many large-scale, human-related problems that are affecting our ecosystems. In a previous chapter, we have already explored how anthropogenic inputs to our atmosphere have dramatically altered the chemistry of precipitation, resulting in acid precipitation, which has caused alarming damage to many aquatic ecosystems (Chapter 7). There is still a third, major human-related alteration occurring in our upper atmosphere, the stratosphere. This is ozone depletion, and the potential consequences are harrowing.

This chapter describes the problem of ozone depletion and the resulting increased penetration of ultraviolet radiation. More importantly, it shows how the above three problems (global warming, acid precipitation, and ozone depletion) may be interrelated. The negative impacts of each of these three stressors are significantly magnified when they occur simultaneously. We refer to such interactions as multiple stressors. There is still relatively little paleolimnological work published on the topic of ultraviolet penetration, and so the few examples we have are still exploratory.

15.2 Stratospheric ozone depletion, ultraviolet radiation, and cancer

Ozone (O_3) is a naturally occurring gas that is present in trace quantities in our atmosphere. It is not distributed evenly throughout the atmosphere, and is most abundant in the stratosphere where it forms the stratospheric ozone layer at an altitude of about 15–40 km, but is most concentrated around 25 km. Ozone, which is continually being formed and broken down, provides a protective shield as it filters out ultraviolet radiation (UV). This includes the high-energy radiation spectrum (280–320 nm) known as ultraviolet B (UV-B), which can have very serious consequences on living organisms, including humans.

In its mildest form, UV-B can cause sunburn. However, it can also damage DNA, depress immune systems, and has been linked to skin cancer and the development of cataracts. Not surprisingly, it has also been linked to a myriad of ecosystem effects in both terrestrial (e.g., reduced crop yields) and aquatic environments (Vincent & Roy 1993).

Although ozone's protective role is so important to life, it is present in minute quantities. If all the ozone present in our atmosphere was compressed into a pure gas at the temperature and pressure of the Earth's surface, its total thickness would be about 3 mm. Without this ozone layer, which filters out about 98% of the incoming UV-B, life as we know it could not have evolved. Alarmingly, there is now solid evidence that this protective shield is diminishing.

An almost continuous record of ozone measurements dates back to 1931, from Arosa, Switzerland. Since that time, instruments have been set up around the world, and satellites have also been used to monitor levels. The data show that ozone had declined in recent decades, and ozone thinning, more commonly referred to in newspaper headlines as the "ozone hole," has been gaining considerable attention over the past decade. Much concern has focused on ozone depletion over the Antarctic, and to a lesser extent the Arctic, but, to varying degrees, this is a global problem. What causes these changes in ozone, a gas that is primarily contained in a diffuse layer high in the stratosphere? Are these changes natural or related to human activities?

As with many environmental problems, the answer is that both natural and human causes are involved. Natural changes occur in ozone levels and the amount of UV radiation reaching the Earth. For example, UV penetration is highest when sunspot activity is high, and vice versa. Humans, however, have been altering the chemistry of the stratosphere, and hence also the penetration of UV.

The first major alarm came in the mid-1980s, when scientists working with the British Antarctic Survey showed that ozone concentrations had declined markedly since the 1960s. These shocking and unpredicted findings were subsequently confirmed. Natural causes were soon ruled out. We have now seen that human activities have altered, to some extent, every ecosystem on this planet. Could human activities have also affected the structure of the stratosphere?

15.3 Refrigerators, air conditioners, aerosol cans, and the ozone layer

Since the mid-1970s, there have been data suggesting that the depletion of the ozone layer has been related to human activities by the release of manufactured chlorine and bromine gaseous compounds. The most common of these are chlorofluorohydrocarbons (CFCs), which were first synthesized in 1928, and whose industrial applications were quickly recognized. For example, they were soon used in air conditioning units, refrigerators, aerosols, plastic foams, fire extinguishers, and other products. CFCs may seem to be innocuous gases, released in small quantities by individuals, but their secondary effects are ominously dangerous. As described in Box 15.1, the same CFCs that were used in our air conditioners

Box 15.1 Chlorofluorohydrocarbons (CFCs) and ozone destruction

CFCs are transparent to sunlight in the visible range of the electromagnetic spectrum, are insoluble in water, and are inert to chemical reactions in the lower atmosphere. These may be desirable properties for many industrial applications, but they are also characteristics that contribute to significant environmental problems. The long residence times of CFCs (e.g., typically from 75 to over 100 years) also add the problem of persistence to these pollutants, similar to other compounds discussed in this book (e.g., Chapter 9).

Ozone (O_3) is continually being formed and broken down naturally in the upper atmosphere by the action of ultraviolet radiation (UV). However, the addition of CFCs has dramatically altered this balance.

When we release CFCs into the atmosphere, they rise into the stratosphere, where UV breaks them down and thus releases chlorine (Fig. 15.1). This free chlorine (Cl) atom finds and reacts with an ozone molecule, and breaks away an oxygen atom (O), forming chlorine

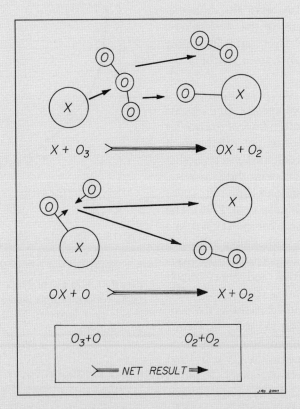

Fig. 15.1 A diagram showing the two major steps that destroy ozone (O_3) molecules in the stratosphere. The catalyst X (which in most cases is a chlorine or bromine atom) cleaves an oxygen (O) atom from an ozone (O_3) molecule, forming OX and an oxygen molecule (O_2). An oxygen atom (O) can then remove the oxygen atom attached to the catalyst, thus releasing the catalyst and allowing it to destroy another O_3, and so the process continues. Modified from Hengeveld (1991).

monoxide (ClO) as a by-product, and leaving an oxygen molecule (O_2). This process, which takes one or two seconds, results in the destruction of an ozone molecule. However, the process continues. A free oxygen atom then removes the oxygen atom attached to the chlorine, thus forming an oxygen molecule, but also freeing the chlorine atom again to destroy yet another ozone molecule. This process continues, until each chlorine atom may destroy over 100,000 ozone molecules! The speed and extent of the destruction of the ozone layer is related to chlorine's extreme efficiency in breaking down ozone molecules. The process only ends when the chlorine atom finally drifts into the lower atmosphere, where it combines with a water molecule and forms hydrochloric acid (HCl).

Although CFCs have received the most media attention, other hydrogen, fluorine, nitrogen, and bromine compounds can also accelerate the rate of ozone destruction.

and deodorant aerosol cans were also dramatically altering the chemistry of the stratosphere, diminishing the ozone layer, and increasing the amount of UV-B penetration reaching the Earth's surface.

Ozone depletion and the resulting increased UV-B penetration should present limnologists and the general public with sufficient reasons for concern, but the situation is potentially much worse. This is due to the combined effects of other stressors simultaneously affecting the ecosystem, in this case acid precipitation (Yan et al. 1996) and climate warming (Schindler et al. 1996). The amount of dissolved organic carbon (DOC) in a lake can be decreased by both adding acidity to precipitation, as well as by decreasing the overall amount of precipitation and hence runoff as a result of global warming (Box 15.2). DOC plays a pre-eminent role in determining the penetration of UV-B radiation (Vincent & Pienitz 1996), because the dissolved humic and fulvic fractions strongly absorb across the UV waveband (Vincent & Roy 1993). For example, Laurion et al. (1997) showed that the water column ratio of UV and photosynthetically active radiation is relatively stable across the DOC range of 4–11 mg of carbon per liter, but increases markedly with decreasing DOC below this level. These colored aromatic compounds can act as a "natural sun screen" in lakes, filtering out

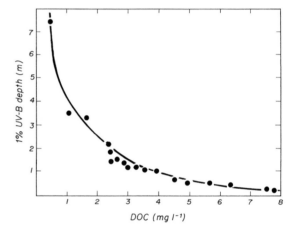

Fig. 15.2 The relationship between measured DOC concentrations in northern and boreal lakes in Canada and the depth of the 1% UV-B penetration isopleth. These data show that as DOC concentrations increase, the depth of UV-B penetration decreases. Modified from Schindler et al. (1996).

harmful UV radiation (Fig. 15.2). If DOC levels are high, the water may take on a tea color.

As shown in Box 15.2, acidic precipitation will result in lower DOC levels (and hence clearer, more UV-transparent waters), due to both direct and indirect results of acidification. Indirect effects predominate, and include enhanced photodegradation at lower pH and increased aggregation and

Box 15.2 The combined effects of stratospheric ozone depletion, acidic precipitation, and climate warming on ultraviolet light penetration in lakes: a three-pronged attack

The penetration of UV radiation, which has increased due to stratospheric ozone depletion, may result in a myriad of problems to aquatic biota. A key factor in this scenario is the amount of dissolved organic carbon (DOC) in the water, as many DOC compounds are colored, and act as a "natural sun screen" to UV penetration.

Under natural conditions, many lakes, such as those that characterize much of the boreal forest region, have naturally high or moderate DOC levels. Much of this DOC is from the catchment, such as from water draining coniferous leaf litter or peat lands. Such lakes are well protected from the damaging effects of UV penetration (Fig. 15.3A).

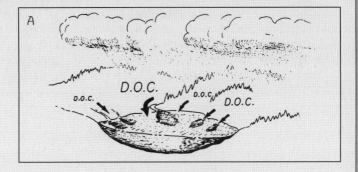

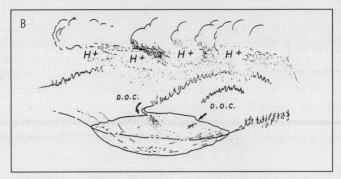

Fig. 15.3 The effects of acidification and warming on dissolved organic carbon (DOC) levels in lakes. (A) The lake in its "natural state," with DOC being delivered to the lake from the catchment. (B) How DOC in the lake is reduced with acidic precipitation. (C) How the supply of DOC to the lake is further diminished under drought conditions, as less DOC is delivered to the lake.

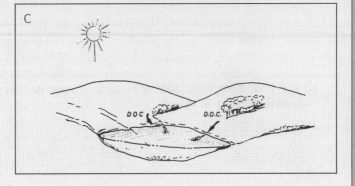

If a lake is receiving acidic precipitation (Fig. 15.3B), DOC levels in the lake will decline, as the photodegradation of DOC is strongly pH dependent. Some additional DOC may be lost from the water as a result of other processes, including the lower pH directly precipitating DOC, or via precipitation of DOC by aluminum that has been mobilized because of catchment acidification (see Chapter 7). This is one reason why many acidic lakes are so clear; the DOC levels have decreased.

With warming, stream and overland inflows to the lake are reduced in some regions (the main vectors for allochthonous DOC transport to the lake), and so DOC levels decline further (Fig. 15.3C). For example, Schindler et al. (1996) showed that DOC concentrations have declined significantly from 1970 to 1990 as a result of decreased stream flow related to climate warming of about 1.6°C, as well as drier conditions, in the Experimental Lakes Area (ELA) of Canada. Over those two decades, DOC in unperturbed ELA reference lakes decreased by an average of 15–20%, allowing for increases of 22–63% in the depth of the UV-B penetration isopleth in the lakes.

Thus, three human-related stressors – stratospheric ozone depletion, acid rain, and climate warming – act in concert to increase the exposure of aquatic organisms to UV radiation. These communities are under a three-pronged attack – a "triple whammy" (Gorham 1996; Schindler 1998), where the combined environmental effects may be more severe than the effects of the individual stressors.

settling with aluminum (released from soils and sediment as a result of acidification; see Chapter 7). Similarly, since much of the DOC entering a lake may be from terrestrial sources, such as leaf litter, and if climate warming results in reduced stream inflows, then the vector for DOC delivery to the lake is removed or minimized. DOC levels drop further, and UV penetrates deeper into the lake, affecting more of the biota. Relatively clear, shallow lakes are most vulnerable, as damaging UV may eventually penetrate through the entire water column.

15.4 Reconstructing past UV penetration in lakes

How does a paleolimnologist track past penetration of radiation of a specific wavelength, such as UV-B, entering a lake? It is certainly not a simple problem. We do not have direct information in sediments of past radiation spectra. New problems present new challenges.

Two main approaches have thus far been used to study the effects of these multiple stressors (Leavitt et al. 2003). One research thrust has been to infer past lakewater DOC (Pienitz et al. 1999), and then reconstruct the past underwater radiation climate (i.e., paleo-optics) and relate this to other changes occurring in the ecosystem, such as terrestrial vegetation changes (e.g., Pienitz & Vincent 2000) or acidic deposition (e.g., Dixit et al. 2001). The second approach is to track past UV-B penetration by studying the biotic response of organisms, in this case pigment composition of blue-green algae (e.g., Leavitt et al. 1997, 1999, 2003), although other biogeochemical techniques may also be useful (e.g., analyses of UV-absorbing sporopollenin monomers in pollen grains and spores; Blokker et al. 2005). Some of these approaches are explored more fully below.

15.5 Paleo-optics: tracking long-term changes in the penetration of ultraviolet radiation in lakes

As noted earlier, several factors can affect the underwater light climate of lakes. Allochthonous input of colored DOC, such as from leaf litter and/or bogs, will certainly affect transparency. This DOC contribution is at least partly related to climate. Once the DOC is in the lake, factors such as acidic precipitation and aluminum inputs can further alter UV-B penetration (Box 15.2). Two examples show how these scenarios can be reconstructed and studied using paleolimnological techniques.

A characteristic feature of many boreal forest lakes, which are surrounded by conifer trees such as spruce, is that DOC levels are relatively high. In fact, many are obviously colored and look like weak tea (and some strongly stained lakes look like strong tea!). Such lakes clearly will not transmit much UV-B into the water column (Fig. 15.2), and the lake biota will be relatively immune to potential damaging effects. In essence, because of colored DOC's ability to attenuate radiation, it acts as an effective filter to UV-B penetration. But how have past climate and vegetation shifts affected UV-B penetration? Are recent increases unprecedented? As UV-B penetration data for lakes have only recently been gathered, we must use paleolimnological inferences.

As discussed in the acidification chapter (Chapter 7), diatom transfer functions can be developed in some lake systems to reconstruct past lakewater DOC. Just as present-day DOC measurements are used to calculate UV-B penetration in lakes (Fig. 15.2), then if past DOC levels can be reconstructed, one can also estimate past UV-B penetration.

A growing volume of paleolimnological research on past DOC concentrations has been accumulating in Arctic treeline regions, as these areas are of considerable interest to paleoclimatologists (Smol & Cumming 2000). The overall reasoning of these studies is that if one can reconstruct past DOC levels, one can also reconstruct the position of the coniferous tree line, as there is a good correlation between conifers in a catchment and DOC levels in lakes (Pienitz et al. 1997a,b; Vincent & Pienitz 1996). For example, a detailed analysis of one treeline lake in the Canadian Sub-Arctic (Queen's Lake) showed marked DOC changes in the past as spruce trees migrated into and out of its catchment during periods of warming and subsequent cooling (MacDonald et al. 1993; Pienitz et al. 1999). Using a diatom–DOC transfer function, the researchers inferred a warming period between approximately 5000 and 3000 ^{14}C yr BP, lakewater DOC increased by up to 5.8 mg l^{-1} in Queen's Lake (Fig. 15.4, upper right graph). As the climate cooled and the trees migrated farther south, leaving behind a treeless tundra catchment, DOC levels dropped precipitously. Pienitz and Vincent (2000) then applied these DOC models to their previously constructed bio-optical models, which were derived from present-day measurements in high-latitude lakes. They showed that the inferred DOC increases were equivalent to about a 100 times decrease in biologically effective UV exposure (Fig. 15.4, bottom graphs), whereas the most recent 3000 years (return to tundra conditions) would have been characterized by a > 50-fold increase in exposure.

These paleolimnological inferences suggest that at or near the Arctic tree line, past UV environments of lakes may have been changed more radically by climatic change (and resulting vegetation shifts) than by recent ozone depletion. If climate warming results in a northward migration of trees, these lakes will likely be less impacted by UV-B radiation, as DOC will increase (in contrast to the more southern boreal forest sites, where warming may result in less runoff, and so less DOC inflow; see Box 15.2). The above scenario is relevant for lakes with significant allochthonous sources of DOC, such as coniferous trees. Likewise, for lakes that are currently located above the tree line, in Arctic tundra environments, climate warming may increase lakewater DOC and decrease UV-B penetration. After surveying 57

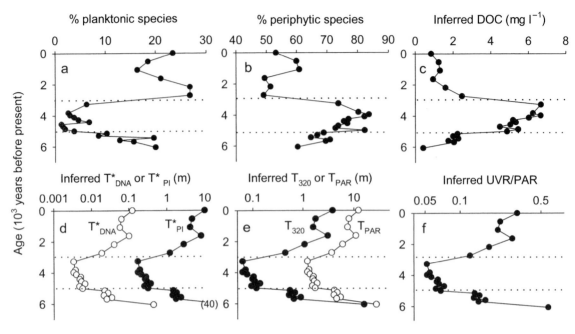

Fig. 15.4 Changes in diatom community structure (percentage of planktonic versus periphytic taxa) and diatom-inferred DOC levels in Queen's Lake. Using equations developed from modern-day lakewater DOC levels and UV penetration, Pienitz and Vincent (2000) used the diatom-inferred DOC values to reconstruct the biologically weighted UV exposure (T^*_{DNA} and T^*_{PI}), as well as the inferred underwater spectral balance (i.e., water column transparency for 320 nm UV radiation (T_{320}), photosynthetically active radiation (T_{PAR}), and the ratio between the two (UV/PAR) over the past 6000 years in Queen's Lake, Northwest Territories, Canada. The dotted lines mark the period of maximum forest cover from c. 5000 to 3000 ^{14}C yr BP, and so highest diatom-inferred DOC and lowest UV and PAR penetration. From Pienitz and Vincent (2000); reprinted by permission from *Nature*, copyright 2000 Macmillan Publishers Ltd.

high-alpine Austrian lakes, which have poorly developed soils and sparse vegetation, Sommaruga et al. (1999) hypothesized that DOC concentrations in these tundra lakes would increase with warming and attenuate more UV radiation. As expected, DOC tended to decline with elevation. Warming in these regions would be expected to promote soil and terrestrial vegetation development, which should increase allochthonous-derived DOC concentrations, which would result in more UV attenuated in the water column. Paleolimnological indicators could potentially be used to track these changes.

Acidification can also result in marked declines in colored DOC and concomitant increases in UV-B penetration (Yan et al. 1996). Using similar

approaches to the treeline example discussed above, Dixit et al. (2001) developed a diatom-inferred DOC transfer function from a calibration set of 80 lakes near Sudbury, Ontario, a region that had received very high levels of acidic precipitation in the past (Chapter 7). Using previously published diatom profiles from three study lakes, they reconstructed past DOC levels in these lakes, from which past UV-B penetration could be calculated. They recorded striking declines in diatom-inferred DOC, and so equally dramatic increases in UV-B penetration, in two of the lakes, which coincided with the diatom-inferred acidification trajectories previously established for these sites. Similarly, with the post-1980s pH recoveries in these lakes, inferred DOC levels increased, and UV-B penetration decreased. Acidification and any subsequent recovery can result in marked changes in underwater spectral properties, confirming previous predictions of the cumulative dangers of multiple stressors (Box 15.2).

15.6 Biological responses to enhanced UV-B exposure: evidence from fossil pigments

The above examples illustrate how diatom species responses have been used to reconstruct past DOC, from which past UV-B penetration can be modeled. A different approach is to reconstruct biological responses that may be directly related to increased UV-B penetration. To date, fossil pigment analyses have taken the lead.

Leavitt et al. (1997, 1999) demonstrated that certain fossil pigments can be used to document historical changes in UV penetration as a result of both climatic changes and acidification. After analyzing the surface sediments from 65 clearwater mountain lakes, they established the distribution of UV-radiation-absorbing sedimentary pigments amongst lakes of differing DOC, and hence UV-B penetration. Although the two primary indicator pigments could not be definitely identified (they have been temporarily named C_a and C_b, pending further analyses), these two UV-B absorbing pigments were very similar to the oxidized and alkaline forms, respectively, of the carotenoid scytonemin, a pigment produced in the sheaths of certain cyanobacteria to protect them from elevated UV-B exposure.

Leavitt et al. (1997, 1999) used these pigment biomarkers to track past UV-B penetration in two different lake systems: alpine lakes subjected to droughts and cooling periods during the late 19th century, as well as an artificially acidified lake in northwestern Ontario. In the two alpine lake sediment cores they examined from Banff National Park (Alberta, Canada), the relative abundances of these UV-radiation-absorbing sedimentary

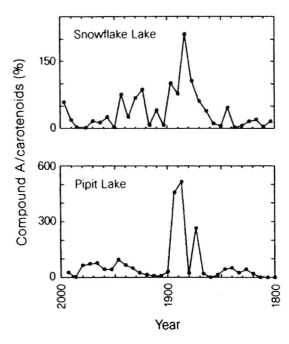

Fig. 15.5 Fossil pigment-inferred increases in the penetration of UV-B radiation in two alpine lake sediment cores from Snowflake Lake (A) and Pipit Lake (B), in Banff National Park, Alberta, Canada. Based on tree-ring data from western Canada, droughts were more common in the late 1800s. These periods coincide with increases in UV-radiation-absorbing pigments (expressed as the percentage of pigment C_a relative to total sedimentary carotenoid. From Leavitt et al. (1997); reprinted by permission from *Nature*, copyright 1997 Macmillan Publishers Ltd.

pigments increased markedly between c. 1850 and 1900, suggesting that UV penetration was high at this time (Fig. 15.5), and much greater than any recent increases that may have been due to ozone depletion. This time period coincided with periods of cool weather and droughts, as inferred from tree-ring data, which likely resulted in decreased lakewater DOC (Fig. 15.3C).

Similarly, C_a and C_b increased strikingly in the post-acidification sediments of Lake 302S, in the Experimental Lakes Area (Fig. 15.6). Lake 302S was artificially acidified in 1980, from pH values of approximately 6.4 to approximately 4.5

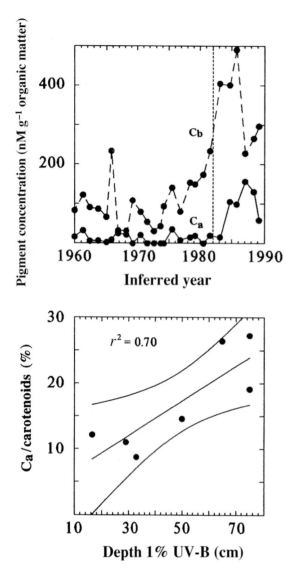

Similar pigment and diatom-based paleo-limnological approaches have been applied to Antarctic lake sediments, where striking changes in past UV penetration have been inferred (Hodgson et al. 2005; Verleyen et al. 2005).

Although these approaches are still very new and have thus far only been applied in a few lakes, they have many potential management applications and implications. For example, they provide insights and benchmarks concerning the natural, background range of variability or pre-anthropogenic changes in UV-B penetration. From the alpine studies, as well as the Arctic treeline studies cited in the previous section, UV-B penetration into lakes has fluctuated dramatically in the past, even before anthropogenic changes to the ozone layer. Similarly, paleolimnological analyses can be used to reconstruct pre- and post-acidification underwater spectral balances, and since many other types of biota can be reconstructed from these same sedimentary profiles, the effects on aquatic food webs can be assessed. By using different statistical approaches, such as variance partitioning, it may potentially be possible to separate the unique effects of some of these multiple stressors, as well as the variance that they have in common.

15.7 Multiple stressors: new combinations of old problems

Fig. 15.6 Concentrations of UV-absorbing carotenoid C_a and C_b in the sediments of Lake 302S, northwestern Ontario. This lake was artificially acidified in 1980, resulting in a decrease in DOC and a corresponding increase in UV-B penetration. From Leavitt et al. (1999); used with permission

by 1987. This acidification resulted in an equally striking decline in lakewater DOC from 7 to 4.5 mg l^{-1} and an estimated eight-fold increase in UV-B penetration (Box 15.2, middle panel). The pigment data faithfully tracked these changes.

The above studies linking climate change, acid rain, and UV-B penetration provide one class of examples showing the ominous effects of multiple stressors. However, many other examples could be cited, but they are very difficult to track and disentangle. In some respects, recent climatic warming may be an overriding factor that will influence and often exacerbate many problems described in this book. For example, with warming, water levels are often decreasing. At its simplest level, this often results in less dilution, and hence a concentrating effect on many pollutants. Many wetlands may also disappear, and with them the loss of biodiversity and the filtering capacities of these natural water purifiers. Lakes will stratify more strongly, and so summer anoxia may become more severe. The biogeochemical processes of lakes will intensify (Schindler 2001), and so on.

These are not simple problems, nor will they have simple solutions. In many respects, we have moved from *relatively* straightforward environmental problems (e.g., eutrophication) to more complex, trans-boundary issues (e.g., acidification, persistent organic pollutants), to more global problems, such as the ones discussed in this chapter. Our vigilance and the development of new techniques must keep pace with these new issues. It is not an easy challenge. Fortunately, paleolimnological approaches are well suited to deal with many of these stressors. At the same time, by reconstructing past biotic changes, paleolimnologists can provide insights on ecosystem responses to these multiple stressors.

15.8 Summary

Ozone depletion, acid rain, and climatic change are environmental problems that have often been discussed in isolation. However, it is now becoming clear that the cumulative effects of these multiple stressors can result in more serious ecosystem repercussions. For example, with the manufacture and release of chlorine and bromine gaseous compounds (e.g., CFCs), the protective stratospheric ozone layer (which filters out harmful ultraviolet-B rays) began to be eroded. Meanwhile, as a result of acidic precipitation, lakewater dissolved organic carbon (DOC) concentrations, which act as natural filters for ultraviolet penetration into lakes, have declined as a result of the drop in pH as well as complexation with aluminum. Moreover, in some regions climate warming is accompanied by reduced stream and overland flows, and therefore less delivery of DOC to lakes. The combined effects of these three stressors can greatly increase UV penetration into lakes. Paleolimnological methods are being developed to track past DOC concentrations in lakes (primarily using diatom transfer functions), whilst fossil pigments can be used to reconstruct some of the biological responses that may be related to changing UV penetration. As new problems are constantly being identified, dealing with multiple stressors will likely pose the greatest challenge facing environmental scientists in the coming years.

16

New problems, new challenges

Why repeat mistakes when there are so many new ones to make?
René Descartes (1596–1650)

16.1 A steadily growing list of environmental problems

This chapter briefly summarizes some of the emerging environmental areas where paleolimnology can be applied. My main goals are to illustrate the versatility of the techniques, and to show how historical perspectives can help managers and scientists in many new settings and applications. The list of topics is by no means complete, as every day we are faced with new environmental challenges.

16.2 Historical perspectives to environmental, habitat, and species conservation

As human populations increase and their activities continue to degrade ecosystems, there are some concerted efforts to conserve and/or restore sections of land and water as parks or reserves. One goal of some of these conservation plans is to set aside areas that are as close as possible to "natural habitats." However, this begs the questions: Which areas represent natural or "pre-

disturbance" conditions? How does one decide (using scientific criteria) which areas should be preserved? Paleoenvironmental techniques can assist conservation efforts by determining which regions represent the most "natural habitats" available (i.e., most closely representing pre-disturbance conditions) and therefore which areas should be prioritized for preservation. Furthermore, paleoecologists can document the historic importance of regions to the life histories of endangered populations, as well as evaluating the cause(s) of habitat deterioration and population dynamics. These approaches can track changes in terrestrial ecosystems by using, for example, pollen grains and macrofossils to characterize vegetation changes, or charcoal and dendroecological techniques to track fire regimes, as well as any changes in aquatic ecosystems using the indicators described in this book. It is hard to imagine a more economical or scientifically sound way to assist with these important decisions.

Conservation efforts sometimes focus on individual species or strains that are in danger of extirpation or extinction. This paleoecological work can range from assisting conservation managers concerned about an endangered charophyte macroalga (Carvalho et al. 2000) or aquatic

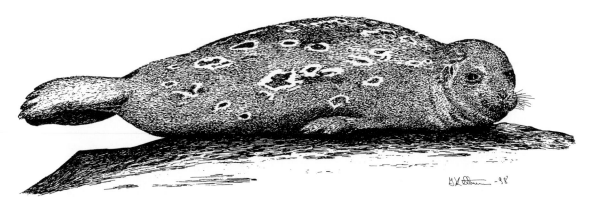

Fig. 16.1 The freshwater seal *Phoca hispida saimensis*, from Lake Pihlajavesi in the Saimaa Lake complex in eastern Finland. Drawing by Iippo Kettunen.

macrophytes (Rasmussen & Anderson 2005), to those working with large mammals. For example, Lake Pihlajavesi in the Saimaa Lake complex in eastern Finland supports one of the world's most endangered endemic seal populations, *Phoca hispida saimensis* (Fig. 16.1). As few as 200 individuals of this subspecies are known to be alive today, with about 50 of them currently living in this lake. Obviously, this region is of critical importance for the survival of this subspecies, and yet the lake has been subjected to increased anthropogenic stressors such as eutrophication since around the 1960s, although nutrient loading has been reduced over the past 20 years or so. Important questions related to the conservation of this population are: What was the "natural habitat" for this freshwater seal before human influences? Are limnological conditions changing to the point that this critical population may be further endangered? Sandman et al. (2000) addressed these questions using a multidisciplinary study of the lake's recent (~ 200-year) history. They showed that, although the effects of eutrophication in the 1960s and 1970s were clearly discernable in the paleolimnological record, limnological conditions have not deteriorated greatly and the lake is close to its "natural state," indicating the seal population is not in imminent danger from water-quality degradation.

Seals are primarily marine mammals, and subspecies that have colonized lakes are rare. In some cases, past fluctuations in marine mammals can be tracked using paleolimnological evidence preserved in coastal lakes where they come ashore. One such example is the Antarctic fur seal, *Arctocephalus gazella*, whose recent population dynamics have concerned conservation managers. Fur seals were almost extirpated from overhunting in the 19th and early 20th centuries, until sealing eventually stopped in 1912. Meanwhile, an intensive whaling industry was established at South Georgia (1902–65), which resulted in the loss of > 90% of the region's baleen whales. This resulted in an increase in the seals' principal food source – krill (euphausids) – as whales were the seals' principal competitors for food. The southern ocean ecosystem is now out of balance. With this extra food availability, it seems that the seal populations have recently been flourishing and expanding in areas such as South Georgia and Signy Island in the South Orkney Islands, and at numerous coastal sites off the west coast of the Antarctic Peninsula (Fig. 16.2).

Although recovery of the fur seal population was an unanticipated by-product of the whaling industry, the recent increases have been so large that they have had significant impacts on the terrestrial and freshwater ecosystems of islands where the seals breed. For example, seal excrement has caused lake eutrophication and the large amount of fur deposition has resulted in some elemental toxicity.

Escalating seal populations have raised important conservation and management concerns

Fig. 16.2 A colony of Antarctic fur seals (*Arctocephalus gazella*) that have come onto the shores of a sub-Antarctic island. Photograph taken by Dominic Hodgson.

(Hodgson et al. 1998a). For example, does the recent increase in seal populations exceed the normal range of variability that has occurred during the past few millennia? Can fragile terrestrial and freshwater Antarctic ecosystems recover from the environmental impacts of such populations? If current populations are, in fact, larger than they have ever been in the past, can the timing of this increase be explained by human influences (i.e., the demise of the sealing industry, coupled with the increase in whaling). How might any measures deemed necessary to limit the impact of the seals be best implemented within the auspices of the Antarctic Treaty?

Hodgson and Johnston (1997) and Hodgson et al. (1998a) addressed the above questions using paleolimnological approaches on seal-impacted lakes from Signy Island. As noted earlier, it is often difficult to directly track vertebrate populations, as they rarely leave reliable proxy indicators. Fur seals are an exception, as the presence, absence, and possibly broad-scale changes in past populations can be estimated by counting the number of fur seal hairs preserved in lake sediments.

By studying dated sediment cores from Sombre Lake, Hodgson et al. showed that fur seals have been visiting the island since the lake's inception (6570 ± 60 ^{14}C yr BP), when the coastline became ice-free. However, population numbers were much lower than current census data, with seal hair deposition never exceeding more than about 25% of current deposition levels. Parallel paleolimnological studies using diatoms tracked the recent eutrophication of the lake as a result of the nutrient fertilization from the seals (Jones & Juggins 1995). The Signy Island ecosystem therefore evolved with the impacts of seals as part of its natural development, although populations were at much lower levels. Any recovery of impacted communities to their former "natural" states of equilibrium and complexity therefore may not be possible with present seal populations.

Hodgson et al. (1998a) also concluded that historical sealing and whaling activities have affected seal populations. The timing of the changes in seal hair concentrations over the past two centuries (Fig. 16.3) matched known sealing and whaling records in the region. Moreover, the

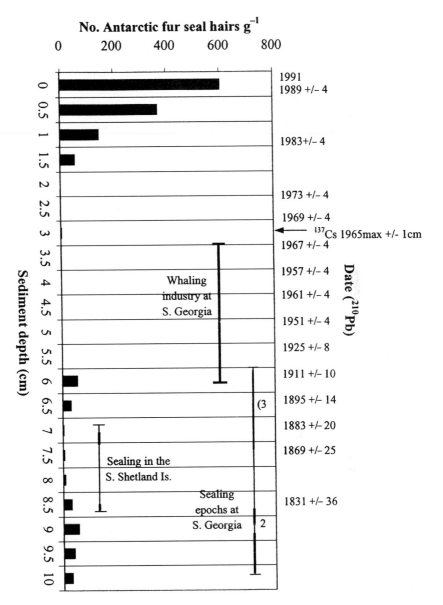

Fig. 16.3 Concentrations of Antarctic fur seal hairs in the high-resolution study of a radiometrically dated sediment core from Sombre Lake, Signy Island. ^{210}Pb and ^{137}Cs dates are shown to the right. Seal hunting ended in 1964. These data were compared to fur seal hair concentrations in a ^{14}C-dated long core collected from the same lake, showing that hair concentrations in the recent sediments were at least four times greater than any time since the lake's inception, over 6000 years ago. From Hodgson et al. (1998a); used with permission.

concentration of seal hairs in the sediments did not appear to be strongly linked to changes in paleo-climatic indicators (Jones et al. 2000), further suggesting that human activities influenced the seal populations. They concluded that some form of management may be required to preserve representative terrestrial and lacustrine ecosystems from further damage from these seal populations which, according to the paleolimnological data, have now expanded to sizes far larger than at any time in the past.

Another application of paleolimnology to conservation science is to assist managers in defining restoration goals for species conservation or reintroductions. About 110 km north of the major mining and smelting installation in Sudbury (Ontario) are several lakes that once supported an endangered strain of brook trout, the aurora trout (*Salvelinus fontinalis*). Aurora trout had declined dramatically and had disappeared entirely from its home range by the late 1960s, presumably due to lake acidification. As a last attempt at saving this taxon from permanent extinction, the Ontario Ministry of Natural Resources preserved this genetic line by maintaining populations in a hatchery, with the goal of eventually reintroducing the aurora trout in "rehabilitated" lakes. With no existing monitoring data, though, how can one determine what the strain's "natural" habitat had been? By the time the problem of imminent extinction was recognized, their home range had already been greatly altered by acidic deposition from the Sudbury smelters. Dixit et al. (1996) used diatom and chrysophyte indicators in ^{210}Pb-dated sediment cores from the aurora trout lakes to determine: (i) the timing, rates, and magnitude of acidification and metal contamination that had occurred in these lakes; (ii) what limnological thresholds were passed that caused the aurora trout populations to decline; (iii) if the pre-impact conditions that existed in these lakes were unique to the Sudbury region, or whether other similar lakes would be reasonable candidates for restocking programs; and (iv) whether these pre-impact conditions could be restored in the original aurora trout lakes. They found that acidification was the root cause of the declines, but that pre-impact conditions were similar to other Sudbury lakes (i.e., the aurora trout lakes did not appear to be limnologically distinct, and so other lakes could likely support this strain, if necessary). Reclamation attempts via liming and reduced emissions have largely been successful – hatchery aurora trout were recently reintroduced and are now naturally reproducing and appear to be doing well – an unfortunately rare happy ending to a very close call at extinction.

16.3 Are collapsing fish stocks due to habitat destruction, over-fishing, natural changes, or all of the above?

The problem of collapsing fish stocks in many regions has been firmly placed on the political, economic, and environmental plate (no pun intended!). Many potential sources of blame exist for declining stocks, ranging from over-fishing to pollution to habitat destruction to possible natural causes such as climate change (although the latter factor now also has anthropogenic overtones, with recent greenhouse gas-induced warming). Once again, the lack of monitoring data of sufficient length and quality hampers the likelihood of clearly and unambiguously identifying the causes of these problems.

As noted earlier in this book, fish are often the "missing factors" in many paleolimnological investigations, as they rarely leave fossils that are present in sufficient quantities. Consequently, as direct inferences of past populations are presently impossible, indirect methods must typically be used (e.g., using invertebrate indicators). However, Finney et al. (2000) demonstrated that by combining paleo-isotope analyses with diatom and invertebrate indicators, long-term trends in past anadromous fish populations, such as Pacific sockeye salmon, can be tracked in Alaskan nursery lakes.

Sockeye salmon (*Oncorhynchus nerka*) is commercially the most important salmon species on the west coast of North America, with significant populations in the western Pacific as well. Sockeye salmon are anadromous, meaning that they reproduce and spend part of their life cycle in lakes, but then spend most of their adult lives in the ocean. For example, the salmon hatch in their nursery lakes where they spend the first 1–3 years of their lives, feeding on invertebrates. They then leave their nursery lake and stream system, and make their way to the ocean, where they then typically spend the next 2–3 years feeding in the North Pacific. It is in this marine environment that they accumulate the majority of their biomass. Sockeye salmon then migrate back, with uncanny accuracy, to their nursery lake–stream system, where they spawn and die.

Given the precarious nature of certain fish stocks, it is critical to know what natural and human-related factors affect fish abundances. Long-term data are needed and, in certain regions, some estimates of sockeye salmon abundance (based on catch records) go back to the late

19th century, which would be considered a long-term record for most of the problems discussed in this book. However, even this record overlaps the period of other human perturbations, such as fishing and habitat destruction. It is important to know how fish populations have fluctuated in the past, prior to human interventions, and then use this information to try and determine the effects of human-related activities.

Although salmon fish fossils are rare, other proxies can be used to track past sockeye salmon abundances. Sockeye salmon accumulate about 99% of their biomass in the North Pacific Ocean, where they feed on plankton, squid, and small fish. Because salmon occupy this high trophic position in the marine food web, they accumulate a high proportion of the heavy nitrogen isotope, ^{15}N, relative to the more common, lighter isotope ^{14}N. The salmon carry this heavy nitrogen signal in their bodies when they return to their natal streams and lakes to spawn (Fig. 16.4). After the arduous upstream journey back to their nursery lakes, the adult sockeye salmon spawn and then die soon afterwards (Fig. 16.5), releasing this heavy nitrogen signal into their nursery lake. This isotope signal is preserved in lake sediments, and so indirectly provides a proxy of past salmon

abundances. The approach, however, requires a careful selection of candidate nursery lakes, as factors such as flushing rates and the relative amounts of allochthonous inputs must all be considered, especially when dealing with relatively low salmon escapement densities (Holtham et al. 2004).

Finney et al. (2000) used nitrogen isotope data to track salmon fluctuations in five Alaskan nursery lakes (as well as two nearby control lakes, where salmon were prevented from entering due to the presence of waterfalls) over the past approximately 300 years (Fig. 16.6). They found that salmon stocks generally matched climate periods of warming and cooling (reconstructed from tree-ring data) in the North Pacific. Salmon populations fluctuated markedly, even before European-style fishers and settlers had arrived in Alaska.

Climate was one part of the story, but commercial fishing also had an effect. Finney et al. (2000) used a multi-proxy approach that included indicators such as diatoms (which were used to infer past nursery lake nutrient levels) and cladoceran concentrations, which tracked past food availability for the young sockeye salmon. Certainly, removing fish before they can spawn

Fig. 16.4 Adult male sockeye salmon struggling up Hansen Creek, Lake Aleknagik, Alaska, to spawn. Photograph by T. Quinn, University of Washington.

Fig. 16.5 Sockeye salmon carcasses along the shore of O'Malley Creek, which drains into Karluk Lake, Kodiak, Alaska. After spawning, the adults die and their carcass-derived nutrients can significantly fertilize nursery lakes. In some lakes, millions of salmon return each year to spawn and die. Photograph by I. Gregory-Eaves, McGill University.

will reduce the carcass-derived nutrient load, and the sedimentary nitrogen signal closely tracked known declines in salmon coincident with the period of commercial fishing. Their paleolimnological data, however, also identified some potentially detrimental limnological repercussions of over-fishing that could also affect young fish stocks. The prolonged 20th-century collapse of the fishery in their primary study lake (Karluk) was also partly driven by reductions

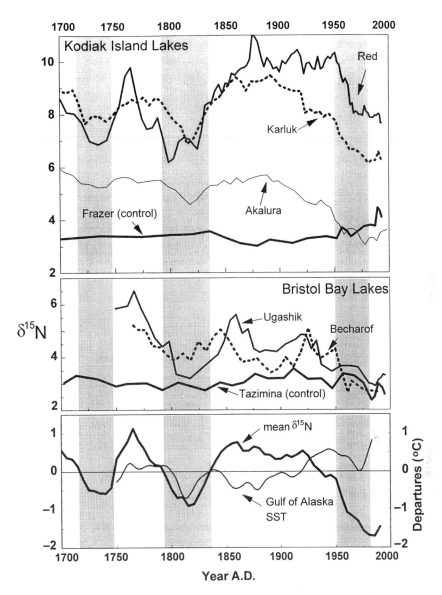

Fig. 16.6 A comparison of sedimentary $\delta^{15}N$ profiles from sockeye salmon nursery and control lakes from the Kodiak Island and Bristol Bay regions of Alaska, over the past approximately 300 years. The sediment cores were dated using ^{210}Pb and tephra layers. The $\delta^{15}N$ profiles track past abundances of sockeye salmon carcasses in the nursery lakes. The control lakes (Frazer and Taziminia) have waterfalls on their outflow streams, thus precluding sockeye salmon returns. However, Frazer Lake had a fish ladder installed in the 1950s, and so the recent increase in $\delta^{15}N$ faithfully tracks the time when sockeye salmon began entering this lake as well. Common patterns in the salmon nursery lakes, as shown by the mean $\delta^{15}N$ in the bottom panel, include lower values in the early 1700s and the early 1800s. These declines corresponded with lower sea-surface temperatures (SST, lower panel), as inferred from tree-ring analyses (D'Arrigo et al. 1999). The more recent declines overlap the periods of commercial harvests in this region. Reprinted from Finney et al. (2000) Impacts of climatic change and fishing on Pacific salmon abundance over the past 300 years. *Science* 290, 795–9. Copyright 2000 American Association for the Advancement of Science.

in lakewater nutrients (inferred from diatom assemblage changes) from over-harvest. A critical source of nutrients to these oligotrophic nursery lakes is, in fact, the dead adult salmon carcasses. Given that salmon can return in very large numbers to some nursery lakes, they can represent an important nutrient subsidy. For example, in Karluk Lake, millions of salmon return in some years, and the decaying carcasses supply over 50% of the lake's nutrient budget.

Higher nutrient levels would result in higher algal abundances, and hence higher zooplankton concentrations (as reflected by the fossil cladoceran data). This, in turn, may provide more food for the juvenile sockeye salmon. A positive feedback system is in place – if more adults return to spawn, there are more eggs and so more young may hatch, and since there are also more adult carcasses, more marine-derived nutrients are recycled to the nursery lake. As a result, more algae and zooplankton can thrive, and so more food is available for the young salmon. By harvesting adult fish, the feedback system may become decoupled, and the fertilization effects of the dead salmon are lost. In short, the nutrients contained in these salmon now go onto our dinner plates, and not back into their nursery lakes. In some systems, this appears to be a critical loss for salmon populations, whereas in others, the trophic dynamics of the ocean are more important (Finney et al. 2000; Schindler et al. 2005).

Reconstructions of salmon populations can also help managers and policy makers to set more realistic catch limits. For example, Finney et al. (2002) used similar paleolimnological techniques to those employed in the original Finney et al. (2000) study, but extended the Karluk Lake sediment records back over 2000 years. They showed that sockeye salmon populations were at amongst their naturally highest levels in over two millennia when the original European fishers first arrived in the Karluk Lake region in the 19th century. Without the benefit of a long "time window" of monitoring made possible via paleolimnological studies, a "snapshot" assessment of very high fish returns would have encouraged overly optimistic estimates of fish stocks (in fact, 13 canneries were built on the Karluk River in the 1880s). Similarly, Schindler et al. (2006) used nitrogen isotope data to help set more realistic harvesting targets for sockeye salmon returning to the Alagnak River in the commercially important Bristol Bay region of Alaska. Salmon runs averaged about 1 million fish from 1995 to 2002, but then returns surged unexpectedly by about five-fold from 2003 to 2005. Schindler et al. (2006) showed that short periods of high salmon abundances did occur about once every century over the past 500 years or so, but that these were transient phenomena that should not be used for setting ecologically realistic catch quotas.

Salmon management must adopt an ecosystem approach, where all aspects of the fish's life cycle must be considered, from nursery lakes, to their natal streams and rivers, to the ocean. The Finney et al. (2000) study was a first attempt to use these paleolimnological approaches, but many other applications are certainly possible at different locations and time scales (e.g., Finney et al. 2002; Gregory-Eaves et al. 2003, 2004; Holtham et al. 2004; Schindler et al. 2005; Selbie et al. 2007), as nursery lakes occur along much of the western coast of the Pacific Northwest of the United States, Canada, and Asia. Each region has different stocks of sockeye salmon, each with different types of stressors (e.g., fishing pressure, dams, habitat alteration, competition from hatchery fish, disease, different climate regimes, etc.). For example, Selbie et al. (2007) used isotope and microfossil evidence to reconstruct variability in the endangered Snake River sockeye salmon returning to Redfish Lake (Idaho). Unprecedented declines in salmon production were inferred over the past approximately 150 years, which could be linked to a spectrum of human interventions, including commercial harvests, sustained interruption of migration access by hydroelectric damming, and non-native fish introductions. Many important research opportunities remain to exploit these paleolimnological records, and thereby also help managers and policy makers to make more scientifically sound decisions to help preserve viable stocks.

As salmon have relatively high fat content, they are also vectors for carrying a variety of bioaccumulated and bioconcentrated pollutants, such as POPs (Chapter 9) and mercury (Chapter 10), from marine systems back into their nursery lakes, where they die. As described earlier (see pp. 162–5), paleolimnological methods can be used to assess the relative importance of these and other biovectors (e.g., seabirds) in contaminant transport (Krümmel et al. 2003, 2005; Blais et al. 2005, 2007; Gregory-Eaves et al. 2007).

16.4 Contrasting the limnological effects of different forestry practices

The limnological repercussions of different forestry practices are generally poorly understood due, at least in part, to the short temporal records available for most studies. For example, a 1997 review on the impacts of forestry practices on lake ecosystems suggests that less than 5% of studies assessed water-quality changes with respect to at least 2 years of pre-disturbance data, and the total duration of most assessments was less than 5 years (Miller et al. 1997). From previous chapters, it is evident that if native forests are cut and not allowed to regrow (e.g., replaced with agriculture), an almost certain repercussion is increased erosion (Chapter 12) and eutrophication of catchment lakes (Chapter 11). However, there is a large gradient of disturbances that may be associated with different forestry practices, such as selective cutting to clear-cutting, to allowing or even encouraging forests to repopulate the landscape. The few paleolimnological studies that specifically addressed forestry effects have similarly resulted in documenting a broad spectrum of responses, from almost no limnological effects to major changes in receiving waters.

Some paleolimnology studies from Finland have recorded marked limnological repercussions to some of their forestry practices, but these included the addition of fertilizers and other watershed changes, in addition to lumbering. For example, forest clear-cutting is often accompanied by plowing the mineral soil to improve growing conditions for new trees. Such postlogging disturbances increase the erosion of materials and loading of nutrients to headwater system. Peatlands have also been drained (up to 80–90% in southern Finland) for forestry purposes, and subsequently fertilized with potassium and phosphorus. Turkia et al. (1998) estimated that more than 5 million ha of peatland (about 50% of Finland's peatlands) have been ditched over the previous 40 years to increase timber production.

As summarized by Simola et al. (1995), it is not surprising that some of these Finnish forestry practices can significantly affect lakes. For example, in a pioneering study addressing peatland drainage and subsequent fertilization from the forestry industry, Simola (1983) tracked diatom and sediment chemistry changes from the annually laminated sediments of Lake Polvijärvi. He showed that the lake experienced marked eutrophication as a result of the nearly complete ditching of virgin peatlands in the drainage basin and their subsequent fertilization. Simola et al. (1995) later found that small headwater lakes were especially affected by these disturbances, with successive small lakes in the system acting as filters, retaining much of the nutrient and particulate input from these intensive forestry activities. Headwater lakes may also originally be more oligotrophic, and so any additional nutrient inputs may be more obvious in these systems. Meanwhile, Turkia et al. (1998) assessed the effects of forestry practices on lakewater quality in six small Finnish lakes using physical, chemical, and diatom-based paleolimnological techniques. Although changes could be related to forestry practices, such as forest ditching and fertilization, in general the changes were surprisingly subtle in all but one lake. The latter study site, Lake Saarijärvi, revealed evidence of marked eutrophication, which was related to anthropogenic activities in the surrounding peatland that was ditched and fertilized several times.

The decreasing sedimentary carbon to nitrogen (C/N) ratio also suggested an increase in algal production. Several of the diatom changes also indicated higher humic concentrations as a result of forestry practices.

Results from North America, where ditching and fertilizing are not common forestry practices, have revealed only subtle and at times ambiguous changes related to past forest cutting, provided that forests were allowed to regenerate. Working in Ontario forest ecosystems, neither Blais et al. (1998a) nor Paterson et al. (1998, 2000) could link any sedimentary changes in physical/chemical variables, nor shifts in past chrysophyte or diatom assemblages, to the removal of trees. Similarly, Laird et al. (2001) and Laird and Cumming (2001) recorded only slight changes in diatom assemblages and sedimentary organic matter content coincident with forest harvesting on Vancouver Island and the central interior of British Columbia, respectively. Meanwhile, Scully et al. (2000) used sediment varves and fossil pigments to track the limnological responses to logging on Long Lake (Michigan) in the late 19th century. They suggested that forest removal resulted in a fundamental change in the physical structure of the lake, as fossil records were consistent with increased wind stress following the removal of trees that led to deeper water column mixing and a reduction in deepwater anoxia.

Although paleolimnological studies specifically addressing the impacts of logging and other forest disturbances are still few in number, it is becoming clear that the degree of disturbance, the procedures followed by logging companies during and after forest removal, the physical and chemical conditions that existed in the lake at the time of disturbance, and the ratio of catchment area to lake size are all important factors that must be considered (Enache & Prairie 2000; Paterson et al. 2002).

The limnological effects of the forestry industry, however, rarely end with the loggers, as often much more serious environmental problems may be the chemicals released as effluents from the pulp and paper mills. Environmental regulations are becoming increasingly strict for these effluents, as research accumulates on their potential toxicological properties. However, in some cases it is not clear which pollutants had been released, nor their concentrations. Similar to the other forms of contaminants described in this book (e.g., Chapter 9), many of these pollutants leave "fingerprints" in lake sediments that can be used to assess and identify a variety of problems and their sources. For example, Leppänen and Oikari (2001) reported on retenes (compounds severely toxic to larval fish) and resin acid concentrations in sediments of a lake where a pulp mill had been closed in 1985. They showed extremely high concentrations in sediments dating from especially about 30 years ago. Concentrations in sediments appeared to faithfully track past activities from the mill (e.g., including what appeared to be a decrease in sedimentary concentrations reflecting a 3-year shutdown of the mill in 1967–70).

Similarly, Meriläinen et al. (2001) used the chemical stratigraphy of resin acids (Fig. 16.7), as well as other geochemical markers, diatoms, and chironomids, to track the limnological history of two basins and a sheltered bay of Lake Päijänne in southern Finland. The lake was subjected to pulp mill effluent since the 1880s. Over the past approximately 30 years, effluents have declined with water protection measures. The concentration of resin acids in the two basins of the lake clearly showed how different pollution profiles and sediment contamination histories can occur in different regions of large lakes (Fig. 16.7). The Lehtiselkä 73 core is near the pulp mill discharge and tracks the effluent history, whilst the Asikkalanselkä basin is about 60 km downstream and shows only negligible changes. The diatom and chironomid assemblages tracked the aquatic ecosystem responses to these disturbances. The core near the effluent outflow showed a dramatic change in benthic chironomid communities, but only a small change on the primarily planktonic diatom flora, coincident with the increases in discharge. With recent declines in discharge over

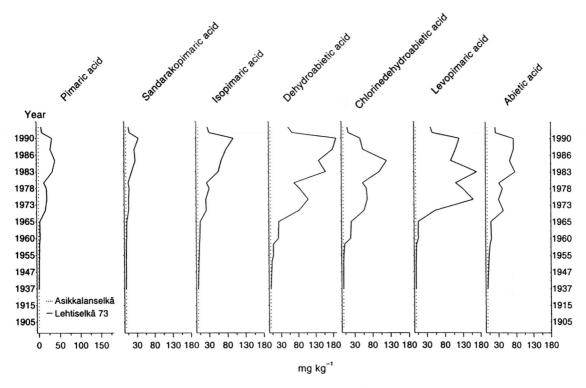

Fig. 16.7 Concentrations of resin acids from pulp and paper mill effluent in Lake Päijänne in southern Finland. The Lehtiselkä 73 core (solid line) is near the pulp mill discharge, and tracks the effluent history, whilst the Asikkalanselkä basin is about 60 km downstream, and shows only negligible changes (dotted line). Recent water-quality improvements are also tracked. From Meriläinen et al. (2001); used with permission.

the past three decades, the benthic chironomid community showed a marked recovery. Interestingly, primary production in the upper waters also appeared to have concurrently increased, even though nutrient loading had decreased. The authors hypothesized that once pulp and paper mill discharges declined, so also was reduced the brown color typical of the released lignins and other colored compounds present in the effluent. Water transparency therefore increased and more photosynthesis was possible. Recent declines in resin acid concentrations, which may have previously inhibited primary production, may also have been a factor. Meanwhile, biological and chemical changes were slight in the relatively unaffected Asikkalanselkä basin.

16.5 Creating new exposure-risk models for human health: a case study reconstructing a century of airborne asbestos pollution

Asbestos is a generic name that refers to a variety of naturally occurring, fibrous silicate minerals that were mined extensively in the 20th century for use as thermal insulators and in other industrial applications. However, since about the middle of the 20th century, it had become increasingly evident that exposure to these fibers could result in very serious health effects, such as the scarring of lung tissues (i.e., asbestosis), lung cancer, and other pulmonary diseases. As airborne asbestos fibers are insoluble in water and accumulate in lake sediments, they can be used to reconstruct past exposure scenarios for different geographical regions, and so provide the long-term environmental data required to develop more realistic exposure-risk models for human health. Webber et al. (2004) developed this approach by reconstructing a century of airborne asbestos pollution from a significant source of fibers: the talc mining industry in northern New York State, appropriately named Talcville. Mining began in the late 19th century, and by the middle of the 20th century over 100,000 tons of talc were being produced annually. However, the human health effects of the talc mining industry had been controversial. Dust collectors were installed by the talc mills in the 1930s and 1940s. Prior to the development of advanced transmission electron microscopic (TEM) techniques in the 1980s, airborne asbestos concentrations were measured using phase-contrast microscopy (PCM). PCM, however, does not have sufficient optical resolution to identify the smaller asbestos particles (e.g., < 2.5 μm diameter), which are the most dangerous from a human health perspective, as they are the ones that reach the lungs' alveoli. Therefore, the available long-term monitoring data were insufficient for developing accurate exposure-risk models.

Webber et al. (2004) documented the concentrations of asbestos fibers in the sediments of two lakes: Clear Lake, which is approximately 8 km downwind of the talc mines and processing facilities; and Sixberry Lake, which is approximately 25 km upwind of the facilities, and so was used as a control site to register ambient asbestos concentrations. As expected, fiber concentrations of the most common asbestos type, chrysotile, increased dramatically in the control lake by the mid-20th century, coincident with the widespread industrial use of asbestos. A steep decline in the last quarter of the century tracked the legislatively mandated reductions in asbestos use. Meanwhile, reconstructed airborne concentrations of anthophyllite fibers, an uncommon asbestos type that was a significant component of the local talc

ore, increased over five-fold from the middle of the 19th century to c. 1967. The paleolimnological records faithfully matched the annual talc production.

The Webber et al. (2004) study was important from a number of perspectives. First, it showed that asbestos fibers can be studied using standard paleolimnological methods. However, it also demonstrated how newer and more sensitive detection methods (in this case TEM) could be used retrospectively, decades after the periods of asbestos release, to correct and provide more accurate data for developing risk-exposure models for human health.

16.6 An anti-fouling paint, recreational boating, and a whole-scale ecosystem shift in shallow English lakes

Tributylin (TBT) is a highly toxic anti-fouling agent that was developed primarily as a paint additive for the hulls of large ocean-going vessels. TBT was very effective at reducing the growth of barnacles, and thereby improving on the speed and efficiency of marine transport. However, following an aggressive advertising campaign in the 1960s, TBT began to be used on smaller boats as well, including recreational vessels used on lakes and rivers. By the 1980s, it was becoming increasingly clear that the highly toxic properties of TBT were having negative impacts on non-target organisms, and international measures began to be implemented to limit its use. In 1987, TBT usage was banned in the United Kingdom.

One inland area where TBT was used extensively was the Norfolk Broads (UK), a network of mostly navigable rivers and shallow lakes (known locally as broads) in eastern England that are popular tourist destinations for recreational boaters. Between 1960 and 1970, the broads experienced a major shift in alternative stable states (Box 11.2), from clearwater macrophyte-dominated lakes to turbid, phytoplankton-dominated systems. The cause for this ecological regime shift was highly debated (e.g., Moss 2001). Although cultural eutrophication would be a likely candidate driving such a change, Sayer et al. (2006) proposed that TBT contamination may have been a key trigger.

Sayer et al. (2006) explored the possible role that TBT played in the loss of submerged macrophytes in the broads. A wide spectrum of paleolimnological indicators were analyzed from radiometrically dated sediment cores collected from two lakes: freshwater Wroxham Broad and brackish Hickling Broad. Both shallow lakes have at least one boatyard along their shores. Submerged macrophytes are currently absent from Wroxham Broad. Meanwhile, charophyte vegetation, which was extirpated in the 1970s from Hickling Broad, has made a modest recovery since the late 1980s. Sayer et al. (2006) documented changes in TBT concentrations in the two sediment cores, as well as shifts in diatoms, cladocerans, fish scales, and mollusks. In addition, charophyte oospores were identified and enumerated from the Hickling Broad sediments.

Sayer et al. (2006) acknowledged that cause and effect are difficult to unravel in a study such as this, and that a variety of factors may have contributed to the demise of macrophytes. However, the timing and nature of the paleolimnological changes implicated TBT as a possible factor in the shift from a clearwater to a turbid state in the broads. Sayer et al. (2006) suggested that TBT contamination resulted in a chronic reduction in grazers, such as cladocerans and mollusks, thus contributing to the shift in alternative stable states. As TBT contamination has occurred in many waterways around the world, the full-scale ecological repercussions of this biocide have yet to be fully assessed.

16.7 Military dump sites: forensic investigations using sediments

Polluters rarely wish to publicize their activities. This is not surprising, as many industrial and government agencies may wish to avoid responsibility for negative repercussions, as well as any associated clean-up costs. As a result, the sources of many contaminants are often hotly contested. Direct documentation of dumping is often not available, and frequently such activities have occurred before any monitoring activities were under way. Paleoenvironmental techniques can be used to clarify the origin, pathway, and fate of various pollutants, such as chemical dump sites.

Some of the most insidious pollutants originate from military operations. An example (although not from a freshwater environment, but from a marine setting) illustrates how sediments can be used, in a forensic fashion, to track down past chemical weapons dump sites. Similar approaches could be used in freshwater environments.

As summarized by Siegel et al. (2000), chemical weapons such as aerial bombs, mines, and artillery and rocket shells were dumped into the Arctic Kara Sea by the former Soviet Union. The locations of such dump sites are considered secret. The chemical weapons believed to have been dumped include Tabun (a nerve agent), as well as mustard gas and Lewisite (chlorovinyl dichloroarsine, $C_2H_2AsCl_3$), two agents that severely burn or blister the skin and may cause permanent damage to the lungs when inhaled. Lewisite, which contains about 36% arsenic (As), can persist in various forms in sea water for months and so can be dispersed by currents. Siegel et al. reasoned that if Lewisite was dumped in the Kara Sea, then As concentrations in sediment cores may help to "fingerprint" the location and composition of the dump site, and to track its contaminant plume.

Siegel et al. collected 31 sediment cores along an area of the Kara Sea where dumping was suspected to have occurred in the late 1940s and the 1950s. Twelve of the cores had As concentrations higher than 150 ppm in the near-surface samples, as compared to baseline concentrations of about 23 ppm. Four of these cores had near-surface-sediment concentrations of 310, 710, 380, and 330 ppm As; these sites occurred in a sequence along an approximately 150 km stretch of sea bed that followed the south-flowing bottom current on the west side of the St Anna Trough. By mapping the sediment concentrations, and coupling these paleoenvironmental data with knowledge of the currents, Siegel et al. proposed a possible location for the dump site.

Pinpointing the sources of pollutants is a major environmental and legal issue in lake, reservoir, and river management. Similar forensic techniques to those used by Siegel et al. can be adopted by paleolimnologists in freshwater settings.

16.8 Nuclear pollution and radioactive waste management

The use of atomic weapons, beginning with the first ground testing in the New Mexico (USA) desert in 1945, ushered in the nuclear age. This era introduced a suite of new environmental issues and problems, many of which lend themselves to discussions based on sedimentary analyses of lacustrine and wetland sediments. These environmental issues and problems commonly fall into two broad categories: (i) dispersal into the environment; and (ii) long-term storage of radionuclides in a safe and secure setting.

Dispersal of radionuclides into the environment commonly occurs in two ways. The first avenue is more general and widespread, typically through the atmosphere, but oceanic dispersal may also occur. The second avenue is more specific and local in nature, typically originating from a human source such as a nuclear facility or a test range, but also from natural sites, such as uranium mines and groundwater reactions with radioactive minerals (e.g., those in granitic regions).

As has been discussed throughout this book, the release of radionuclides, such as ^{137}Cs or ^{241}Am, has been exploited by paleolimnologists as a dating tool (see Chapter 4). Other fallout radionuclides, such as ^{239}Pu, ^{240}Pu, ^{60}Co, ^{131}I, ^{90}Sr, ^{155}Eu, and ^{207}Bi, can also be studied (Joshi &

McNeely 1988). In addition to providing geochronological control, these data can be used to track the geographical extent and degree of nuclear contamination by measuring the distributions and activities of various isotopes on different spatial and temporal scales. The two major types of applications are tracking past depositional patterns from nuclear weapons testing, and reconstructing fallout from the inadvertent release of radionuclides, most notably the Chernobyl (Ukraine) nuclear accident.

As described in the geochronological part of this book (Chapter 4), deposition of isotopes such as ^{137}Cs, which are released from nuclear weapons testing, reached significant levels in the atmosphere by 1954, and peaked in 1963 (Pennington et al. 1973). Thereafter, following the ratification of treaties banning atmospheric nuclear bomb testing, the deposition of ^{137}Cs dropped precipitously. This resultant sedimentary signature (see Fig. 4.7) of a rise in ^{137}Cs dated to the mid-1950s, peaking at 1963, and then dropping again, has been recorded in a large number of sediment profiles.

In some regions, sediments record another increase in ^{137}Cs and related radionuclides originating from the April 1986 Chernobyl nuclear reactor accident and fire. This accident released about 2×10^{18} Bq of radioactivity into the atmosphere. The fallout from Chernobyl primarily affected parts of the former Soviet Union, Europe, and Turkey. Nonetheless, even in regions such as the United Kingdom, several thousand kilometers from the site of the accident, ^{137}Cs deposition from the Chernobyl reactor fire was comparable to or exceeded that from weapons fallout (Appleby et al. 1990). As some of this radioactivity becomes rapidly incorporated into sedimentary profiles, paleolimnologists can use this information to track the radioactive plume in both time and space. For example, Kansanen et al. (1991) used sediment traps and cores to determine the sedimentation rates of fallout nuclides on a lake in southern Finland, as well as the spatial distribution of radionuclides on the lake bottom. Their findings included the clear tendency for sediment inventories to increase with water depth. These patterns could then be used as baseline data for future investigations on this lake.

Similarly, Johnson-Pyrtle et al. (2000) tracked ^{137}Cs distributions in cores from the Lena River drainage in Siberia, and showed the effect of the Chernobyl accident on these tundra lake and pond sediment profiles. They also demonstrated that, although the highest concentrations of ^{137}Cs were now buried in the sediments, the average ^{137}Cs levels in the surface sediments were still higher than those of nearby estuarine and marine sediments. Hence, these lakes and ponds may represent potential ^{137}Cs contamination sources to the adjacent aquatic environment.

Radionuclides can also be released by other activities, such as ore extraction. For example, uranium mining and milling near Elliot Lake (Ontario, Canada) has significantly increased the release of the naturally occurring radioisotopes of the ^{238}U and ^{232}Th decay chains to the local environment, primarily from the tailings. Profiles of ^{238}U and ^{232}Th decay chain radionuclides in Quirke Lake sediment cores were enriched one to three orders of magnitude in the more recent sediments relative to background levels measured in sediments deposited before the mining and milling operations were under way (McKee et al. 1987). The profiles also recorded modest declines in these isotopes in the most recent sediments, apparently tracking recent improvements in wastewater management. As regulators and the general public become more concerned about the release and eventual fate of radionuclides, sedimentary approaches can be used to assess and advise on these emerging problems.

Another application of paleolimnology to nuclear pollution is to screen potential sites for long-term storage of nuclear waste. An unfortunate by-product of nuclear power and some other nuclear-based activities is the accumulation of radioactive wastes, which require long-term (e.g., potentially on millennial scales) storage in isolated repositories. Storage of radionuclides in safe and secure settings involves some understanding of future environmental change on various time scales, depending on the radionuclide. Short-lived or low concentration radionuclides, such as those used in medical facilities (commonly referred to as low-level nuclear waste) are readily stored in standard engineered facilities, because such waste, although dangerous, is not viewed as being hazardous. Conversely, high-level nuclear waste, commonly those radionuclides derived from nuclear power plants and military activities, which are both highly radioactive and have very long half-lives, require storage facilities that must survive tens to hundreds of thousands of years with only minimal or no loss of the materials to the external environment. In order to assess the suitability of potential long-term repositories for high-level nuclear waste storage, a number of engineering, geological, hydrological, climate, and societal issues must be considered. Analysis of lacustrine sedimentary records that extend back in time for several glacial periods and that are in close proximity to the proposed repository offer detailed insights into the full range of possible climates during one or more long-eccentricity cycles. Understanding the variability of the past then provides a basis to estimate potential variability in the future, at least within the future time frame of modern tectonic and oceanographic settings.

A critical factor that must be assessed before a site can be designated as a long-term waste storage facility is the stability of its hydrology, as water–waste interactions may release radioactive materials to the external environment. As described by Forester and Smith (1992), the reconstruction of a site's hydrological history using paleolimnological techniques provides fundamental information on its probable isolation through some specified future period (i.e., regions that have been hydrologically isolated over the past few millennia would more likely be hydrologically stable and isolated in the future). Any evaluation using models of hydrological responses to possible future climatic changes is more credible if realistic boundary conditions and estimates of natural variability can first be derived from paleoenvironmental reconstructions.

For example, a considerable amount of pale-olimnological and related work has been completed in the Yucca Mountain region of Nevada (Forester et al. 1999). This region was being considered a potential repository for all US high-level nuclear and military waste. Long-term perspectives on climatic and hydrological change provide some "boundaries" or "envelopes" of possible future changes, which can be used to evaluate whether future changes of comparable magnitude might compromise the integrity of the proposed waste burial site.

16.9 Taste and odor problems in water supplies

The most frequent complaint put to many lake and reservoir managers is that the water tastes or smells "bad," with the offending odor or taste often compared to cucumbers or rotting fish. There are several potential sources for these tastes and odors – not least of which may be rotting fish! However, the source can more often be related to extracellular organic compounds produced by phytoplankton blooms, with certain chrysophytes (e.g., *Synura*, *Uroglena*) and some diatoms the most likely culprits in oligotrophic systems, and cyanobacterial taxa in more productive lakes.

Although taste and odor problems have rarely been considered using paleolimnological approaches, there are certainly many potential applications. Like many environmental problems, little or no historical data are available to determine whether offensive blooms are increasing in frequency and, if so, which environmental factors are triggering these problems. Scientists and managers are reporting that complaints from consumers appear to be increasing; however, without long-term records, it is impossible to answer many crucial questions. For example, are such blooms influenced by human activities, such as acidification, eutrophication, and/or accelerated climate change? Or have such algae

bloomed naturally in these lakes long before humans began influencing limnological characteristics? Only by more fully understanding the reason(s) for these blooms can managers begin to treat the problem.

Paleolimnology can provide many of these answers. For example, the two dominant taste- and odor-causing chrysophytes, *Synura petersenii* and *Uroglena americana*, can be identified in the sedimentary records by their siliceous scales and cysts, respectively (Figs. 5.5 & 5.6). Past cyanobacterial populations can be tracked using fossil pigments. Although we cannot directly reconstruct (not yet at least) the organic compounds causing the problems, we can reconstruct the populations that lead to these problems.

Complaints regarding taste and odor episodes in oligotrophic Ontario (Canada) lakes are commonly related to *Synura petersenii* (Nicholls & Gerrath 1985). The frequency of complaints appears to be increasing, but records range from spotty to non-existent. Paterson et al. (2004) addressed this problem on a regional scale by examining scaled chrysophytes in sediment cores from a broad spectrum of Canadian Shield lakes. By comparing chrysophyte scale assemblages in recent (present-day) and pre-industrial sediments (i.e., a top/bottom paleolimnological approach; see Box 4.1), they showed that *S. petersenii* had increased in relative abundance in almost every lake over the past approximately 150 years (Fig. 16.8). More detailed paleolimnological analyses of two lakes indicated that these increases began about 60–80 years ago. This trend suggests that the threat of taste and odor events has in fact increased. However, statistical analyses found no correlation between the percentage abundance of this chrysophyte and acidification or eutrophication-related variables at a regional scale. The authors speculated that recent increases in ultraviolet-B (UV-B) penetration may have been a factor, as these motile, colonial chrysophytes can retreat to form deep-water populations, protected from the damaging effects of UV-B radiation. Such studies need to be completed in other regions, but the potential

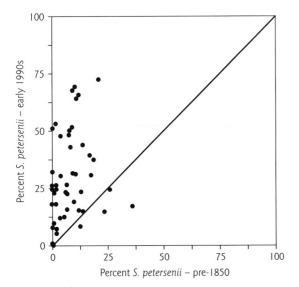

Fig. 16.8 A comparison of the relative abundances of *Synura petersenii* chrysophyte scales from surface sediments (i.e., the early 1990s; the years in which the cores were collected) and from pre-industrial (i.e., pre-1850) sediments from a broad spectrum of boreal shield lakes in Ontario, Canada. The diagonal line represents the 1 : 1 line. Any points plotting above the 1 : 1 line indicate that, in these lakes, the relative proportion of *S. petersenii* scales has increased since pre-industrial times. From Paterson et al. (2004); used with permission of Blackwell Publishing.

certainly exists to clarify many of these relationships using paleolimnological approaches.

16.10 Salinization of water supplies

Salinization of water is a general term that refers to the increase in salinity, which may arise from a number of sources. It is a serious water-quality issue because, as the salinity of a water supply increases, its value as a source of drinking, agriculture, or industrial water decreases, with significant economic and ecological costs (Williams 1987, 1999). In this section, salinization is used to describe increased concentrations of dissolved salts as a direct result of human activities. In closed-basin lakes in semi-arid regions, salinity levels may be closely linked to climatic changes, and typically increase during drought periods, when evaporation exceeds precipitation (Fritz et al. 1999). The reconstruction of such trends has important climatic considerations, but it is not the focus of this discussion, although one could argue that increased salinity of surface waters due to warming induced by greenhouse gases is a form of pollution.

Road salt is commonly used as a de-icing agent, especially on North American roads. A strong correlation often exists between the concentration of sodium and chloride in surface waters and the use of winter road de-icing salts (Mattson & Godfrey 1994). Although there are now widespread concerns that surface-water salinity is increasing, little or no historical data exist. Fortunately, many biological indicators can be used to track past salinity trends. For example, seepage from a road salt storage facility established in 1953 adjacent to Fonda Lake (Michigan) was tracked using diatom (Tuchman et al. 1984) and chrysophyte (Zeeb & Smol 1991) indicators. The paleolimnological data suggested that the lake's salinity began to increase in the 1960s, and especially from about 1968 to about 1972. However, between roughly 1974 and 1980, salinity levels appear to have decreased, likely due to the construction of an asphalt pad for the salt storage facility in the early 1970s.

As part of the EMAP biomonitoring program (see pp. 207–9), Dixit et al. (1999) developed transfer functions to infer overall trends in lake-water chloride concentrations in northeastern US lakes, based on diatom assemblages. Although this transfer function was weaker than the ones developed for pH and total phosphorus, overall trends in lakewater salinity could probably be inferred. Using the "top/bottom" approach for 257 lakes and reservoirs, they showed that, in support of recent concerns that surface waters located near highways were increasing in salinity, diatom-inferred chloride concentrations have increased in many systems. Most of the sites that showed the largest chloride increases currently had relatively high lakewater levels and were located near roads and in areas of high population densities. Such approaches can also be used in detailed stratigraphic analyses of individual lakes (A.S. Dixit et al. 2000).

Runoff from urbanized or residential areas tends to have higher salinities than runoff from forested areas (Prowse 1987). Siver et al. (1999) explored the influences of land-use changes on 23 Connecticut water bodies, using diatom and chrysophyte paleolimnological analyses. There was no evidence of acidification is this lake set, but about 25% of the sites recorded increases in inferred specific conductivities, a trend that was most pronounced in watersheds that were highly residential in nature. Lakes situated in catchments that remained at least 80% forested rarely increased in conductivity; however, once over 25% of a catchment was deforested (which coincided with more road construction), significant changes were evident. Such data are useful to lake managers who are charged with setting land-use guidelines.

Certain mining and industrial activities can also release salts, either directly or indirectly, into water systems. Once again, historical records are rarely, if ever, maintained. Siver (1993), using a transfer function that related chrysophyte scales to lakewater specific conductivity in Connecticut (USA) lakes, hindcasted conductivity changes over the past approximately 170 years in Long Pond. Although absolute changes were not large, the inferred specific conductivity had more than tripled over the past century. The rise in ion concentrations could be correlated with known changes in the catchment, such as lakeside cottage development, but also the establishment of a surface silica mine.

With increased pumping of groundwater in many regions of the world, the rate of withdrawal may well exceed the rate of water renewal (Williams 1987). This deficit may be compensated for by the inflow of lateral water, which may have a higher salt content. By using indicators at

groundwater springs (see Section 16.12), these problems can also be assessed using paleolimnological approaches.

16.11 Water diversions: lakes and rivers dying the death of irrigation canals and impoundment

For millennia, humans have been very effective at altering the flow of water to meet their agricultural and industrial needs. These actions have not been without environmental repercussions. One of the most dramatic examples of lake degradation due to water diversion is the recent tragedy of the Aral Sea, which borders on Kazakhstan, Uzbekistan, and Turkmenistan. A once thriving fishing and resort area, it was, until recently, the world's fourth largest inland body of water. However, as part of the former Soviet Union's poorly conceived plans for economic expansion, large river diversion projects were initiated in the 1950s to allow for wide-scale cotton production in this naturally arid environment. With much of its freshwater supply diverted, the Aral Sea began to disappear and increase in salinity. From 1960 to 1990, its surface area and volume decreased by 40% and 60%, respectively, and its salinity rose from 9 to 30 mg l (approaching the salinity of the oceans). The resort facilities, which were popular tourist attractions in the 1950s, are now approximately 120 km away from the current waterfront! None of the 20 indigenous fish species have survived, and the thousands of people who were employed in the fisheries industry are now without work (Fig. 16.9). A large number of ecological, health, and social problems have been linked to this near total destruction of an entire ecosystem.

As alarming as the 20th-century changes in the Aral Sea have been, paleolimnological and archeological studies show that this has just been one of a number of fluctuations in past hydrological changes (Boomer et al. 2000). Prior to the establishment of agriculture and the need for

Fig. 16.9 Part of the former Aral Sea fishing fleet, now abandoned, about 20 km from the water's edge. In some areas, the Aral coastline has receded over 150 km since 1960. Photograph taken near Akespe, by Ian Boomer.

irrigation canals, the size and salinity of the Aral Sea was primarily controlled by regional climatic factors, with lower water levels and higher salinities coeval with warmer, drier conditions. However, over the past approximately 6000 years, the sedimentary data record a long history of water supply mismanagement, mainly due to past attempts at irrigation, which began about 5000 years ago, but also due to water diversions being employed as a weapon during times of war, as well as to meet other economic and political objectives. These data help put the current Aral Sea tragedy in context; although, as pointed out by Boomer et al. (2000), the present environmental catastrophe is in many respects far worse, as it is compounded by the use of chemical fertilizers, pesticides, and defoliants.

The potential repercussions of diverting water for agricultural purposes are a key issue in many other regions, including the dryland areas of East Africa. These regions are highly sensitive to inter-annual and decadal trends in rainfall, and so it is not always clear if water diversions are markedly affecting lake systems, or whether such changes are within the range of natural variability. One lake of considerable interest with regard to these questions is Lake Naivasha, in central Kenya. The river flowing into the lake is currently a major source of water for cash crop agriculture, with about 15% of the inflow directly pumped and diverted. How significantly has this practice affected the salinity and other limnological characteristics of Lake Naivasha, and in particular of Lake Oloidien, its hydrologically closed satellite basin? It is known that lake levels and salinity have been closely linked to past droughts, such as those that occurred in the 1920s and again in the 1940s and 1950s. But were these droughts truly "exceptional"? If not, then the additional strain of the recent water diversion projects may make this already sensitive ecosystem much more prone to ecological disaster if subjected to future droughts.

Verschuren et al. (1999) addressed the above questions by using a spectrum of paleolimnological techniques to track the past two centuries of limnological change in Lake Oloidien. Algal, cyanobacterial, and invertebrate community successional changes in a dated sediment core were used to track past changes in lake level and related salinity shifts. Their study revealed a highly dynamic lake that fluctuated from freshwater to saline conditions, depending on local climates. For example, during much of the 19th century and between c. 1940 and 1991, the lake was shallow and saline. However, during wetter periods, such as between c. 1890 and the late 1930s, water levels increased to the point that Lake Oloidien became confluent with nearby Lake Naivasha, and created a freshwater system with a papyrus swamp, well-developed submerged macrophyte beds, and reduced algal production. Interestingly, this period coincides with the period of early ranching and dairy farming by the white colonists. Conditions for dairy farming were remarkably good at this time, and in fact considerably out of balance with the long-term patterns of lake levels and salinity for this lake. With subsequent droughts, which the paleolimnological data showed were not exceptional (as previously believed) but typical for this region, additional water had to be diverted from the lake to sustain the local agriculture and industry.

Verschuren et al. (1999) showed that Lake Oloidien is a highly sensitive ecosystem, with low water levels and high salinity a natural aspect of its long-term development. Equally, studies such as this warn that any additional stresses (such as water diversions) can seriously affect these ecosystems, as they may already be at their ecological extremes due to climatic changes. As Verschuren et al. concluded, sustainable development in dryland regions must recognize that water is not a fixed resource, but may vary strongly at time scales far longer than the few years on which most management decisions are based.

Some of the effects of damming can be documented using other indicators. Kowalewski et al. (2000) tracked the demise of mollusk populations in the Colorado River (USA and Mexico) ecosystem by dams and irrigation projects using shells that had accumulated over centuries in the river's delta. The impoundments and diversions of the Colorado River began in the 1930s, but little is known about the natural river flow. Kowalewski et al. (2000), using radiocarbon techniques to date 125 mollusk shells collected from the Colorado River Delta (Gulf of California), showed that these shells ranged in age from AD 950 to 1950. Conservative estimates of pre-diversion mollusk population sizes were approximately 6×10^9 bivalve mollusks, or about 50 individuals per square meter. In contrast, the present mollusk population is about 94% lower. There can be little doubt that benthic productivity of this river ecosystem was drastically altered by decreased flows of freshwater and nutrients to the delta.

Intuitively, a decrease in the supply of fresh water is recognized by limnologists as a potential environmental problem. Although far less common, human-caused disruptions in the natural flow of salt water may also have negative impacts. One such example is from southwest Tasmania (Australia), where the development of a large hydroelectric facility and dam had severely altered the ecology of the only three known meromictic lakes in the temperate rainforests of the Southern Hemisphere. In 1977, the Gordon Power Development built a dam across the Gordon River, which significantly altered the hydrology of the region. From its inception, high discharges from the dam have exceeded natural river water flows and restricted the degree of salt wedge penetration in the estuary. Surveys of the region affected by the dam identified three meromictic lakes in the floodplain of the lower reaches of the river (Tyler & Bowling 1990). Meromictic stability calculations indicated that the penetration of the salt wedge up the estuary was critical in delivering sufficient salts to maintain chemical stratification in these lakes. Denied the penetration of brackish water up the river by the new dam, the chemoclines of these three unique lakes were being driven progressively deeper, and eventually chemical stratification would break down totally and, in the case of the two lakes, their water columns would mix. Tyler and Vyverman (1995) summarized the microbiological changes resulting from these changes in salinity, and hence chemical stability.

The potential loss of the only three known meromictic lakes from this region elicited considerable public and scientific concerns. However, the limnological data only extended back to the time after the construction of the dam, with very little evidence to show that it was in fact the dam that was causing the disruption of meromictic stability. The history of past chemical stratification, however, could be gleaned from the sedimentary record.

Hodgson et al. (1997, 1998b) removed a 17 m long core from Lake Fidler, one of the meromictic lakes, to provide an approximately 8000-year perspective on the lake's development and to determine the impact, if any, of the 1977 dam. Diatom analyses showed the development of the site from a former channel of the Gordon River, and then to an open backwater, until an alluvial sand bar was likely deposited and the lake proper was formed (Hodgson et al. 1997). A background signal of estuarine and marine diatoms in the lake's sediments showed that periodic inputs of brackish water from the riverine salt wedge were a common phenomenon in the lake's long-term development. From 2070 yr BP, the lake's meromixis stabilized, with a distinct community of freshwater diatoms characterizing its mixolimnion (Hodgson et al. 1997). Fossil pigment analyses allowed for the further refinement of these interpretations, as chlorophylls and carotenoids can be used to track microbiological communities specifically associated with meromixis (Hodgson et al. 1998b). For example, the bacteriochlorophylls characteristic of the anaerobic green phototrophic sulfur bacteria *Chlorobium limicola* and *Chlorochromatium aggregatum* require specific microhabitats present at the chemoclines of meromictic lakes. The establishment of this bacterial community, coupled with the diatom changes noted above, identified the development of the chemocline and the onset of meromixis about 2070 years ago.

The Fidler Lake paleolimnological data showed that the declining meromictic stability evident after 1977 was inconsistent with the lake's natural development. Moreover, the fossil pigment studies of the microbiological communities during the early stages of chemocline development, when frequent reversions to holomixis occurred, showed that these organisms can quickly re-establish themselves, once meromictic stability is regained. These paleolimnological data are being used as part of the management strategy for the dam and, since 1990, the dam has been used less intensively, which has allowed occasional salt wedges to become established higher up the river, thus replenishing the meromictic stability of the floodplain lakes. The three meromictic lakes are now part of a World Heritage Area.

16.12 Environmental histories at the interface of groundwater and streams: the paleoecology of springs

The pollution of surface waters is clearly the focus of this book. Nonetheless, as mentioned in the introductory chapter, the amount of groundwater dwarfs the volume of water held in our lakes and rivers. Some estimates suggest that if all of our planet's groundwater was pumped to the surface, it would cover the globe to a depth of 120 m! In contrast, the volume of fresh water contained in our surface reservoirs (i.e., lakes, rivers, swamps, etc.) would only cover the Earth's surface to a depth of about 25 cm.

Because groundwater is out of sight, it is easily taken for granted. Nonetheless, the deterioration of groundwater quality is a cause of serious concern. Common problems include contamination from both point and diffuse sources, such as leaky pipelines and septic systems, industrial wastes and landfills, livestock and agricultural wastes, pesticides and other contaminants, salt intrusion, as well as many other impacts. Because groundwater may move very slowly (e.g., the residence time of water underground may range from weeks to millennia), once an aquifer is contaminated, it may remain unusable for decades. Yet, our dependence on groundwater is increasing. As populations grow, and the quantity and quality of easily accessible surface waters become inadequate, groundwater reserves are often exploited. For many regions, groundwater is the sole source of fresh water.

Although there is growing concern about the deterioration of groundwater quality, even less long-term data are typically available for these freshwater stores than for surface waters. Several different lines of research are under way to track groundwater quality, including chemical and isotopic techniques. My goal is not to review these approaches (a separate volume would be necessary), but to summarize some studies that use the same paleolimnological techniques described for lakes and rivers to track long-term changes in groundwater by studying stratigraphic deposits at springs.

Groundwater does not stay underground forever, as it eventually reappears at the surface in regions called discharge areas (as opposed to recharge areas, where surface waters seep underground to become part of the groundwater). Springs are typical discharge sites, representing the transition zones between groundwater and streams.

Tracking long-term changes in groundwater quality poses some challenges to the paleolimnologist, as sediments do not accumulate when the water is underground. However, when groundwater discharges to the surface, stratigraphic information on the past history of source aquifers may accumulate in the sediments of discharge pools and other water bodies. Williams and Williams (1997) and Reavie et al. (2001) used this approach to track the history of groundwater flow and quality in a southern Ontario spring over the past 160 years or so from the caddisfly, chironomid, pollen, diatom, chrysophyte, testate amoeba, and phytolith indicators preserved in the pool sediments downstream of a spring outflow. Their paleolimnological data tracked a gradual transition from a slow river environment in the early 1800s to a spring-fed pond environment by

the late 19th century. Marked changes in species assemblages could be linked to natural events, such as Hurricane Hazel in 1954, as well as to human-related factors such as agriculture and urban development. The increase in salinity-tolerant diatoms reflected possible groundwater salinification from road salt seepage.

As the authors acknowledged, these studies are still preliminary and exploratory, but such approaches may present water-quality managers with important tools in assessing trends in groundwater quality, provided that suitable stratigraphic sequences can be located near discharge zones. The environmental optima and tolerances of indicators characteristic of spring outflows can be quantified using similar techniques developed for surface-sediment calibration sets. For example, Sabater and Roca (1990) demonstrated that diatom assemblages in 28 mountain outflows reflected water quality. Similar calibration studies should be conducted in other groundwater discharge sites, preferably with the inclusion of other indicators, such as ostracods (Forester 1991) and the chemical signals preserved in their shells (Wansard & Mezquita 2001). Moreover, depending on the problems being investigated, other types of proxy data can be used. For example, if pesticides or other types of groundwater contamination were potential problems, then these variables could be measured and tracked over long time frames using standard paleolimnological techniques.

16.13 Summary

Human activities are constantly altering ecosystems and posing new problems to environmental managers. Paleolimnology has demonstrated its strength and flexibility by responding to many of these new challenges. Some recent examples include problems related to habitat and species conservation, identifying the provenance of pollutants, assessing the relative impacts of different forestry practices, tracking nuclear pollution plumes and assisting managers with decisions regarding the location of nuclear storage facilities, taste and odor problems in water supplies, salinity changes, water diversions, and tracking long-term changes in groundwater quantity and quality. The list of potential applications is increasing steadily. Fortunately, the development of creative new approaches, as well as the constant improvement and refinement of established techniques, is allowing paleolimnologists to meet these challenges.

17

Paleolimnology: a window on the past, a key to our future

Retrospection is introspection.
Jean-Paul Sartre (1905–80)

17.1 Some personal perspectives and reflections

Historians have frequently observed that people pay more attention to history when they are concerned about the future. As shown by the examples used in this book, there is much to worry about. It is therefore not surprising that paleolimnology has been enjoying so much interest and activity over recent years.

Looking back at my career thus far as a paleolimnologist, I am certainly encouraged by progress in the scientific development of my field and how these approaches have been applied to environmental issues. However, this optimism is tempered by a growing realization of the magnitude of the environmental issues we are facing, and the new and progressively more complex problems that society is continuing to produce. The original outline for the first edition of this book was for a shorter, simpler volume. In the intervening years that I spent writing, the list of problems and applications grew steadily, requiring considerable expansion of many sections and the addition of one new chapter. With this second edition, a similar expansion was required, as new

environmental problems were recognized. There is no shortage of work for applied paleolimnologists.

When I was first asked to write this book, some of the original reviewers suggested that I conclude with some personal perspectives. I am keenly aware that much of what I write here is not overly original, and instead is an amalgamation of my readings and discussions with many colleagues. Perhaps, though, some of these points warrant repeating.

It is now clear that no part of our planet is safe from environmental degradation of some kind. The human "footprint" is now virtually everywhere, from the polar regions to the tallest mountains, to the tropical forests. Water bodies are downhill from land and so are at least partial sinks for disturbances occurring in their watersheds. We must now also acknowledge that many pollutants may be transported in the atmosphere, by the wind, and so we must consider inputs from the airshed. Paleolimnology has shown that even isolated, seemingly pristine water bodies have been impacted by human activities. The rates may be very slow in some cases, perhaps imperceptible using conventional, short-term monitoring programs, but even remote ecosystems are changing due to human activities.

Paleolimnological data have shown that we almost always underestimate the magnitude of our actions. I am reminded of the quote used to open Chapter 9, which was from a review of Rachel Carson's book *Silent Spring*, which warned us of the potential negative effects of pesticides. We have a tremendously poor record of predicting the environmental repercussions of our actions, and typically conjure up overly optimistic scenarios of the consequences. The levels of disturbance we are causing are often staggering. For example, ever since humans learned to use fire, we have been altering the composition of our atmosphere, with little foresight as to where this uncontrolled experiment will take us. With increases in sulfuric and nitric oxide emissions, we have been conducting a large and unintentional titration of our water bodies with strong acids. With the development of freons, we even began eroding away the Earth's protective shield, the ozone layer, many kilometers above us. The problems are confounded because we tend to look at environmental issues in isolation, and rarely take into account synergistic reactions or the actions of multiple stressors occurring simultaneously. Unfortunately, the combined effects are often more negative than the actions of pollutants in isolation. The long-term environmental consequences of action or inaction are another major factor that is rarely considered. Once again, we are often tempted to underestimate the magnitude of our problems.

17.2 The tasks before us

The tasks before us are not simple. We must never underestimate the resources and influences of the vested interest groups in environmental debates. As I look back on our work during the acid rain debates, I am still dismayed at how difficult it was to get action. Even though about 95% of the scientific data clearly showed that acid rain was the problem, the remaining 5% of "doubt" received a totally disproportionate amount of media and political attention. What we learned during the acid rain discussions is that delay and obfuscation can often translate into large financial savings to polluters, at a high cost to the environment. Little has changed as we continue to debate current environmental problems, such as climatic change.

As I personally become more involved at the interface between science and policy, and spend more time arguing environmental cases before politicians, I am sometimes reminded of the speech given by Tolkien's (1954) Galadriel, who referred to their struggle as "the long defeat"; the elves in Middle-earth continued to soldier on, yet seemingly always losing ground to deterioration. However, the fact that I and others still rage when it comes to fighting for environmental concerns means that we still have hope. To paraphrase H.G. Wells' sentiments, the odds are against humanity, but are still worth fighting for.

Science is not a democracy where all opinions are given equal value. Assertions must be weighed, based on the data they can provide to support their claims, must be verifiable by independent analysis, and must be subjected to peer review. Scientific hypotheses are not weighted by the glossiness of the paper used in print advertisement, nor by the level of polish and spin put on "sound bites" for the media. Thankfully, history shows us that science is self-correcting. In the end, truth does emerge. But if the truth is stalled, how much more environmental damage is caused?

Science must become a bigger part of the decision-making process. As Charles Dickens (1843) reminds us in his classic novel *A Christmas Carol*, the ghost presents Scrooge with the allegorical twins Want and Ignorance. The spirit warns Scrooge to beware them both, but above all, to beware Ignorance. Science must become a larger player in the decision-making process. Education must play a key role. Scientific findings must also be communicated more effectively to the public and politicians, to counterbalance the assaults from vested interest groups who have little concern for the effects of environmental degradation. In many respects, environmental scientists (and certainly paleolimnologists) are new to policy. And policy makers are new to scientists. This can no longer continue; we can no longer afford to be apart. Constructive interactions are required. Paleolimnological approaches can be used to identify, clarify, and help mitigate environmental problems. With cutbacks in many government agencies, and the associated declines in standard monitoring programs, the high-quality long-term data that can be provided using paleolimnological approaches are especially relevant.

I think the problems facing us can be illustrated by a conversation that I had with a businessman – an investment banker – on an airplane. He asked me what I did, and we began to talk about acid rain and other problems. He seemed skeptical. He then said "Well, for you scientists, your level of certainty is fine. But in business, we need a lot more certainty than you scientists are ready to offer to make these expensive decisions, such as cutbacks in emissions and so on." I sat there wondering: What actually are "investment bankers"? Some people might equate them to professional gamblers! What percentages are they basing their decisions on for advising someone to sell their stocks because they anticipate they will go down, whilst they sell the same stocks to someone else who speculates they will go up in value? The man I spoke to seemed to know a lot about the costs of mitigation. I wonder if he ever estimated the long-term costs of doing nothing? I suppose he would have been flying in First Class, and not in the cheap seats with me, if he had better data!

There may be some hope. After the acid rain debates, and as the world's attention shifted to other environmental problems, many countries, on paper at least, agreed to Principle 15 of the Rio Declaration from the 1992 UN Conference on Environment and Development (UNCED), commonly known as the *Earth Summit*. This principle states that "lack of scientific certainty will not prevent action." I wonder if this will make a difference in dealing with our future problems?

Have we learned any lessons from the past? I wonder and worry what will happen with new, major problems such as greenhouse gases and climatic warming. Will the "spin-doctors of obfuscation" again be weaving their webs of delay and using claims of insufficient data to excuse inaction? Probably. They already are. What are the costs of doing nothing? Or delaying action? History tells us they are high. One of our biggest enemies is complacency; constructive interventions are required.

17.3 Concluding comments

Paleolimnology has constantly been finding new applications. From its beginnings as a more esoteric and descriptive scientific discipline, a significant portion of paleolimnological research has shifted seamlessly to applied fields (e.g., eutrophication, acidification, ultraviolet penetration, climatic warming). One could argue that these topic changes were simply a matter of chasing available research funding. This cynical analysis may be partly correct, but only to a point.

I would more vigorously argue that the strength of a science is its breadth and depth, and its flexibility to address the changing concerns and questions of other scientists and the public at large. The precision and accuracy of paleolimnological inference techniques have been improving rapidly. Constant change, even if it brings improvements, can cause some frustration, but I prefer this frustration over the worry I would feel working in a scientific field that has become static, not always improving and honing its tools to face new problems. Change is challenging, but if it comes with constant improvements, change is also good. Paleolimnology has shown again and again that it has this breadth, this flexibility, and this strength.

Some paleolimnological approaches are still quite rudimentary. However, very often they provide the *only* data available. And any data are better than no data at all. They provide some answer, or at least some guidance, to scientists and decision makers. If one is going into battle, a blunt battle axe is better than no weapon at all (Håkansson & Peters 1995). Conversely, one sometimes hears the complaint that paleolimnologists are using "overly complex statistical approaches." If such statistics are not required, then there is good reason for complaint. One should always use the most straightforward (and most defensible) approach to explain one's data. There are numerous paleolimnological examples, as shown in this book, where complex statistics are not required. However, there is no need to apologize for using *appropriate* statistical approaches. Complex data sets (such as the ones often used in biological calibration sets) often require "complex" statistical treatments (and "complex" is a relative term). As Albert Einstein cautioned, "Everything should be as simple as possible, but no simpler." It is satisfying to see that many of the statistical and other approaches developed in the paleolimnological context (many of them initially criticized) are now being applied to neolimnological and other ecological problems.

The applied aspects of both neolimnology and paleolimnology have allowed more scientists to be involved in these fields. Many fundamental scientific questions can be studied and answered using applied studies. Unplanned interventions can be treated as unintentional experiments, which can now be studied using retrospective techniques. G. Harris (1994) reminds us: "As pure science becomes harder to justify and fund we must make every effort to derive general principles from the study of applied problems. Ecologists should not be afraid of applied problems, they can tell us much about general principles."

These are exciting times for paleolimnology. For example, at the *International Paleolimnology Symposium* held in Duluth (Minnesota) in June 2006, the formation of the *International Paleolimnology Association* was unanimously approved. As I write this concluding chapter (2007), the *Journal of Paleolimnology* is celebrating its 20th year of publication. Expectations for paleolimnology are now very high, but this is justified. It is fair to propose an ambitious agenda for an ambitious science. There is nothing wrong with "pushing the envelope" in our applications, as long as we simultaneously appreciate and clearly explain the limitations and assumptions of our work. Paleolimnology is steadily coalescing into a rigorous and potent science. I am tremendously delighted and proud of the accomplishments of my colleagues. I can hardly wait to see what they will do in the coming years!

Glossary

accelerator mass spectrometry (AMS)   Used by paleolimnologists primarily for radiocarbon (^{14}C) dating. Carbon atoms are ionized and accelerated to high velocities; they are then subjected to strong magnetic fields that separate the isotopes into individual ion beams that can be measured. The amount of ^{14}C relative to ^{12}C in the sample is used to estimate the age of the material.

accumulation rate   The rate at which sediments or constituents of sediments (such as diatom valves or other indicators) accumulates at the sediment surface. It is typically expressed as a quantity (mass or number) per unit of surface area (e.g., cm^2) of sediment per year. Essentially, it is the concentration multiplied by the sedimentation rate. It is referred to as influx data in some studies.

adnate   Joined closely together. For example, adnate diatoms are periphytic diatoms that are closely associated with their substrates.

afforestation   The planting or natural regrowth of trees on land that was deforested for at least 50 years.

albedo   The measure of the reflectivity of a surface. Snow, for example, has a relatively high albedo, whereas darkly colored soil has a relatively low albedo.

alkenones   Highly resistant organic compounds. The ratios of some alkenones are being used in climate reconstructions.

allochthonous   Derived from outside the lake or water body. Also called **allogenic**.

allogenic   See **allochthonous**.

alluvial sediments   Sediments that have been deposited as a result of flowing water (e.g., a river).

alternative equilibria   In limnology, this refers to the two main equilibrium states that shallow lakes often exhibit. One is a relatively clearwater, macrophyte-dominated state, and the second is a turbid, phytoplankton-dominated state, which precludes extensive rooted macrophyte growth.

analogue techniques   See **modern analogue techniques**.

anemophilous pollen   Pollen grains dependent on wind currents for transport.

Anthropocene   A new geological epoch, proposed by Crutzen (2002), representing the period of human-dominated Earth's history. It was proposed that the Anthropocene could have started in the latter part of the 18th century, when polar ice core data indicate the beginning of higher concentrations of greenhouse gases, which also coincides with the development of the steam engine. However, Ruddiman (2003) has argued that the Anthropocene started several thousand years ago.

anoxia   This literally means the absence of oxygen. However, in the limnological literature, it is often considered to be waters with dissolved oxygen concentrations of less than 1 mg l^{-1}. See also **hypoxia** and **oxic waters**.

artificial neural network   A network of simple processors that works by creating connections, which to some extent imitate a biological neural network, adapting and learning from past patterns. It is a method employed by some paleolimnologists to develop transfer functions.

asbestosis   A serious lung disease linked to inhaling asbestos fibers, which irritate and cause scarring of lung tissues.

athalassic lakes   Inland saline lakes (i.e., saline lakes that are not affected by marine influences).

autochthonous   Derived from inside the lake or water body. Also called **authigenic**.

authigenic   See **autochthonous**.

Bayesian statistics   Statistics that incorporate prior knowledge into probability calculations. Such approaches are currently being explored by some paleolimnologists in the development of transfer functions.

BP   Before present. The abbreviation is used in radiocarbon dating, where "present" is conventionally defined as AD 1950.

benthos   Organisms living on and in the sediments, as well as on other substrates.

bight   A bend in a coastline that forms an open bay.

billabong   An Australian term used to refer to a water body that either occupies an elongated basin in the bed of a temporary stream or a cutoff bend in a river (i.e., an oxbow lake).

bioaccumulation   The net accumulation of a contaminant in and on an organism from all sources (e.g., water, air, and solid phases in the environment).

bioconcentration   A more restrictive term than bioaccumulation, in that it refers to the net accumulation in and on an organism of a contaminant from water only.

biomagnification   The cumulative, increased chemical concentrations of contaminants in tissues of organisms that are magnified at higher trophic levels.

bioturbation   The mixing of sediments by organisms (e.g., burrowers).

black water   See **gray water**.

bottom-up controls   A term used especially in studies of food web interactions. Systems are said to have strong "bottom-up" controls when physical and chemical variables (e.g., nutrients) are strongly influencing the composition and abundance of organisms. See also **top-down controls** and **cascading trophic interactions**.

brackish water   Water with salinity levels intermediate to those found in fresh water and sea water, such as water in estuaries.

calendar years   See **radiocarbon years**.

calibrated years   See **radiocarbon years**.

carapace   The chitinous shield that covers part of the body of crustaceans (e.g., Cladocera). In ostracodes, the carapace is calcified and consists of two halves.

carcinogen   A substance that has the potential to cause cancer.

cascading trophic interactions   The effects of predators and herbivores cascade down through the food web from one trophic level to another. For example, a large piscivore population may depress the number of herbivorous fish, which may then allow zooplankton to thrive, which may graze heavily on phytoplankton. See **top-down controls**.

cellulose   A long-chain polysaccharide composed of glucose units. A fundamental constituent of plant cell walls.

chemocline   See **meromictic lake**.

chitin   A polysaccharide compound found in the exoskeletons of many invertebrates (e.g., Cladocera), as well as some other biota. Chitin is resistant to deterioration when buried in sediments. It is sometimes referred to as "nature's plastic," as it occurs in many taxa and is resistant to degradation.

clastic   Composed of fragments of pre-existing sediment or rock.

condensation   The conversion of a substance from the vapor phase to a condensed phase (liquid or solid). For example, water vapor condenses to liquid water as part of the hydrological cycle. Some contaminants are also volatilized, enter an airborne phase, but then condense back to Earth.

critical load   The highest load of a pollutant that will not cause changes leading to long-term harmful effects on the most sensitive systems.

cryptogenic taxon   A taxon of unknown origin.

delta   An alluvial sediment deposit located at the interface between a river system and a body of water (e.g., a lake or ocean) resulting from loss of velocity of river flow.

demersal   Typically found at or near the bottom of lakes and oceans, such as demersal fish communities.

dendroecology   Using the annual growth rings of trees to decipher past environmental changes.

destructive techniques   Laboratory techniques that destroy the sample (e.g., an acid digestion of sediment is a destructive technique).

diagenesis   The physical, chemical, and biological changes undergone by a sediment after it is deposited.

dimictic   A thermally stratified lake that mixes twice a year, typically in spring and autumn.

drainage lake   A lake that typically has an outlet and an inlet, and receives most of its hydrological input from streams (i.e., an open basin).

ecdysis   Moulting (shedding of the outer layer of integument), such as occurs with several invertebrate groups (e.g., cladocerans, chironomid larvae) when they shed their exoskeletons.

ecotone   The boundary between two distinct communities or ecosystems, such as the ecotone between Arctic treeline and tundra ecozones.

emerging POPs   Newly developed persistent organic pollutants (POPs), such as polybrominated diphenyl ethers (PBDEs), polychlorinated naphthalenes (PCNs), perfluorooctanesulfonate (PFOS), and perfluorinated compounds (PFCs). See **legacy POPs**.

endemic   A species thought to occur in only one area. For example, Lake Baikal is known to contain many endemic species, which are only found in that lake.

enrichment factor   Used primarily in geochemical studies of lake sediments where, for example, the concentration of a metal or a contaminant is enriched in the recent sediments, with respect to pre-disturbance (i.e., background) sediments.

ephippium   The sexual resting stage of certain invertebrates, such as Cladocera. The plural is "ephippia."

epilimnion   The warmer (less dense) surface water layer of a thermally stratified lake.

epilithon   Living on the surfaces of rocks and stones.

epineuston   A community of organisms that inhabit the top of the surface water layer.

epipelon   Living associated with the sediments.

epiphyton   Living on the surfaces of plants.

epipsammon   Living on the surfaces of sand grains.

epizooic   Living on the surfaces of animals.

eurythermal   An organism with broadly defined tolerances for temperature.

eurytopic   An organism with broadly defined environmental tolerances.

eutrophic lake   A nutrient-rich, highly productive lake.

eutrophication   Nutrient enrichment. Cultural eutrophication refers to nutrient enrichment as a result of human activities, such as sewage discharge or agricultural runoff.

exotic species   An introduced species.

fluvial lake   A widening in a river system where flow rates slow to the point that the river takes on lacustrine features.

fractionation   See **isotopic fractionation**.

frustule   The siliceous cell wall of a diatom, composed of two valves.

gray water   Domestic waste water that was typically used for cleaning. This would include water that goes down the drains in kitchen and bathroom sinks, as well as from baths and showers. In some cases, gray water may have some limited uses without pre-treatment, such as irrigation. In contrast, water that is flushed down the toilet, and so contaminated with sewage, is often referred to as **black water**.

greenhouse effect   Greater heat retention in the lower atmosphere due to the absorption and re-radiation by clouds and greenhouse gases of longwave terrestrial radiation.

greenhouse gases   Those gases that are contributing to the greenhouse effect, including water vapor, carbon dioxide, tropospheric ozone, nitrous oxide, and methane. Carbon dioxide (CO_2) often receives most attention, as its concentrations have increased markedly as a result of industrialization. Increased concentrations of greenhouse gases are believed responsible for accelerated warming of the planet due to an enhanced greenhouse effect.

guano   Feces, as in bird guano.

gullying   Formation of ditches or gullies due to erosion by water.

Haber–Bosch process   The principal commercial method of producing ammonia (a key component of fertilizers) by direct combination of nitrogen and hydrogen under high pressure in the presence of a catalyst, often iron.

half-life   The length of time for a process to run 50% to completion. Often used in reference to the decay of radioactive isotopes, it is the period of time required for half of a given quantity of the parent nuclide to decay into daughter products. The term is also used to describe half the length of time that a compound is in one form before it is converted into another form.

herbivore   An organism that eats plant material. For example, an invertebrate that eats algae would be considered a herbivore.

hindcast   The opposite of forecast.

Holocene   The last epoch within the Quaternary period. It is traditionally defined to have begun 10,000 ^{14}C years ago, although some now place the boundary at 10,800 ^{14}C years ago. In calendar years, the start of the Holocene is usually placed about 11,550 calendar years BP (based on radiocarbon and ice-core records).

holomictic lake   See **meromictic lake**.

hypolimnion   The colder (and hence denser), deepwater layer of a thermally stratified lake.

hyponeuston   The community of organisms that live just under the surface water layer.

hypoxia   The term refers to low-oxygen conditions; however, the threshold oxygen concentrations that designate hypoxic conditions are not formally defined. For example, in some North American aquatic ecosystems where lake trout is considered a keystone species, hypoxia is often defined as waters with dissolved oxygen concentrations lower than 4 mg l^{-1}. Meanwhile, some marine ecologists set the hypoxia boundary at 2–3 mg l^{-1}. See **anoxia** and **oxic waters**.

internal loading   Used primarily in eutrophication work, the term refers to the release of phosphorus from lake sediments under conditions of anoxia.

isotopic fractionation   Processes that result in a change in the isotopic ratio of a compound. For example, with evaporation water molecules containing the lighter isotope of oxygen (i.e., ^{16}O) will tend to be evaporated preferentially compared to water molecules with the heavier isotope of oxygen (i.e., ^{18}O).

lacustrine   An adjective referring to lakes. For example, lake sediments may be referred to as lacustrine sediments.

late glacial   The time of transition between glacial conditions and the establishment of the following interglacial conditions; generally between about 10,000 and 14,000 ^{14}C years ago.

Law of Superposition   Younger sedimentary sequences are deposited on top of older sequences. Therefore, in an undisturbed sedimentary sequence, the deepest deposits are the oldest; the shallowest are the youngest.

legacy POPs   Persistent organic pollutants (POPs) such as DDT or PCBs that were developed some time ago but, despite recent bans on their production in most areas, still persist in the environment. See **emerging POPs**.

Liebig's Law of the Minimum   Under steady state conditions, the growth of an organism is dependent on the amount of essential material that is available in least supply relative to demand.

limnology   The study of inland waters (e.g., lakes, rivers, reservoirs).

littoral zone   The shallower parts of a water body along the shore; often defined as the area where rooted aquatic macrophytes can grow.

LOI   See **loss-on-ignition**.

loss-on-ignition (LOI)   The weight loss of dried sediment after igniting it to about 550°C in a muffle furnace, for about 2 hours (the temperatures and times used vary slightly between laboratories). It provides an estimate of the organic matter of sediments.

lentic   The adjective for still waters, such as lakes (as opposed to **lotic**, which refers to running waters, such as in streams).

lotic   The adjective for running waters, such as streams (as opposed to **lentic**, which refers to still waters, such as in lakes).

macrophytes   In the limnological literature, the term refers to aquatic angiosperms (i.e., higher plants that live in water). It may include submerged or emergent forms, as well as those with floating leaves, such as water lilies.

magnetic susceptibility   The degree to which a material can be magnetized by an external magnetic field. Typically, iron-bearing minerals exhibit higher susceptibility than minerals with little or no iron. The latter include water, most organic matter, calcite, and many evaporite minerals.

mandible   Either of a pair of mouthparts in arthropod invertebrates. The lower jaw of an insect.

meromictic lake A partially mixing lake (i.e., a lake that is chemically stratified, which has a deepwater layer of dense, high-conductivity water that does not mix with the upper water region during, for example, spring and fall overturn, as in other thermally stratified lakes). The deepwater, non-mixing layer is called the **monimolimnion**; the upper layer of water that can be more easily mixed is called the **mixolimnion**. The two zones are separated by a chemical (and hence density) gradient known as the **chemocline**. Lakes whose water columns mix completely are referred to as **holomictic lakes**.

mesotrophic lake A lake with intermediate levels of nutrient concentrations and production.

messenger A small, free-falling weight that is often used in conjunction with cable-operated devices, such as many gravity corers. The messenger is clipped or otherwise attached to the cable in such a way that it can be released from the surface and then slides down the cable to the device, and thereby triggers a mechanism (e.g., in gravity corers, the messenger is used to trigger closure of the top of the core tube).

metalimnion The middle layer of water, which exhibits a rapid temperature change, between the **epilimnion** and **hypolimnion** in a thermally stratified lake. The zone of most rapid temperature (and hence density) change is called the **thermocline**.

microtome An instrument used to cut materials into thin sections for microscopic examination. Most microtomes are used to cut organic tissues; however, paleolimnologists have also used microtomes for fine-resolution sectioning of sediment cores.

mixolimnion See **meromictic lake**.

modern analogue techniques A method of environmental reconstruction that is based on matching fossil assemblages to similar modern, surface-sediment assemblages.

monimolimnion See **meromictic lake**.

mutagenic Causing mutations in genes.

nekton Animals with relatively strong locomotory controls (as opposed to **zooplankton**).

neolimnology The study of present-day limnological conditions and processes.

neural networks See **artificial neural networks**.

neuston The community of organisms that inhabit the surface of the water.

non-destructive techniques Laboratory techniques that do not destroy the sediment (i.e., the sediment can typically then be used for other analyses). For example, magnetic susceptibility is a non-destructive technique.

nursery lake A lake where young are reared. When dealing with sockeye salmon, nursery lakes are the sites where mature sockeye salmon spawn and die, eggs hatch, and the young live for 1–3 years before they migrate to the Pacific Ocean.

oligotrophic lake A lake with low levels of nutrients and production.

omnivore An organism with a varied diet (e.g., eating both plant and animal matter).

ornithogenic Influenced by birds. For example, sediments collected from lakes and ponds influenced by bird colonies (e.g., via **guano**) may be referred to as ornithogenic sediments.

orthophosphate The form of phosphorus (PO_4^{3-}) that is readily available for algal growth.

otolith A calcium carbonate concretion from the inner ear chambers of a fish. The daily and annual growth rings of otoliths contain information about the environment that the fish experienced during its life.

oxbow lake An abandoned meander of a river that has become isolated from the main stream channel.

oxic waters   Water containing dissolved oxygen. The term can also be used, for example, to describe "oxic sediments." See **anoxia** and **hypoxia**.

paleolimnology   The multidisciplinary science that uses the physical, chemical, and biological information preserved in sedimentary profiles to reconstruct past environmental conditions in inland aquatic systems. In its broadest sense, paleolimnology also includes studies of long-term changes in basin configuration and shoreline geomorphology.

paludification   The process of water-logging mineral soils and the development of fen or bog ecosystems that form peat.

parthenogenesis   Reproduction without fertilization, whereby females produce other female offspring, without male gametes. This occurs for part of the life cycle of, for example, Cladocera and rotifers.

periglacial   Near an ice sheet.

periphyton   Microbial growth upon substrates, such as periphytic algae.

persistent organic pollutants (POPs)   A group of organic chemicals exhibiting the combined properties of persistence, bioaccumulation, toxicity, and long-range environmental transport. See **legacy POPs** and **emerging POPs**.

photic zone   The depth of the water column where sufficient light penetrates so that photosynthesis is greater than respiration. It is also called the euphotic zone and the trophogenic zone (as opposed to the aphotic and tropholytic zone).

piscicide   A chemical used to kill fish.

piscivore   An organism that eats fish.

planktivore   An organism that eats plankton.

plankton   A term used to describe organisms that have little locomotor control, and whose positions in the water column are largely at the mercy of physical factors (e.g., currents).

phytoplankton   Plankton with the ability to photosynthesize (e.g., algae found in the open water).

polymictic lake   A typically shallow lake whose entire water column mixes frequently (i.e., is not thermally stratified for extended periods of time).

pond   A water body that is relatively shallow. It is sometimes delineated as a water body that is shallow enough so that its entire bottom can support the growth of rooted aquatic macrophytes (i.e., the entire pond is technically a littoral zone).

postglacial   The period following the retreat of the last Pleistocene ice sheet.

potline   A term used in the aluminum processing industry. A pot is where the electrolyte reduction process occurs. An aluminum plant may have many of these pots (e.g., up to about 200) electrically connected to one another in long rows called potlines.

proportional counter   A radiation detector, used in some laboratories for radiocarbon (^{14}C) dating, consisting of a gas-filled tube in which the intensity of the discharge pulses is proportional to the energy of the ionizing particles.

proxy data   Data obtained by scientists such as paleolimnologists to reconstruct past environmental conditions. For example, diatom species composition can be used as a proxy for lakewater pH.

pyritization   A measure of the amount of pyrite (FeS_2) that is formed in sediments. The degree of pyritization has been used as an index of **anoxia**.

radiocarbon years   Ages based on radiocarbon (^{14}C) measurements, but not corrected for past variations in the ^{14}C concentration in the Earth's atmosphere. By convention, expressed as years before AD 1950 (see BP, before present). These dates can then be converted to calibrated years (equivalent to calendar years, expressed as AD, BC, or BP), using a calibration that accounts for past variations in atmospheric ^{14}C.

redox potential   A composite measure of the overall intensity of a system's oxidizing or reducing potential.

reservoir   A water body used for the storage, regulation, and/or control of water. In the examples used in this book, most reservoirs are artificial and are formed by impounding rivers or otherwise raising water levels.

residence time   In limnology, this term is often used to describe the water renewal rate. For example, a river would typically have a shorter residence time than a lake, and a rapidly flushing lake would have a shorter residence time than a slowly flushing system. The term is also used for the length of time a chemical is in a system (e.g., the residence time of a contaminant in the atmosphere or in a water body).

rift lake   A lake formed by tectonic activity, such as the large rift lakes in Eastern Africa (e.g., Lake Tanganyika).

saline lakes   Lakes with relatively high salinity; often defined by biologists as lakes containing water with a salinity of greater than 1 g l^{-1}, but higher levels are also cited. For example, geologists often define saline lakes as those with salinity levels greater than 3 g l^{-1}.

sclerochronology   The science that uses calcified structures, such as fish otoliths, mollusk shells, and corals, to reconstruct environmental conditions.

scintillation counter   An instrument used in some labs for radiocarbon (^{14}C) dating that measures ionizing radiation that interacts with a phosphor to emit light pulses (i.e., scintillations).

scrubber   One of many different types of pollution control devices used on smoke stacks or other point sources of pollution. For example, scrubbers are used by some industries to decrease acid rain emissions.

sediment focusing   A process causing sediments to accumulate at different rates in different parts of a basin through time. The term also refers to the process by which sediments are deposited in marginal zones of a basin and then redeposited in deeper water zones.

sedimentation rate   The rate of sediment accumulation, typically expressed as millimeters per year. See also **accumulation rate**.

seepage lakes   Lakes that do not have surface water inlets and outlets, and receive most of their water from direct precipitation and/or groundwater seepage.

Shelford's Law of Tolerance   Different organisms can survive within a range of environmental conditions, but have an ecological minimum and maximum, which represents their limits of tolerance.

siliciclastic   A term pertaining to clastic sediment composed of silicon-bearing minerals such as quartz, feldspars, clay minerals, and so forth.

specific heat   The amount of heat needed to change the temperature of 1 g of a substance by 1°C.

stained lake   A lake that is colored. The term is often used to refer to lakes with high levels of colored dissolved organic carbon.

statoblast   Buds produced by freshwater Bryozoa; they are important means of reproduction, dispersal, and survival under unfavorable conditions.

statospore   See **stomatocyst**.

stenothermal   An organism with narrowly defined tolerances for temperature.

stomatocyst The siliceous, endogenously formed resting stage of a chrysophyte alga. These cysts were often called **statospores** in the older literature.

supported ^{210}Pb ^{210}Pb that is derived from the *in situ* decay of the parent radionuclide ^{226}Ra (see **unsupported ^{210}Pb**).

surface-sediment calibration set A set of surface-sediment samples from a range of lakes (or other water bodies), used to help define the ecological optima and tolerances of indicators, or in some other way calibrate indicators to environmental conditions. The overall goal is often to develop some form of quantitative transfer function that will relate indicators to environmental variables, so that these proxy data can be used in down-core reconstructions of past environmental conditions. Also called a **training set**.

surface tension A measure of the strength of a liquid's surface film.

taphonomy The study of the death and decay of organisms, including the processes involved in the preservation and decomposition of fossils.

target load This term uses similar criteria as critical loads of pollutants, but it allows for other factors, such as financial and social considerations.

tephra Volcanic ash. Tephra layers in lake sediments can be used to date profiles if the date of the volcanic eruption is known.

teratological forms Morphologically deformed organisms.

thermocline The layer of maximum temperature (and hence density) change in a thermally stratified lake, separating the **epilimnion** from the **hypolimnion**.

thermophilic Preferring higher temperatures.

top/bottom approach A "before-and-after" type of paleolimnological assessment that compares inferred limnological conditions that occurred before a putative impact, and then compared to inferred present-day conditions. Indicators are examined in the surface (i.e., the "top" sample) sediments, representing an integrated sample of present-day conditions, and then compared to inferences made from sediments deposited before anthropogenic changes were believed to have taken place (i.e., the "bottom" sample). For example, if one was examining the effects of acidic deposition in North America, samples from sediments deposited in the early 1800s would represent pre-impact, "bottom" samples. This approach allows for a larger number of lakes to be sampled than would be possible if full sediment cores were analyzed, and provides regional assessments of overall limnological changes.

top-down controls A term used especially in studies of food web interactions. Systems are said to have strong "top-down" controls when activities such as grazing or predation are strongly influencing the composition and abundance of organisms lower in the food web. For example, grazing by herbivorous zooplankton affecting phytoplankton numbers would be an example of a top-down control. See also **bottom-up controls** and **cascading trophic interactions**.

training set See **surface-sediment calibration set**.

transfer function A mathematical model, also called an inference model, that describes the relationship between biological taxa (or other proxy data) and environmental data. It is used to infer past values of environmental variables from fossil assemblages. See **surface-sediment calibration set**.

unsupported ^{210}Pb ^{210}Pb that is derived primarily from the atmospheric flux. Unsupported ^{210}Pb can be calculated by subtracting supported activity from the total activity (see **supported ^{210}Pb**).

valve In diatoms, two siliceous valves comprise one frustule (i.e., half of a diatom's cell wall). The term is also used with other organisms, such as the valves of mollusks and other invertebrates.

vapor pressure   The pressure of the vapor of a substance in equilibrium with its pure condensed phase (solid or liquid).

varve   A discernible sedimentary structure (couplet), representing a single year of deposition.

void ratio   When referring to sediments, the fraction of total space not occupied or taken up by the sediments themselves.

winterkills   This typically refers to fish kills that occur under ice cover, as a result of oxygen depletion.

X-ray diffraction (also **X-ray diffractometry, XRD**)   A commonly used method for analysis of the mineralogical composition and crystallography of a sediment sample; the diffraction of a beam of X-rays by the three-dimensional array of atoms in a crystal.

zooplankton   Animals suspended in the water column, with limited locomotor control (e.g., Cladocera).

References

Aaby, B. and Berglund, B.E. (1986) Characterization of peat and lake deposits. In B.E. Berglund (ed.), *Handbook of Holocene Palaeoecology and Palaeohydrology*. Chichester: John Wiley, pp. 231–46.

Abbott, M.B. and Wolfe, A.P. (2003) Intensive pre-Incan metallurgy recorded by lake sediments from the Bolivian Andes. *Science*, 301, 1893–5.

Abbott, M.B., Binford, M.W., Brenner, M., and Kelts, K.R. (1997) A 3500 ^{14}C yr high-resolution record of water-level changes in Lake Titicaca, Bolivia/Peru. *Quaternary Research*, 47, 169–80.

Allan, R.J. and Kenney, B.C. (1978) Rehabilitation of eutrophic prairie lakes. *Verhandlungen der Internationale Vereinigung von Limnologie*, 20, 214–24.

Alverson, K., Oldfield, F., and Bradley, R.S. (eds.) (1999) *Past Global Changes and Their Significance for the Future*. Amsterdam: Elsevier. [The chapters in this book were subsequently published as journal papers in a dedicated issue of *Quaternary Science Reviews*, 19, 1–479.]

American Fisheries Society (2003) *Acid Rain: Are the Problems Solved?* Trends in Fisheries Science and Management 2, Bethesda, MD: American Fisheries Society.

Amsinck, S.L., Jeppesen, E., and Ryves, D. (2003) Cladoceran stratigraphy in two shallow brackish lakes with special reference to changes in salinity, macrophyte abundance and fish predation. *Journal of Paleolimnology*, 29, 495–507.

Anderson, N.J. and Battarbee, R.W. (1994) Aquatic community persistence and variability: A palaeolimnological perspective. In P.S. Giller, A.G. Hildrew, and D.G. Raffaelli (eds.), *Aquatic Ecology: Scale, Pattern and Process*. Oxford: Blackwell Scientific, pp. 233–59.

Anderson, N.J. and Korsman, T. (1990) Land-use change and lake acidification: Iron Age de-settlement in northern Sweden as a pre-industrial analogue. *Philosophical Transactions of the Royal Society of London*, B 327, 373–6.

Anderson, N.J. and Odgaard, B.V. (1994) Recent palaeolimnology of three shallow Danish lakes. *Hydrobiologia*, 275/276, 411–22.

Anderson, T.W. (1974) The chestnut pollen decline as a time horizon in lake sediments in eastern North America. *Canadian Journal of Earth Sciences*, 11, 678–85.

Antoniades, D. and Douglas, M.S.V. (2002) Characterization of High Arctic stream diatom assemblages from Cornwallis Island, Nunavut, Canada. *Canadian Journal of Botany*, 80, 50–8.

Appelberg, M. and Svenson, T. (2001) Long-term ecological effects of liming – the ISELAW Programme. *Water, Air, and Soil Pollution*, 130, 1745–50.

Appleby, P.G. (1993) Forward to the lead-210 dating anniversary series. *Journal of Paleolimnology*, 9, 155–60.

Appleby, P.G. (2001) Chronostratigraphic techniques in recent sediments. In W.M. Last and J.P. Smol (eds.), *Tracking Environmental Change Using Lake Sediments*. Volume 1: *Basin Analysis, Coring, and Chronological Techniques*. Dordrecht: Kluwer, 171–203.

Appleby, P.G. and Oldfield, F. (1978) The calculation of ^{210}Pb dates assuming a constant rate of supply of unsupported ^{210}Pb to the sediment. *Catena*, 5, 1–8.

Appleby, P.G., Richardson, N., and Nolan, P.J. (1991) ^{241}Am dating of lake sediments. *Hydrobiologia*, 214, 35–42.

Appleby, P.G., Richardson, N., Nolan, P.J., and Oldfield, F. (1990) Radiometric dating of the United Kingdom SWAP sites. *Philosophical Transactions of the Royal Society of London*, B 327, 233–8.

Apsimon, H., Thornton, I., Fyfe, W., et al. (1990) Anthropogenically induced global change – Report of Working Group 3, IUGS Workshop on Global Change Past and Present. *Palaeogeography, Palaeoclimatology, Palaeoecology*, 82, 97–111.

Arctic Monitoring and Assessment Programme (1998) *AMAP Assessment Report: Arctic Pollution Issues*. Oslo: AMAP.

Armitage, P., Cranston, P.S., and Pinder, L.C.V. (eds.) (1995) *The Chironomidae: The Biology and Ecology of Non-Biting Midges*. London: Chapman and Hall.

Atkinson, K.M. and Haworth, E.Y. (1990) Devoke Water and Loch Sionascaig: recent environmental changes and the post-glacial overview. *Philosophical Transactions of the Royal Society of London*, B 327, 349–55.

Attenborough, D. (1984) *The Living Planet*. London: Collins/British Broadcasting Corporation.

Ball, P. (1999) *Life's Matrix: A Biography of Water*. New York: Farrar Straus & Giroux.

Baron, J.S., Norton, S.A., Beeson, D.R., and Herrman, R. (1986) Sediment diatom and metal stratigraphy from Rocky Mountain lakes with special reference to atmospheric deposition. *Canadian Journal of Fisheries and Aquatic Sciences*, 43, 1350–62.

Baron, J.S., Rueth, H.M., Wolfe, A.M., et al. (2000) Ecosystem response to nitrogen deposition in the Colorado Front Range. *Ecosystems*, 3, 352–68.

Barrie, L., Gregor, D., Hargrave, B., et al. (1992) Arctic contaminants: sources, occurrence and pathways. *The Science of the Total Environment*, 122, 1–74.

Battarbee, R.W. (2000) Palaeolimnological approaches to climate change, with special regard to the biological record. *Quaternary Science Reviews*, 19, 107–24.

Battarbee, R.W., Cronberg, G., and Lowry, S. (1980) Observation of the occurrence of scales and bristles of *Mallomonas* spp. (Chrysophyceae) in the micro-laminated sediments of a small lake in Finnish North Karelia. *Hydrobiologia*, 71, 225–32.

Battarbee, R.W., Gasse F., and Stickley, C.E. (eds.) (2004) *Past Climate Variability through Europe and Africa*. Dordrecht: Springer.

Battarbee, R.W., Smol, J.P., and Meriläinen, J. (1986) Diatoms as indicators of pH: An historical review. In J.P. Smol, R.W. Battarbee, R.B. Davis, and J. Meriläinen (eds.), *Diatoms and Lake Acidity*. Dordrecht: Dr W. Junk Publishers, pp. 5–14.

Battarbee, R.W., Anderson, N.J., Jeppesen, E., and Leavitt, P.R. (2005) Combining palaeolimnological and limnological approaches in assessing lake ecosystem response to nutrient reduction. *Freshwater Biology*, 50, 1772–80.

Battarbee, R.W., Charles, D.F., Dixit, S.S., and Renberg, I. (1999) Diatoms as indicators of surface water acidity. In E.F. Stoermer and J.P. Smol (eds.), *The Diatoms: Applications for the Environmental and Earth Sciences*. Cambridge: Cambridge University Press, pp. 85–127.

Battarbee, R.W., Mason, J., Renberg, I., and Talling, J.F. (eds.) (1990) *Palaeolimnology and Lake Acidification*. London: The Royal Society.

Battarbee, R.W., Allott, T.E.H., Juggins, S., Keiser, A.M., Curtis, C., and Harriman, R. (1996) Critical loads of acidity to surface waters: An empirical diatom-based palaeolimnological model. *Ambio*, 25, 366–9.

Battarbee, R.W., Jones, V.J., Flower, R.J., et al. (2001) Diatoms. In J.P. Smol, H.J.B. Birks, and W.M. Last (eds.), *Tracking Environmental Change Using Lake Sediments*. Volume 3: *Terrestrial, Algal, and Siliceous Indicators*. Dordrecht: Kluwer, pp. 155–202.

Beamish, J. and Harvey, H.H. (1972) Acidification of LaCloche Mountain lakes, Ontario, and resulting fish mortalities. *Journal of the Fisheries Research Board of Canada*, 29, 1131–43.

Belis, C. (1997) Palaeoenvironmental reconstruction of Lago di Albano (Central Italy) during the late Pleistocene using fossil ostracod assemblages. *Water, Air, and Soil Pollution*, 99, 593–600.

Belzile, N. and Morris, J.R. (1995) Lake sediments: Sources or sinks of industrially mobilized elements. In J.M. Gunn (ed.), *Restoration and Recovery of an Industrial Region*. New York: Springer-Verlag, pp. 183–93.

Bennett, K.D. and Parducci, L. (2006) DNA from pollen: principles and potential. *The Holocene*, 16, 1031–4.

Bennett, K.D. and Willis, K.J. (2001) Pollen. In J.P. Smol, H.J.B. Birks, and W.M. Last (eds.), *Tracking Environmental Change Using Lake Sediments*. Volume 3: *Terrestrial, Algal, and Siliceous Indicators*. Dordrecht: Kluwer, Dordrecht, pp. 5–32.

Bennion, H. and Battarbee, R.W. (2007) The European Union Water Framework Directive: opportunities for palaeolimnology. *Journal of Paleolimnology*, 38, 285–95.

Bennion, H., Appleby, P.G., and Phillips, G.L. (2001) Reconstructing nutrient histories in the Norfolk Broads, UK: implications for the role of diatom-total phosphorus transfer functions in shallow lake management. *Journal of Paleolimnology*, 26, 181–204.

Bennion, H., Fluin, J., and Simpson, G.L. (2004) Assessing eutrophication and reference conditions for Scottish freshwater lochs using subfossil diatoms. *Journal of Applied Ecology*, 41, 124–38.

Bennion, H., Wunsam, S., and Schmidt, R. (1995) The validation of diatom-phosphorus transfer functions: An example from Mondsee, Austria. *Freshwater Biology*, 34, 271–83.

Berglund, B. (1986) *Handbook of Holocene Palaeoecology and Palaeohydrology*. Chichester: John Wiley.

Betts-Piper, A.A. (2001) Chrysophyte stomatocyst-based paleolimnological investigations of environmental changes in Arctic and alpine environments. M.Sc. thesis, Queen's University. Department of Biology, Kingston.

Beyens, L. and Meisterfeld, R. (2001) Protozoa: Testate amoebae. In J.P. Smol, H.J.B. Birks, and W.M. Last (eds.), *Tracking Environmental Change Using Lake Sediments*. Volume 3: *Terrestrial, Algal, and Siliceous Indicators*. Dordrecht: Kluwer, pp. 121–53.

Bianchi, T.S., Engelhaupt, E., Westman, P., Andrén, T., Roff, C., and Elmgren, R. (2000) Cyanobacterial blooms in the Baltic Sea: Natural or human induced? *Limnology and Oceanography*, 45, 716–26.

Bindler, R. (2006) Mired in the past – looking to the future: Geochemistry of peat and the analysis of past environmental changes. *Global and Planetary Change*, 53, 209–21.

Binford, M.W., Brenner, M., Whitmore, T.J., Higuera-Gundy, A., Deevey, E.S., and Leyden, B. (1987) Ecosystems, paleoecology and human disturbance in subtropical and tropical America. *Quaternary Science Reviews*, 6, 115–28.

Birks, H.H. (2001) Plant macrofossils. In J.P. Smol, H.J.B. Birks, and W.M. Last (eds.), *Tracking Environmental Change Using Lake Sediments*. Volume 3: *Terrestrial, Algal, and Siliceous Indicators*. Dordrecht: Kluwer, pp. 49–74.

Birks, H.H. (2002) The recent extinction of *Azolla nilotica* in the Nile Delta, Egypt. *Acta Palaeobotanica*, 42, 203–13.

Birks, H.H. and Birks, H.J.B. (2006) Multi-proxy studies in palaeolimnology. *Vegetation History and Archaebotany*, 15, 235–51.

Birks, H.H., Battarbee, R.W., and Birks, H.J.B. (2000) The development of the aquatic ecosystem at Kråkenes Lake, western Norway, during the late-glacial and early-Holocene – a synthesis. *Journal of Paleolimnology*, 23, 91–114.

Birks, H.H., Peglar, S.M., Boomer, I., et al. (2001) Palaeolimnological responses of nine North African lakes in the CASSARINA Project to recent environmental changes and human impact detected by plant macro-fossil, pollen, and faunal analyses. *Aquatic Ecology*, 35, 405–30.

Birks, H.J.B. (1995) Quantitative palaeoenvironmental reconstructions. In D. Maddy and J.S. Brew (eds.), *Statistical Modelling of Quaternary Science Data*. Cambridge: Quaternary Research Association, pp. 161–254.

Birks, H.J.B. (1998) Numerical tools in palaeolimnology – Progress, potentialities, and problems. *Journal of Paleolimnology*, 20, 307–32.

Birks, H.J.B., Berge, F., Boyle, J.F., and Cumming, B.F. (1990a) A palaeoecological test of the land-use hypothesis for recent lake acidification in South-West Norway using hill-top lakes. *Journal of Paleolimnology*, 4, 69–85.

Birks, H.J.B., Juggins, S., Lotter, A., and Smol, J.P. (eds.) (in preparation) *Tracking Environmental Change Using Lake Sediments*. Volume 5: *Data Handling and Statistical Techniques*. Dordrecht: Springer.

Birks, H.J.B., Line, J.M., Juggins, S., Stevenson, A.C., and ter Braak, C.J.F. (1990b) Diatoms and pH reconstruction. *Philosophical Transactions of the Royal Society of London*, B 327, 263–78.

Bissett, A., Gibson, J.A.E., Jarman, S.N., Swaddling, K.M., and Cromer, L. (2005) Isolation, amplification, and identification of ancient copepod DNA from lake sediments. *Limnology and Oceanography: Methods*, 3, 533–42.

Björck, S. and Wohlfarth, B. (2001) ^{14}C chronostratigraphic techniques in paleolimnology. In W.M. Last and J.P. Smol (eds.), *Tracking Environmental Change Using Lake Sediments*. Volume 1: *Basin Analysis, Coring, and Chronological Techniques*. Dordrecht: Kluwer, pp. 205–45.

Blais, J.M. (1996) Using isotopic tracers in lake sediments to assess atmospheric transport of lead in eastern Canada. *Water, Air, and Soil Pollution*, 92, 329–42.

Blais, J.M. and Kalff, J. (1993) Atmospheric loading of Zn, Cu, Ni, Cr, and Pb to lake sediments: The role of catchment, lake morphometry, and physico-chemical properties of the elements. *Biogeochemistry*, 23, 1–22.

Blais, J.M. and Muir, D.C.G. (2001) Paleolimnological methods and applications for persistent organic pollutants. In W.M. Last and J.P. Smol (eds.), *Tracking Environmental Change Using Lake Sediments*. Volume 2: *Physical and Geochemical Methods*. Dordrecht: Kluwer, pp. 271–98.

Blais, J.M., Duff, K.E., Laing, T.E., and Smol, J.P. (1999) Regional contamination in lakes from the Noril'sk Region in Siberia, Russia. *Water, Air, and Soil Pollution*, 110, 389–404.

Blais, J.M., France, R.L., Kimpe, L.E., and Cornett, R.J. (1998a) Climatic changes have had greater effect on erosion and sediment accumulation than logging and fire: Evidence from ^{210}Pb chronology in lake sediments. *Biogeochemistry*, 43, 235–52.

Blais, J.M., Kalff, J., Cornett, R.J., and Evans, R.D. (1995) Evaluation of ^{210}Pb dating in lake sediments using stable Pb, *Ambrosia* pollen, and [137]Cs. *Journal of Paleolimnology*, 13, 169–78.

Blais, J.M., Macdonald, R.W., Mackay, D., Webster, E., Harvey, C., and Smol, J.P. (2007) Biologically mediated transport of contaminants to aquatic ecosystems. *Environmental Science & Technology*, 41, 1075–84.

Blais, J.M., Schindler, D.W., Muir, D., Kimpe, L., Donald, D., and Rosenberg, B. (1998b) Accumulation of persistent organochlorine compounds in mountains of western Canada. *Nature*, 395, 585–8.

Blais, J.M., Kimpe, L.E., McMahon, D., et al. (2005) Arctic seabirds transport marine-derived contaminants. *Science*, 309, 445.

Blais, J.M., Schindler, D.W., Muir, D.C.G., et al. (2001) Melting glaciers: A major source of persistent organochlorines to subalpine Bow Lake in Banff National Park, Canada. *Ambio*, 30, 410–15.

Blenckner, T., Malmaeus, K., and Petterson, K. (2006) Climatic change and the risk of eutrophication. *Verhandlungen der Internationale Vereinigung von Limnologie*, 29, 1837–40.

Blokker, P., Yeloff, D., Boelen, P., Boekman, R.A., and Rozema, J. (2005) Development of a proxy for past surface UV-B irradiation: A thermally assisted hydrolysis and methylation of py-GC/MS method for the analysis of pollen and spores. *Analytical Chemistry*, 77, 6021–31.

Bonacina, C., Bonomi, G., and Monti, C. (1986) Oligochaete cocoon remains as evidence of past lake pollution. *Hydrobiologia*, 143, 395–400.

Boomer, I., Aladin, N., Plotnikov, I., and Whatley, R. (2000) The palaeolimnology of the Aral Sea: a review. *Quaternary Science Reviews*, 19, 1259–78.

Borbely, J. (2001) Modelling the spread of the spiny waterflea (*Bythotrephes longimanus*) in inland lakes in Ontario using gravity models and GIS. MSc thesis, University of Windsor, Windsor, Ontario.

Borcard, D., Legendre, P., and Drapeau, P. (1992) Partially out the spatial component of ecological variation. *Ecology*, 73, 1045–55.

Bos, D. and Cumming, B.F. (2003) Sedimentary cladoceran remains and their relationship to nutrients and other limnological variables in 53 lakes from central British Columbia, Canada. *Canadian Journal of Fisheries and Aquatic Sciences*, 60, 1177–89.

Bos, D., Cumming, B., Watters, C., and Smol, J.P. (1996) The relationship between zooplankton, conductivity and lake-water ionic composition in 111 lakes from the Interior of British Columbia, Canada. *International Journal of Salt Lake Research*, 5, 1–15.

Boudreau, B.P. (1999) Metals and models: Diagenetic modeling in freshwater lacustrine sediments. *Journal of Paleolimnology*, 22, 227–51.

Bouillon, D.F. (1995) Developments in emission control technologies/strategies: A case study. In J.M. Gunn (ed.), *Restoration and Recovery of an Industrial Region*. New York: Springer-Verlag, pp. 275–85.

Bourbonniere, R.A. and Meyers, P.A. (1996) Sedimentary geolipid records of historical changes in the watersheds and productivities of Lakes Ontario and Erie. *Limnology and Oceaonography*, 41, 352–9.

Boyle, J.F. (1995) A simple closure mechanism for a compact, large-diameter, gravity corer. *Journal of Paleolimnology*, 13, 85–7.

Boyle, J.F. (2001) Inorganic geochemical methods in paleolimnology. In W.M. Last and J.P. Smol (eds.), *Tracking Environmental Change Using Lake Sediments. Volume 2: Physical and Geochemical Methods*. Dordrecht: Kluwer, pp. 83–141.

Bradbury, J.P. (1988) A climatic-limnologic model of diatom succession for paleolimnological interpretation of varved sediments at Elk Lake, Minnesota. *Journal of Paleolimnology*, 1, 115–31.

Bradbury, J.P. and Dean, W.E. (eds.) (1993) *Elk Lake, Minnesota: Evidence for Rapid Climate Change in the North-Central United States*. Special Paper 276. U.S. Geological Survey, Denver, CO.

Bradley, R.S. (1999) *Paleoclimatology: Reconstructing Climates of the Quaternary*. San Diego: Academic Press.

Bradshaw, E.G. and Anderson, N.J. (2001) Validation of a diatom–phosphorus calibration set for Sweden. *Freshwater Biology*, 46, 1035–48.

Bradshaw, E.G., Rasmussen, P., and Odgaard, B.V. (2005) Mid- to late-Holocene land-use change and lake development at Dallund Sø, Denmark: Synthesis of multiproxy data, linking land and lake. *The Holocene*, 15, 1152–62.

Brännvall, M.-L., Bindler, R., Emteryd, O., and Renberg, I. (2001) Four thousand years of atmospheric lead pollution in northern Europe: a summary from Swedish lake sediments. *Journal of Paleolimnology*, 25, 421–35.

Brännvall, M.-L., Bindler, R., Emteryd, O., Nilsson, M., and Renberg, I. (1997) Stable isotope and concentration records of atmospheric lead pollution in peat and lake sediments in Sweden. *Water, Air, and Soil Pollution*, 100, 243–52.

Brännvall, M.-L., Bindler, R., Renberg, I., Emeteryd, O., Bartnicki, J., and Billström, K. (1999) The medieval metal industry was the cradle of modern large-scale atmospheric lead pollution in northern Europe. *Environmental Science & Technology*, 33, 4391–5.

Brenner, M., Whitmore, T., Flannery, M.S., and Binford, M.W. (1993) Paleolimnological methods for defining target conditions in lake restoration: Florida case studies. *Lake and Reservoir Management*, 7, 209–17.

Brenner, M., Whitmore, T.J., Lasi, M.A., Cable, J.E., and Cable, P.H. (1999) A multi-proxy trophic state reconstruction for shallow Orange Lake, Florida, USA: Possible influence of macrophytes on limnetic concentrations. *Journal of Paleolimnology*, 21, 215–33.

Brinkhurst, R.O. (1974) *The Benthos of Lakes*. London: Macmillan.

Brinkhurst, R.O., Chua, K.E., and Batoosingh, E. (1969) Modifications in sampling procedures as applied to studies on the bacteria and tubificid oligochaetes inhabiting aquatic sediments. *Journal of the Fisheries Research Board of Canada*, 26, 2581–93.

Brodersen, K.P. and Quinlan, R. (2006) Midges as palaeoindicators of lake productivity, eutrophication, and hypolimnetic oxygen. *Quaternary Science Reviews*, 25, 1995–2012.

Brodersen, K.P., Whiteside, M.C., and Lindegaard, C. (1998) Reconstruction of trophic state in Danish lakes using subfossil chydorid (Cladocera) assemblages. *Canadian Journal of Fisheries and Aquatic Sciences*, 55, 1093–103.

Brodersen, K.P., Odgaard, B.V., Vestergaard, O., and Anderson, N.J. (2001) Chironomid stratigraphy in the shallow and eutrophic Lake Søbygaard, Denmark: Chironomid–macrophyte co-occurrence. *Freshwater Biology*, 46, 253–67.

Brooks, S.J. (2006) Fossil midges (Diptera: Chironomidae) as palaeoclimatic indicators for the Eurasian region. *Quaternary Science Reviews*, 25, 1894–910.

Brooks, S.J., Bennion, H., and Birks, H.J.B. (2001) Tracing lake trophic history with a chironomid-total phosphorus inference model. *Freshwater Biology*, 46, 513–33.

Brown, K., Zeeb, B., Smol, J.P., and Pienitz, R. (1997) Taxonomy and ecological characterization of chrysophyte stomatocysts from northwestern Canada. *Canadian Journal of Botany*, 75, 842–63.

Brown, S.R. (1968) Bacterial carotenoids from freshwater sediments. *Limnology and Oceanography*, 13, 233–41.

Brown, S.R., McIntosh, H.J., and Smol, J.P. (1984) Recent paleolimnology of a meromictic lake: fossil pigments of photosynthetic bacteria. *Verhandlungen der Internationalen Vereinigung von Limnologen*, 22, 1357–60.

Brugam, R.B. and Lusk, M. (1986) Diatom evidence for neutralization in acid surface mine lakes. In J.P. Smol, R.W. Battarbee, R.B. Davis, R.B., and J. Meriläinen (eds.), *Diatoms and Lake Acidity*. Dordrecht: Kluwer, pp. 115–29.

Brunel, J. (1956) Addition du *Stephanodiscus binderanus* á la flore diatomique de l'Amérique du Nord. *Le Naturaliste Canadien*, 83, 89–95.

Callender, E. and Van Metre, P.C. (1997) Reservoir sediment cores show U.S. lead declines. *Environmental Science & Technology*, 31, 424A–428A.

Cameron, E., Prévost, C.L., McCurdy, M., Hall, G., and Doidge, B. (1998) Recent (1930) natural acidification and fish-kill in a lake that was an important food source for an Inuit community in Northern Québec, Canada. *Journal of Geochemical Exploration*, 64, 197–213.

Campana, S.E. (2005) Otolith science entering the 21st century. *Marine and Freshwater Research*, 56, 485–95.

Campana, S.E. and Thorrold, S.R. (2001) Otoliths, increments, and elements: keys to a comprehensive understanding of fish populations? *Canadian Journal of Fisheries and Aquatic Sciences*, 58, 30–8.

Campbell, L.M., Hecky, R.E., Muggide, R., Dixon, G.G., and Ramlal, P.S. (2003) Variation and distribution of total mercury in water, sediment and soil from northern Lake Victoria, East Africa. *Biogeochemistry*, 65, 195–211.

Carignan, R. and Nriagu, J.O. (1985) Trace metal deposition and mobility in the sediments of two lakes near Sudbury, Ontario. *Geochimica et Cosmochimica Acta*, 49, 1753–64.

Carignan, R. and Tessier, A. (1985) Zinc deposition in acid lakes: The role of diffusion. *Science*, 228, 1524–6.

Carignan, R., Lorrain, S., and Lum, K. (1994) A 50-year record of pollution by nutrients, trace metals and organic chemicals in the St. Lawrence River. *Canadian Journal of Fisheries and Aquatic Sciences*, 51, 1088–100.

Carlton, J.T. (1996) Biological invasions and cryptogenic species. *Ecology*, 77, 1653–5.

Carmichael, W.W. (1994) The toxins of Cyanobacteria. *Scientific American*, 270, 78–86.

Carmichael, W.W. (1997) The cyanotoxins. *Advances in Botanical Research*, 27, 211–56.

Carpenter, S.R. and Cottingham, K.L. (1997) Resilience and restoration of lakes. *Conservation Ecology*, 1(1), 2; online journal, http://www.consecol.org/vol1/iss1/art2

Carpenter, S.R., Kitchell, J.F., and Hodgson, J.R. (1985) Cascading trophic interactions and lake productivity. *BioScience*, 35, 634–9.

Carson, R. (1962) *Silent Spring*. Boston: Houghton Mifflin.

Carvalho, L., Bingham, N., and Martin, A. (2000) Lagoonal charophyte conservation: a palaeoecological approach. *Verhandlungen der Internationalen Vereinigung von Limnologen*, 27, 884–6.

Cattaneo, A., Asioli, A., Comoli, P., and Manca, M. (1998) Organisms response in a chronically polluted lake supports hypothesized link between stress and size. *Limnology and Oceanography*, 43, 1938–43.

Cattaneo, A., Couillard, Y., Wunsam, S., and Courcelles, M. (2004) Diatom taxonomic and morphological changes as indicators of metal pollution and recovery in Lac Dufault (Québec, Canada). *Journal of Paleolimnology*, 32, 163–75.

Charles, D.F. (1990) A checklist for describing and documenting diatom and chrysophyte calibration data sets and equations for inferring water chemistry. *Journal of Paleolimnology*, 3, 175–8.

Charles, D.F. and Smol, J.P. (1990) The PIRLA II project: regional assessment of lake acidification trends. *Verhandlungen der Internationalen Vereinigung von Limnologen*, 24, 474–80.

Charles, D.F. and Whitehead, D.R. (1986) The PIRLA project: Paleoecological Investigation of Recent Lake Acidification. *Hydrobiologia*, 143, 13–20.

Charles, D.F., Dixit, S.S., Cumming, B.F., and Smol, J.P. (1991) Variability in diatom and chrysophyte assemblages and inferred pH: paleolimnological studies of Big Moose L., N.Y. *Journal of Paleolimnology*, 5, 267–84.

Charles, D.F., Battarbee, R.W., Renberg I., van Dam, H., and Smol, J.P. (1989) Paleoecological analysis of lake acidification trends in North America and Europe using diatoms and chrysophytes. In S.A. Norton, S.E. Lindberg, and A.L. Page (eds.), *Acid Precipitation*. Volume 4: *Soils, Aquatic Processes, and Lake Acidification*. Stuttgart: Springer-Verlag, pp. 207–76.

Charles, D.F., Binford, M.W., Fry, B.D., et al. (1990) Paleoecological investigation of recent lake acidification in the Adirondack Mountains, N.Y. *Journal of Paleolimnology*, 3, 195–241.

Chivas, A.R. and Holmes, J.A. (eds.) (2002) *The Ostracoda: Applications in Quaternary Research.* Geophysical Monograph Series 131. Washington, DC: American Geophysical Union, pp. 1–313.

Christensen, E. and Zhang, X. (1993) Sources of polycyclic aromatic hydrocarbons to Lake Michigan determined from sediment records. *Environmental Science & Technology*, 27, 139–46.

Christensen, J.R., MacDuffee, M., Macdonald, R.W., Whiticar, M., and Ross P.S. (2005) Pacific salmon deliver persistent organic pollutants to British Columbia grizzly bears. *Environmental Science & Technology*, 39, 6952–60.

Christie, C.E. and Smol, J.P. (1996) Limnological effects of 19th century canal construction and other disturbances on the trophic state history of Upper Rideau Lake, Ontario. *Lake and Reservoir Management*, 12, 78–90.

Clerk, S.A. (2001) Fossil chironomids as indicators of water quality impacts from aquaculture activities. M.Sc. thesis, Queen's University, Kingston.

Clerk, S., Selbie, D.T., and Smol, J.P. (2004) Cage aquaculture and water quality changes in the LaCloche Channel, Lake Huron, Canada: A paleolimnological assessment. *Canadian Journal of Fisheries and Aquatic Sciences*, 61, 1691–701.

Cocquyt, C. and Israël, Y. (2004) A microtome for sectioning lake sediment cores at a very high resolution. *Journal of Paleolimnology*, 32, 301–4.

Cohen, A.S. (2003) *Paleolimnology: The History and Evolution of Lake Systems.* Oxford: Oxford University Press.

Cohen, A.S., Palacios-Fest, M.R., McGill, J., et al. (2005a) Paleolimnological investigations of anthropogenic environmental change in Lake Tanganyika: I. An introduction to the project. *Journal of Paleolimnology*, 34, 1–18.

Cohen, A.S., Palacios-Fest, M.R., Msaky, E.S., et al. (2005b) Paleolimnological investigations of anthropogenic environmental change in Lake Tanganyika: IX. Summary of paleorecords of environmental change and catchment deforestation at Lake Tanganyika and impacts on the Lake Tanganyika ecosystem. *Journal of Paleolimnology*, 34, 125–45.

Conley, D.J. and Schelske, C.L. (2001) Biogenic silica. In J.P. Smol, H.J.B. Birks, and W.M. Last (eds.), *Tracking Environmental Change Using Lake Sediments. Volume 3: Terrestrial, Algal, and Siliceous Indicators.* Dordrecht: Kluwer, pp. 281–93.

Coolen, M.J.L., Muyzer, G., Rijpstra, W.I.C., Schouten, S., Volkman, J.K., and Sinninghe Damsté, J.S. (2004) Combined DNA and lipid analyses of sediments reveal changes in Holocene haptophyte and diatom populations in an Antarctic lake. *Earth and Planetary Science Letters*, 223, 225–39.

Cooper, S.R. (1995) Chesapeake Bay watershed historical land use: Impact on water quality and diatom communities. *Ecological Applications*, 5, 703–23.

Cooper, S.R. (1999) Estuarine paleoenvironmental reconstructions using diatoms. In E.F. Stoermer and J.P. Smol (eds.), *The Diatoms: Applications to the Environmental and Earth Sciences.* Cambridge: Cambridge University Press, pp. 352–73.

Cooper, S.R. (2000) *The History of Water Quality in North Carolina Estuarine Waters as Documented in the Stratigraphic Record.* Water Resources Research Institute of the University of North Carolina, Raleigh. Report No. 2000–327.

Cooper, S.R. and Brush, G. (1991) Long-term history of Chesapeake Bay anoxia. *Science*, 254, 992–6.

Cooper, S.R. and Brush, G.S. (1993) A 2500-year history of anoxia and eutrophication in Chesapeake Bay. *Estuaries*, 16, 617–26.

Cosby, B.J., Hornberger, J.N., Galloway, J.N., and Wright, R.F. (1985) Freshwater acidification from atmospheric deposition of sulphuric acid: a quantitative model. *Environmental Science & Technology*, 19, 1144–9.

Cottingham, K., Rusak, J., and Leavitt, P.R. (2000) Increased ecosystem variability and reduced predictability following fertilization: Evidence from palaeolimnology. *Ecology Letters*, 3, 340–8.

Cranwell, P.A. (1984) Organic geochemistry of lacustrine sediments: triterpenoids of higher-plant origin reflecting post-glacial vegetational succession. In E.Y. Haworth and J.W.G. Lund (eds.), *Lake Sediments and Environmental History.* Minneapolis: University of Minnesota Press, pp. 69–92.

Cromer, L., Gibson, J.A.E., Swadling, K., and Ritz, D.A. (2005) Faunal microfossils: Indicators of Holocene ecological change in a saline Antarctic lake. *Palaeogeography, Palaeoclimatology, Palaeoecology,* 221, 83–97.

Cronberg, G. and Sandgren, C.D. (1986) A proposal for the development of standardized nomenclature and terminology for chrysophycean statospores. In J. Kristiansen and R.A. Andersen (eds.), *Chrysophytes: Aspects and Problems.* Cambridge: Cambridge University Press, pp. 317–28.

Crutzen, P.J. (2002) Geology of mankind. *Nature,* 415, 23.

Cumming, B.F., Smol, J.P., and Birks, H.J.B. (1992a) Scaled chrysophytes (Chrysophyceae and Synurophyceae) from Adirondack (N.Y., USA) drainage lakes and their relationship to measured environmental variables, with special reference to lakewater pH and labile monomeric aluminum. *Journal of Phycology,* 28, 162–78.

Cumming, B.F., Davey, K., Smol, J.P., and Birks, H.J. (1994) When did Adirondack Mountain lakes begin to acidify and are they still acidifying? *Canadian Journal of Fisheries and Aquatic Sciences,* 51, 1550–68.

Cumming, B.F., Glew, J.R., Smol, J.P., and Norton, S.A. (1993) Comment on "Core compression and surficial sediment loss of lake sediments of high porosity caused by gravity coring" (Crusius and Anderson). *Limnology and Oceanography,* 38, 695–9.

Cumming, B.F., Smol, J.P., Kingston, J.C., et al., (1992b) How much acidification has occurred in Adirondack region (New York, USA) lakes since pre-industrial times? *Canadian Journal of Fisheries and Aquatic Sciences,* 49, 128–41.

Cunningham, L., Raymond, B., Snape, I., and Riddle, M.J. (2005) Benthic diatom communities as indicators of anthropogenic metal contamination at Casey Station, Antarctica. *Journal of Paleolimnology,* 33, 499–513.

Currie, D.C. and Walker, I.R. (1992) Recognition and palaeohydrological significance of fossil black fly larvae, with a key to Nearctic genera (Diptera: Simuliidae). *Journal of Paleolimnology,* 7, 23–35.

Curry, B.B. (1998) An environmental tolerance index for ostracodes as indicators of physical and chemical factors in aquatic habitats. *Palaeogeography, Palaeoclimatology, Palaeoecology,* 148, 51–63.

Curry, B. and Delorme, D. (2003) Ostracode-based reconstruction from 23,300 to about 20,250 cal yr BP of climate, and paleohydrology of a groundwater-fed pond near St. Louis, Missouri. *Journal of Paleolimnology,* 29, 199–207.

D'Arrigo, R., Wiles, G., Jacoby, G., and Villalba, R. (1999) North Pacific sea surface temperatures: past variations inferred from tree rings. *Geophysical Research Letters,* 26, 2757–60.

Damon, E., Donahue, D.J., Gore, B.H., et al. (1989) Radiocarbon dating of the Shroud of Turin. *Nature,* 337, 611–14.

Daniels, R.A. (1996) *Guide to the Identification of Scales of Inland Fishes of Northeastern North America.* New York State Museum Bulletin 488, pp. 1–97.

Daniels, R.A. and Peteet, D. (1998) Fish scale evidence for rapid post-glacial colonization of an Atlantic coastal pond. *Global Ecology and Biogeography Letters,* 7, 467–76.

Dapples, F., Lotter, A.F., van Leeuwen, J.F.N., van der Knaap, W.O., Dimitriadis, S., and Oswald, D. (2002) Paleolimnological evidence for increased landslide activity due to forest clearing and land-use since 3600 cal BP in the western Swiss Alps. *Journal of Paleolimnology,* 27, 239–48.

Dauvalter, V. (1992) Concentration of heavy metals in superficial lake sediments of Pechenga District, Murmansk Region, Russia. *Vatten,* 2, 141–5.

Dauvalter, V. (1994) Heavy metals in lake sediments of the Kola Peninsula, Russia. *The Science of the Total Environment,* 158, 51–61.

Dauvalter, V. (1997a) Heavy metal concentrations in lake sediments as an index of freshwater ecosystem pollution. In R.M.M. Crawford (ed.), *Disturbances and Recovery in Arctic Lands: An Ecological Perspective.* Dordrecht: Kluwer, pp. 333–51.

Dauvalter, V. (1997b) Metal concentrations in sediments in acidifying lakes in Finnish Lapland. *Boreal Environment Research,* 2, 369–79.

Davidson, G.A. (1988) A modified tape-peel technique for preparing permanent qualitative microfossil slides. *Journal of Paleolimnology,* 1, 229–34.

Davidson, T.A., Sayer, C.D., Perrow, M.R., and Tomlinson, M.L. (2003) Representation of fish communities by scale sub-fossils in shallow lakes: implications for inferring percid–cyprinid shifts. *Journal of Paleolimnology*, 30, 441–9.

Davis, M.B. (1976) Erosion rates and land use history in southern Michigan. *Environmental Conservation*, 3, 139–48.

Davis, R.B., Anderson, D.S., Norton, S.A., and Whiting, M.C. (1994) Acidity of twelve northern New England (U.S.A.) lakes in recent centuries. *Journal of Paleolimnology*, 12, 103–54.

Davis, R.B., Hess, C.T., Norton, S.A., Hanson, D.W., Hoagland, K.D., and Andreson, D.S. (1984) ^{137}Cs and ^{210}Pb dating of sediments from soft-water lakes in New England (U.S.A.) and Scandinavia, a failure of ^{137}Cs dating. *Chemical Geology*, 44, 151–85.

DeDeckker, P. (1988) An account of the techniques using ostracodes in palaeolimnology in Australia. *Palaeogeography, Palaeoclimatology, Palaeoecology*, 62, 463–75.

DeDeckker, P., Colin, J.P., and Peypouquet, J.P. (eds.) (1988) *Ostracoda in the Earth Sciences*. Amsterdam: Elsevier.

de Wit, C.A., Alaee, M., and Muir, D.C.G. (2006) Levels and trends of brominated flame retardants in the Arctic. *Chemosphere*, 64, 209–33.

de Wolf, H. and Cleveringa, P. (1999) The impact of the beer industry on medieval water quality. In S. Mayama, M. Idei, and I. Koizumi (eds.), *Proceedings of the Fourteenth International Diatom Symposium*. Koenigstein: Koeltz Scientific Books, pp. 511–22.

Dean, W.E., Jr. (1974) Determination of carbonate and organic matter in calcareous sediments and sedimentary rocks by loss on ignition: Comparison with other methods. *Journal of Sedimentary Petrology*, 44, 242–8.

Dearing, J.A. (1992) Sediment yields and sources in a Welsh upland lake-catchment during the past 800 years. *Earth Surface Processes and Landforms*, 17, 1–22.

Dearing, J.A. (1994) Reconstructing the history of soil erosion. In N. Roberts (ed.), *The Changing Global Environment*. Oxford: Blackwell, pp. 242–61.

Dearing, J.A. (1999) Holocene environmental change from magnetic proxies in lake sediments. In B.A. Maher and R. Thompson (eds.), *Quaternary Climates, Environments and Magnetism*. Cambridge: Cambridge University Press, pp. 231–78.

Dearing, J.A. and Foster, I.D.L. (1993) Lake sediments and geomorphological processes: Some thoughts. In J. McManus and R.W. Duck (eds.), *Geomorphology and Sedimentology of Lakes and Reservoirs*. Chichester: John Wiley, pp. 5–14.

Dearing, J.A. and Jones, R.T. (2003) Coupling temporal and spatial dimensions of global sediment flux through lake and marine sediment records. *Global and Planetary Change*, 39, 147–68.

Dearing, J.A., Alström, K., Bergman, A., Regnell, J., and Sandgren, P. (1990) Recent and long-term records of soil erosion from southern Sweden. In J. Boardman, I.D.L. Foster, and J.A. Dearing (eds.), *Soil Erosion on Agricultural Land*. Chichester: John Wiley, pp. 173–91.

Dearing, J.A., Håkansson, H., Liedberg-Jönsson, B., et al. (1987) Lake sediments used to quantify the erosional response to land use change in southern Sweden. *Oikos*, 50, 60–78.

Dearing, J.A., Jones, R.T., Shen, J., et al. (2007) Using multiple archives of past climate–human–environment interactions to inform future landscape sustainability: the lake Erhai catchment, Yunnan Province, China. *Journal of Paleolimnology* (in press).

Deevey, E.S., Jr. (1969) Coaxing history to conduct experiments. *BioScience*, 19, 40–3.

Delorme, D. (1982) Lake Erie oxygen: The prehistoric record. *Canadian Journal of Fisheries and Aquatic Sciences*, 39, 1021–9.

Detenbeck, N.E., Johnston, C.A., and Niemi, G.E. (1993) Wetland effects of lake quality in the Minneapolis/St. Paul metropolitan area. *Landscape Ecology*, 8, 39–61.

Diamond, J. (1995) Easter's end. *Discover*, August 1995, 63–9.

Dickens, C. (1843) *A Christmas Carol*. London: Elliott Stock.

Dillon, P.J. and Rigler, F.H. (1974) The phosphorus–chlorophyll relationship in lakes. *Limnology and Oceanography*, 19, 767–73.

Dillon, P.J., Evans, H.E., and Scholer, P.J. (1988) The effects of acidification on metal budgets of lakes and catchments. *Biogeochemistry*, 5, 201–20.

Dillon, P.J., Reid, R.A., and de Grosbois, E. (1987) The rate of acidification of aquatic ecosystems in Ontario, Canada. *Nature*, 329, 45–8.

Dillon, P.J., Nicholls, K.H., Scheider, W.A., Yan, N.D., and Jeffries, D.S. (1986) *Lakeshore Capacity Study, Trophic Status*. Research and Special Projects Branch, Ontario Ministry of Municipal Affairs and Housing. Queen's Printer for Ontario, Toronto.

Dixit, A.S., Dixit, S.S., and Smol, J.P. (1992) Algal microfossils provide high temporal resolution of environmental trends. *Water, Air, and Soil Pollution*, 62, 75–87.

Dixit, A.S., Dixit, S.S., and Smol, J.P. (1996) Long-term trends in limnological characteristics in the Aurora trout lakes, Sudbury, Canada. *Hydrobiologia*, 335, 171–81.

Dixit, A.S., Hall, R., Leavitt, P., Quinlan, R., and Smol, J.P. (2000) Effects of sequential depositional basins on lake response to urban and agricultural pollution: A palaeoecological analysis of the Qu'Applelle Valley, Saskatchewan, Canada. *Freshwater Biology*, 43, 319–38.

Dixit, S.S. and Smol, J.P. (1994) Diatoms as environmental indicators in the Environmental Monitoring and Assessment – Surface Waters (EMAP–SW) program. *Environmental Monitoring and Assessment*, 31, 275–306.

Dixit, S.S. and Smol, J.P. (1995) Diatom evidence of past water quality changes in Adirondack seepage lakes (New York, U.S.A.). *Diatom Research*, 10, 113–29.

Dixit, S.S., Dixit, A.S., and Smol, J.P. (1989) Relationship between chrysophyte assemblages and environmental variables in 72 Sudbury lakes as examined by canonical correspondence analysis (CCA). *Canadian Journal of Fisheries and Aquatic Sciences*, 46, 1667–76.

Dixit, S.S., Dixit, A.S., and Smol, J.P. (1991) Multivariable environmental inferences based on diatom assemblages from Sudbury (Canada) lakes. *Freshwater Biology*, 26, 251–65.

Dixit, S.S., Dixit, A.S., and Smol, J.P. (1992) Assessment of changes in lake water chemistry in Sudbury area lakes since preindustrial times. *Canadian Journal of Fisheries and Aquatic Sciences*, 49 (Suppl. 1), 8–16.

Dixit, S.S., Dixit, A.S., Smol, J.P., and Keller, W. (1995) Reading the records stored in the lake sediments: A method of examining the history and extent of industrial damage to lakes. In J.M. Gunn (ed.), *Restoration and Recovery of an Industrial Region*. New York: Springer-Verlag, pp. 33–44.

Dixit, S.S., Keller, W., Dixit, A.S., and Smol, J.P. (2001) Diatom-inferred dissolved organic carbon reconstructions provide assessments of past UV-B penetration in Canadian shield lakes. *Canadian Journal of Fisheries and Aquatic Sciences*, 58, 543–50.

Dixit, S.S., Dixit, A., Smol, J.P., Hughes, R., and Paulsen, S. (2000) Water quality changes from human activities in three northeastern USA lakes. *Lake and Reservoir Management*, 16, 305–21.

Dixit, S.S., Smol, J.P., Charles, D.F., Hughes, R.M., Paulsen, S.G., and Collins, G.B. (1999) Assessing water quality changes in the lakes of the Northeastern United States using sediment diatoms. *Canadian Journal of Fisheries and Aquatic Sciences*, 56, 131–52.

Dixit, S.S., Cumming, B.F., Kingston, J.C., et al. (1993) Diatom assemblages from Adirondack lakes (N.Y., U.S.A.) and the development of inference models for retrospective environmental assessment. *Journal of Paleolimnology*, 8, 27–47.

Dobson, J.E., Rush, R.M., and Peplies, R.W. (1990) Forest blowdown and lake acidification. *Annals of the American Association of Geographers*, 80, 343–61.

Donar, C.M., Neely, R.K., and Stoermer, E.F. (1996) Diatom succession in an urban reservoir system. *Journal of Paleolimnology*, 15, 237–43.

Donovan, J.J., Smith, A.J., Panek, V.A., Engstrom, D.R., and Ito, E. (2002) Climate-driven hydrologic transients in lake sediment records: calibration of groundwater conditions using 20th century drought. *Quaternary Science Reviews*, 21, 605–24.

Doubleday, N.C. and Smol, J.P. (2005) Atlas and classification scheme of Arctic combustion particles suitable for paleoenvironmental work. *Journal of Paleolimnology*, 33, 393–431.

Douglas, M.S.V. and Smol, J.P. (1999) Freshwater diatoms as indicators of environmental change in the High Arctic. In E.F. Stoermer and J.P. Smol (eds.), *The Diatoms: Applications for the Environmental and Earth Sciences*. Cambridge: Cambridge University Press, pp. 227–44.

Douglas, M. and Smol, J.P. (2000) Eutrophication and recovery in the High Arctic: Meretta Lake revisited. *Hydrobiologia*, 431, 193–204.

Douglas, M.S.V. and Smol, J.P. (2001) Siliceous protozoan plates and scales. In J.P. Smol, H.J.B. Birks, and W.M. Last (eds.), *Tracking Environmental Change Using Lake Sediments. Volume 3: Terrestrial, Algal, and Siliceous Indicators*. Dordrecht: Kluwer, pp. 265–79.

Douglas, M.S.V., Smol, J.P., and Blake, W., Jr. (1994) Marked post-18th century environmental change in High Arctic ecosystems. *Science*, 266, 416–19.

Douglas, M.S.V., Smol, J.P., Savelle, J.M., and Blais, J.M. (2004) Prehistoric Inuit whalers affected Arctic freshwater ecosystems. *Proceedings of the National Academy of Sciences*, 101, 1613–17.

Duff, K.E. and Smol, J.P. (1995) Chrysophycean cyst assemblages and their relationship to water chemistry in 71 Adirondack Park (New York, U.S.A.) lakes. *Archiv für Hydrobiologie*, 134, 307–36.

Duff, K., Zeeb, B., and Smol, J.P. (1995) *Atlas of Chrysophycean Cysts*. Dordrecht: Kluwer.

Duff, K.E., Laing, T.E., Smol, J.P., and Lean, D. (1999) Limnological characteristics of lakes located across the treeline zone of northern Siberia. *Hydrobiologia*, 391, 203–20.

Duffy, M.A., Perry, L.J., Kearns, C.M., Weider L.J., and Hairston, Jr., N.G. (2000) Paleogenetic evidence for a past invasion of Onondaga Lake, New York, by exotic *Daphnia curvirostris* using mtDNA from dormant eggs. *Limnology and Oceanography*, 45, 1409–14.

Dumont, H.J., Cocquyt, C., Fontugne, M., et al. (1998) The end of moai quarrying and its effect of Lake Rano Raraku, Easter Island. *Journal of Paleolimnology*, 20, 409–22.

Edlund, M.B. and Stoermer, E.F. (2000) A 200,000-year, high-resolution record of diatom productivity and community makeup from Lake Baikal shows high correspondence to the marine oxygen-isotope record of climate change. *Limnology and Oceanography*, 45, 948–62.

Edlund, M.B., Taylor, C.M., Schelske, C.L., and Stoermer, E.F. (2000) *Thalassiosira baltica* (Bacillariophyta), a new exotic species in the Great Lakes. *Canadian Journal of Fisheries and Aquatic Sciences*, 54, 610–15.

Eggermont, H., Heiri, O., and Verschuren, D. (2006) Fossil Chironomidae (Insecta: Diptera) as quantitative indicators of past salinity in African lakes. *Quaternary Science Reviews*, 25, 1966–94.

Einarsson, A., Óskarsson, H., and Haflidason, H. (1993) Stratigraphy of fossil pigments and *Cladophora* and its relationship with deposition of tephra in Lake Myvatn, Iceland. *Journal of Paleolimnology*, 8, 15–26.

Eisenreich, S., Capel, P., Robbins, J., and Bourbonniere, R. (1989) Accumulation and diagenesis of chlorinated hydrocarbons in lacustrine sediments. *Environmental Science & Technology*, 23, 1116–26.

Ek, A.S. and Renberg, I. (2001) Heavy metal pollution and lake acidity changes caused by one thousand years of copper mining at Falun, central Sweden. *Journal of Paleolimnology*, 26, 89–107.

Ekdahl, E., Teranes, J., Wittkop, C., Stoermer, E., Reavie, E., and Smol, J.P. (2007) Diatom assemblage response following Iroquoian and Euro-Canadian eutrophication of Crawford Lake, Ontario, Canada. *Journal of Paleolimnology*, 37, 233–46.

Elias, S.A. (2001) Coleoptera and Trichoptera. In J.P. Smol, H.J.B. Birks, and W.M. Last (eds.), *Tracking Environmental Change Using Lake Sediments. Volume 4: Zoological Indicators*. Dordrecht: Kluwer, pp. 67–80.

Elton, C.S. (2000) *The Ecology of Invasions by Animals and Plants*. Chicago: The University of Chicago Press.

Enache, M. and Prairie, Y.T. (2000) Paleolimnological reconstruction of forest fire induced changes in lake biogeochemistry (Lac Francis, Abitibi, Quebec, Canada). *Canadian Journal of Fisheries and Aquatic Sciences*, 57 (Suppl. 2), 146–54.

Engstrom, D. and Wright, H.E., Jr. (1984) Chemical stratigraphy of lake sediments. In E. Haworth and J. Lund (eds.), *Lake Sediments and Environmental History*. Minneapolis: University of Minnesota Press, pp. 11–67.

Engstrom, D.R., Whitlock, C., Fritz, S.C., and Wright, H.E., Jr. (1991) Recent environmental changes inferred from the sediments of small lakes in Yellowstone's northern range. *Journal of Paleolimnology*, 5, 139–74.

Engstrom, D.R., Whitlock, C., Fritz, S.C., and Wright, H.E., Jr. (1994) Reinventing erosion in Yellowstone's northern range. *Journal of Paleolimnology*, 10, 159–61.

Environment Canada and the U.S. Environmental Protection Agency (1995) *The Great Lakes: An Environmental Atlas and Resource Book*. Toronto: Environment Canada/Chicago: U.S. Environmental Protection Agency.

Erästö, P. and Holmström, L. (2006) Selection of prior distributions and multiscale analysis in Bayesian temperature reconstructions based on fossil assemblages. *Journal of Paleolimnology*, 36, 69–80.

Esposito, K. (1998) The metal that slipped away. *Wisconsin Natural Resources Magazine*, 21(1), 19–25.

European Commission (2000) Directive 2000/60/EC of the European Parliament and on the Council of 23 October 2000 establishing a framework for Community action in the field of water policy. *Official Journal L*, 327/1.

European Commission (2003) *Common Implementation Strategy for the Water Framework Directive (2000/60/EC), Guidance Document No. 10, Rivers and Lakes – Typology, Reference Conditions and Classification Systems*. Produced by Working Group 2.3 – REFCOND. Luxembourg: Office for Official Publications of the European Communities.

Evenset, A., Christensen, G.N., and Kallenborn, R. (2005) Selected chlorobornanes, polychlorinated naphthalenes and brominated flame retardants in Bjørnøya (Bear Island) freshwater biota. *Environmental Pollution*, 136, 419–30.

Evenset, A., Christensen, G.N., Skotvold, T., et al. (2004) A comparison of organic contaminants in two High Arctic lake ecosystems, Bjørnøya (Bear Island), Norway. *Science of the Total Environment*, 318, 125–41.

Fallu, M., Pienitz, R., Walker, I.R., and Overpeck, J. (2004) AMS ^{14}C dating of tundra lake sediments using chironomid head capsules. *Journal of Paleolimnology*, 31, 11–22.

Fernández, P., Vilanova, R.M., Martínez, C., Appleby, P., and Grimalt, J.O. (2000) The historical record of atmospheric pyrolytic pollution over Europe in the sedimentary PAH from remote mountain lakes. *Environmental Science & Technology*, 34, 1906–13.

Finney, B.P., Gregory-Eaves, I., Douglas, M.S.V., and Smol, J.P. (2002) Fisheries productivity in the northeastern Pacific Ocean over the past 2200 years. *Nature*, 416, 729–33.

Finney, B.P., Gregory-Eaves, I., Sweetman, J., Douglas, M.D., and Smol, J.P. (2000) Impacts of climatic change and fishing on Pacific salmon abundance over the past 300 years. *Science*, 290, 795–9.

Fisher, T.G. (2004) Vibracoring from lake ice with a lightweight monopod and piston coring apparatus. *Journal of Paleolimnology*, 31, 377–82.

Fitzgerald, W.F., Engstrom, D.E., Mason, R.P., and Nater, E.A. (1998) The case for atmospheric mercury contamination in remote areas. *Environmental Science & Technology*, 32, 1–7.

Flannery, T. (2005) *The Weather Makers: How Man Is Changing the Climate and What It Means for Life on Earth*. New York: Atlantic Monthly Press.

Flenley, J.R., King, S.M., Jackson, J., Chew, C., Teller, J.T., and Prentice, M.E. (1991) The late-Quaternary vegetational climatic history of Easter Island. *Journal of Quaternary Science*, 6, 85–115.

Flower, R. (1998) Paleolimnology and recent environmental change in Lake Baikal: an introduction and overview of interrelated concurrent studies. *Journal of Paleolimnology*, 20, 107–17.

Ford, M.S. (1990) A 10,000-yr history of natural ecosystem acidification. *Ecological Monographs*, 60, 57–89.

Forester, R.M. (1991) Ostracode assemblages from springs in the Western United States: Implications for paleohydrology. *Memoirs of the Entomological Society of Canada*, 155, 181–201.

Forester, R.M. and Smith, A.J. (1992) Microfossils as indicators of paleohydrology and paleoclimate. In Nuclear Energy Agency, *Paleohydrological Methods and Applications*. Paris: Organization for Economic Co-operation and Development, pp. 39–57.

Forester, R.M., Bradbury, J.P., Carter, C., et al. (1999) *The Climatic and Hydrologic History of Southern Nevada During the Late Quaternary*. Open File Report 98-635. Denver, CO: US Geological Survey.

Forman, M.R. and Whiteside, M.C. (2000) Occurrence of *Bythotrephes cederstroemi* in inland lakes in northeastern Minnesota as indicated from sediment records. *Verhandlungen der Internationalen Vereinigung von Limnologen*, 27, 1552–5.

Forrest, F., Reavie, E.D., and Smol, J.P. (2002) Comparing limnological changes associated with 19th century canal construction and other catchment disturbances in four lakes within the Rideau Canal system, Ontario, Canada. *Journal of Limnology*, 61, 183–97.

Förstner, U. and Wittmann, G.T.W. (1981) *Metal Pollution in the Aquatic Environment*. Berlin: Springer-Verlag.

Foster, D.L., Mighall, T.M., Proffitt, H., Walling, D.E., and Owens, P.N. (2006) Post-depositional ^{137}Cs mobility in the sediments of three shallow coastal Lagoons, SW England. *Journal of Paleolimnology*, 35, 881–95.

Foster, I.D.L. and Walling, D.E. (1994) Using reservoir deposits to reconstruct changing sediment yields and sources in the catchment of the Old Mill Reservoir, South Devon, UK, over the past 50 years. *Hydrological Sciences*, 39, 347–68.

Fox, H.M. (1965) Ostracod Crustacea from ricefields in Italy. *Memoire dell'Istituto Italiano di Idrobiologia*, 18, 205–14.

Francis, D.R. (1997) Bryozoan statoblasts in the recent sediments of Douglas Lake, Michigan. *Journal of Paleolimnology*, 17, 255–61.

Francis, D.R. (2001) Bryozoan statoblasts. In J.P. Smol, H.J.B. Birks, and W.M. Last (eds.), *Tracking Environmental Change Using Lake Sediments*. Volume 4: *Zoological Indicators*. Dordrecht: Kluwer, pp. 105–23.

Francis, G. (1878) Poisonous Australian lakes. *Nature*, 18, 11–12.

Francus, P. (ed.) (2004) *Image Analysis, Sediments and Paleoenvironments*. Dordrecht: Springer.

Francus, P., Bradley, R.S., Abbott, M.B., Patridge, W., and Keimig, F. (2002) Paleoclimate studies of minerogenic sediments using annually resolved textural parameters. *Geophysical Research Letters*, 29; doi:10.1029/2002GL015082.

Frazier, B., Wiener, J., Rada, R., and Engstrom, D. (2000) Stratigraphy and historic accumulation of mercury in recent depositional sediments in the Sudbury River, Massachusetts, U.S.A. *Canadian Journal of Fisheries and Aquatic Sciences*, 57, 1062–72.

Frey, D.G. (1964) Remains of animals in Quaternary lake and bog sediments and their interpretation. *Archiv für Hydrobiologie, Ergebnisse der Limnologie*, 2, 1–114.

Frey, D.G. (1974) Paleolimnology. *Mitteilungen Internationale Vereinigung Limnologia*, 20, 95–123.

Fritz, S. (1989) Lake development and limnological response to prehistoric and historic land-use changes in Diss, Norfolk, U.K. *Journal of Ecology*, 77, 182–202.

Fritz, S. and Carlson, R.E. (1982) Stratigraphic diatom and chemical evidence for acid strip-mine lake recovery. *Water, Air, and Soil Pollution*, 17, 151–63.

Fritz, S.F., Cumming, B.F., Gasse, F., and Laird, K. (1999) Diatoms as indicators of hydrologic and climatic change in saline lakes. In E.F. Stoermer and J.P. Smol (eds.), *The Diatoms: Applications for the Environmental and Earth Sciences*. Cambridge: Cambridge University Press, pp. 41–72.

Frost, T.M. (2001) Freshwater sponges. In J.P. Smol, H.J.B. Birks, and W.M. Last (eds.), *Tracking Environmental Change Using Lake Sediments*. Volume 3: *Terrestrial, Algal, and Siliceous Indicators*. Dordrecht: Kluwer, pp. 253–63.

Fry, B. (1986) Stable sulfur isotopic distribution and sulfate reduction in lake sediments of the Adirondack Mountains, New York. *Biogeochemistry*, 2, 329–43.

Galassi, S., Valsecchi, S., and Tartari, G. (1997) The distribution of PCB's and chlorinated pesticides in two connected Himalayan lakes. *Water, Air, and Soil Pollution*, 99, 717–25.

Gandouin, E., Maasri, A., Van Vliet-Lanoë, B., and Franquet, E. (2006) Chironomid (Insecta: Diptera) assemblages from a gradient of lotic and lentic waterbodies in river floodplains of France: A methodological tool for paleoecological applications. *Journal of Paleolimnology*, 35, 149–66.

García, A. (1994) Charophyta: their use in paleolimnology. *Journal of Paleolimnology*, 10, 43–52.

Ginn, B., Cumming, B.F., and Smol, J.P. (2007) Tracking pH changes since pre-industrial times in 51 low-alkalinity lakes in Nova Scotia, Canada. *Canadian Journal of Fisheries and Aquatic Sciences* (in press).

Glew, J. (1988) A portable extruding device for close interval sectioning of unconsolidated core samples. *Journal of Paleolimnology*, 1, 235–9.

Glew, J. (1989) A new trigger mechanism for sediment samplers. *Journal of Paleolimnology*, 2, 241–3.

Glew, J. (1991) Miniature gravity corer for recovering short sediment cores. *Journal of Paleolimnology*, 5, 285–7.

Glew, J. (1995) Conversion of shallow water gravity coring equipment for deep water operation. *Journal of Paleolimnology*, 14, 83–8.

Glew, J.R., Smol, J.P., and Last, W.M. (2001) Sediment core collection and extrusion. In W.M. Last and J.P. Smol (eds.), *Tracking Environmental Change Using Lake Sediments. Volume 1: Basin Analysis, Coring, and Chronological Techniques*. Dordrecht: Kluwer, pp. 73–105.

Goldberg, E.D. (1985) *Black Carbon in the Environment: Properties and Distribution*. New York: Wiley Interscience.

Gomez, A. and Carvalho, G.R. (2000) Sex, parthenogenesis and genetic structure of rotifers: microsatellite analysis of contemporary and resting egg bank populations. *Molecular Ecology*, 9, 203–14.

Goosse, H., Renssen, H., Timmermann, A., and Bradley, R.S. (2005) Internal and forced climate variability during the last millennium: a model-data comparison using ensemble simulations. *Quaternary Science Reviews*, 24, 1345–60.

Goosse, H., Renssen, H., Timmermann, A., Bradley, R.S., and Mann, M.E. (2006) Using paleoclimate proxy-data to select optimal realisations in an ensemble of simulations of the climate of the past millennium. *Climate Dynamics*, 27, 165–84.

Gorham, E. (1996) Lakes under a three-pronged attack. *Nature*, 381, 109–10.

Gorham, E. and Gordon, A.G. (1960) The influence of smelter fumes upon the chemical composition of lake waters near Sudbury, Ontario, and upon the surrounding vegetation. *Canadian Journal of Botany*, 38, 477–87.

Gorham, E.G. (1989) Scientific understanding of ecosystem acidification: A historical review. *Ambio*, 18, 150–4.

Granberg, K. (1983) *Closteridium perfringens* (Holland) as an indicator of human effluent in the sediment of Lake Tuomiojärvi, central Finland. *Hydrobiologia*, 103, 181–4.

Greenwood, M.T., Wood, P.J., and Monk, W.A. (2006) The use of fossil caddisfly assemblages in the reconstruction of flow environments from floodplain paleochannels of the River Trent, England. *Journal of Paleolimnology*, 35, 747–61.

Gregory-Eaves, I., Finney, B.P., Douglas, M.S.V., and Smol, J.P. (2004) Inferring sockeye salmon (*Oncorhynchus nerka*) population dynamics and water quality changes in a stained nursery lake over the past ~ 500 years. *Canadian Journal of Fisheries and Aquatic Sciences*, 61, 1235–46.

Gregory-Eaves, I., Smol, J.P., Douglas, M.S.V., and Finney, B.P. (2003) Diatoms and sockeye salmon (*Oncorhynchus nerka*) population dynamics: Reconstructions of salmon-derived nutrients in two lakes from Kodiak Island, Alaska. *Journal of Paleolimnology*, 30, 35–53.

Gregory-Eaves, I., Demers, M.J., Kimpe, L., et al. (2007) Tracing salmon-derived nutrients and contaminants in freshwater food webs across a pronounced spawner density gradient. *Environmental Toxicology and Chemistry*, 26, in press.

Guhrén, M., Bigler, C., and Renberg, I. (2007) Liming placed in a long-term perspective: A paleolimnological study of 12 lakes in the Swedish liming program. *Journal of Paleolimnology*, 37, 247–58.

Guilizzoni, P. and Lami, A. (1988) Sub-fossil pigments as a guide to the phytoplankton history of the acidified Lake Orta (N. Italy). *Verhandlungen der Internationalen Vereinigung von Limnologen*, 23, 874–9.

Guilizzoni, P., Lami, A., Marchetto, A., Appleby, P.G., and Alvisi, F. (2001) Fourteen years of palaeolimnological research of a past industrial polluted lake (L. Orta, Northern Italy): an overview. *Journal of Limnology*, 60, 249–62.

Gunn, J.M. (ed.) (1995) *Restoration and Recovery of an Industrial Region*. New York: Springer-Verlag.

Gupta, B.K.S., Turner, R.E., and Rabalais, N.N. (1996) Seasonal oxygen depletion in continental-shelf waters of Louisiana: Historical record of benthic foraminifers. *Geology*, 24, 227–30.

Haas, J.N. (1999) Charophyte population dynamics during the Late Quaternary at Lake Bibersee, Switzerland. *Australian Journal of Botany*, 47, 315–24.

Hairston, Jr., N.G., Kearns, C.M., Demma, L.P., and Effler, S.W. (2005) Species-specific *Daphnia* phenotypes: A history of industrial pollution and pelagic ecosystem response. *Ecology*, 86, 1669–78.

Hairston, N.G., Jr., Perry, L., Bohonak, A., Fellows, M., Kearns, C., and Engstron, D. (1999) Population biology of a failed invasion: Paleolimnology of *Daphnia exilis* in upstate New York. *Limnology and Oceanography*, 44, 477–86.

Håkanson, L. and Jansson, M. (1983) *Principles of Lake Sedimentology*. Berlin: Springer-Verlag.

Håkanson, L. and Peters, R.H. (1995) *Predictive Limnology*. Amsterdam: SPB Academic Publishing.

Hall, R.I. and Smol, J.P. (1999) Diatoms as indicators of lake eutrophication. In: E.F. Stoermer and J.P. Smol (eds.), *The Diatoms: Applications for the Environmental and Earth Sciences*. Cambridge: Cambridge University Press, pp. 128–68.

Hall, R.I. and Yan, N.D. (1997) Comparing annual population growth of the exotic invader *Bythrotrephes* by using sediment and plankton records. *Limnology and Oceanography*, 42, 112–20.

Hall, R.I., Leavitt, P., Dixit, A., Quinlan, R., and Smol, J.P. (1999a) Limnological succession in reservoirs: A paleolimnological comparison of two methods of reservoir formation. *Canadian Journal of Fisheries and Aquatic Sciences*, 56, 1109–21.

Hall, R.I., Leavitt, P.R., Quinlan, R., Dixit, A., and Smol, J.P. (1999b) Effects of agriculture, urbanization and climate on water quality in the Northern Great Plains. *Limnology and Oceanography*, 44, 739–56.

Hamilton, W.L. (1994) Recent environmental changes inferred from the sediments of small lakes in Yellowstone's northern range (Engstrom et al., 1991). *Journal of Paleolimnology*, 10, 153–7.

Harper, M.A. (1994) Did Europeans introduce *Asterionella formosa* Hassall to New Zealand? In J.P. Kociolek (eds.), *Proceedings of the 11th International Diatom Symposium*, California Academy of Sciences, San Francisco, pp. 479–84.

Harriman, R. and Morrison, B.R.S. (1982) The ecology of streams draining forested and non-forested catchments in an area of central Scotland subject to acidic precipitation. *Hydrobiologia*, 88, 251–63.

Harris, G.P. (1994) Pattern, process and prediction in aquatic ecology. A limnological view of some general ecological problems. *Freshwater Biology*, 32, 143–60.

Harris, M.A., Cumming, B.F., and Smol, J.P. (2006) Assessment of recent environmental changes in New Brunswick (Canada) lakes based on paleolimnological shifts in diatom species assemblages. *Canadian Journal of Botany*, 84, 151–63.

Hausmann, S. and Pienitz, R. (2007) Seasonal climate inferences from high- resolution modern diatom data along a climate gradient: a case study. *Journal of Paleolimnology*, in press.

Hay, M.B., Smol, J.P., Pipke, K., and Lesack, L. (1997) A diatom-based paleohydrological model for the Mackenzie Delta, Northwest Territories, Canada. *Arctic and Alpine Research*, 29, 430–44.

Hecky, R.E. (1993) The eutrophication of Lake Victoria. *Verhandlungen der Internatioanlen Vereinigung von Limnologen*, 25, 39–48.

Heinke, G.W. and Deans, B. (1973) Water supply and waste disposal systems for Arctic communities. *Arctic*, 26, 149–59.

Heinrichs, M.L. and Walker, I.R. (2006) Fossil midges and palaeosalinity: potential as indicators of hydrological balance and sea-level change. *Quaternary Science Reviews*, 25, 1948–65.

Heiri, O., Lotter, A.F., and Lemcke, G. (2001) Loss on ignition as a method for estimating organic and carbonate content in sediments: reproducibility and comparability of results. *Journal of Paleolimnology*, 25, 101–10.

Hendy, C.H. (2000) Late Quaternary lakes in the McMurdo Sound region of Antarctica. *Geografiska Annaler*, 82 A, 411–32.

Hengeveld, H. (1991) *Understanding Atmospheric Change*. State of the Environment Report, Atmospheric Environment Service, Ottawa.

Hermanson, M.K. (1998) Anthropogenic mercury deposition to Arctic lake sediments. *Water, Air, and Soil Pollution*, 101, 309–21.

Hermanson, M.K. and Brozowski, J.R. (2005) History of Inuit community exposure to lead, cadmium, and mercury in sewage lake sediments. *Environmental Health Perspectives*, 113, 1308–12.

Hickman, M. and Schweger, C. (1991) Oscillathanthin and myxoxanthophyll in two cores from Lake Wabamum, Alberta, Canada. *Journal of Paleolimnology*, 5, 127–37.

Hidore, J.J. (1996) *Global Environmental Change – Its Nature and Impact*. Upper Saddle River, NJ: Prenctice Hall.

Hodell, D. and Schelske, C. (1998) Production, sedimentation, and isotopic composition of organic matter in Lake Ontario. *Limnology and Oceanography*, 43, 200–14.

Hodell, D., Schelske, C., Fahnenstiel, G., and Robbins, L. (1998) Biologically induced calcite and its isotopic composition in Lake Ontario. *Limnology and Oceanography*, 43, 187–99.

Hodgson, D.A. and Johnston, N.M. (1997) Inferring seal populations from lake sediments. *Nature*, 387, 30–1.

Hodgson, D.A., Tyler, P., and Vyverman, W. (1997) The palaeolimnology of Lake Fidler, a meromictic lake in south-west Tasmania and the significance of recent human impact. *Journal of Paleolimnology*, 18, 312–33.

Hodgson, D.A., Johnston, N.M., Caulkett, A.P., and Jones, V.J. (1998a) Palaeolimnology of Antarctic fur seal *Arctocephalus gazella* populations and implications for Antarctic management. *Biological Conservation*, 83, 145–54.

Hodgson, D.A., Vyverman, W., Chepstow-Lusty, A., and Tyler, P.A. (2000) From rainforest to wasteland in 100 years: The limnological legacy of the Queenston mines, Western Tasmania. *Archiv für Hydrobiologie*, 149, 153–76.

Hodgson, D.A., Wright, S.W., Tyler, P.A., and Davies, N. (1998b) Analysis of fossil pigments from algae and bacteria in meromictic Lake Fidler, Tasmania, and its application to lake management. *Journal of Paleolimnology*, 19, 1–22.

Hodgson, D.A., Vyverman, W., Verleyen, E., et al. (2005) Late Pleistocene record of elevated UV radiation in an Antarctic lake. *Earth and Planetary Science Letters*, 236, 765–72.

Holmes, J.A. (1996) Trace-element and stable-isotope geochemistry of non-marine ostracod shells in Quaternary palaeoenvironmental reconstruction. *Journal of Paleolimnology*, 15, 223–35.

Holmes, J.A. (2001) Ostracoda. In J.P. Smol, H.J.B. Birks, and W.M. Last (eds.), *Tracking Environmental Change Using Lake Sediments*. Volume 4: *Zoological Indicators*. Dordrecht: Kluwer, pp. 125–51.

Holtham, A.J., Gregory-Eaves, I., Pellatt, M., et al. (2004) Reconstructing sockeye salmon (*Oncorhynchus nerka*) dynamics in Alaska and British Columbia using paleolimnology: The influence of flushing rates, terrestrial input and low salmon escapement densities. *Journal of Paleolimnology*, 32, 255–71.

Huang, Y., Shuman, B., Wang Y., and Webb, T. (2004) Hydrogen isotope ratios of individual lipids in lake sediments as novel tracers of climatic and environmental change: A surface sediment test. *Journal of Paleolimnology*, 31, 363–75.

Husar, R.B., Sullivan, T.J., and Charles, D.F. (1991) Methods for assessing long-term trends in atmospheric deposition and surface water chemistry. In D.F. Charles (ed.), *Acidic Deposition and Aquatic Ecosystems: Regional Case Studies*. New York: Springer-Verlag, pp. 65–82.

Hustedt, F. (1937–9) Systematische und ökologische Untersuchungen über der Diatomeen-Flora von Java, Bali, Sumatra. *Archiv für Hydrobiologie* (Suppl.), 15 & 16.

Hutchinson, G.E. (1957) *A Treatise on Limnology. I. Geography, Physics, and Chemistry*. New York: John Wiley.

Hutchinson, G.E. (1958) Concluding remarks. *Cold Spring Harbor Symposium on Quantitative Biology*, 22, 415–27.

Hutchinson, G.E. and Cowgill, U. (1970) XII. The history of the lake: A synthesis. *Transactions of the American Philosophical Society*, 60, 163–70.

Hutchinson, N.J., Neary, B.P., and Dillon, P.J. (1991) Validation and use of Ontario's Trophic Status Model for establishing lake development guidelines. *Lake and Reservoir Management*, 7, 13–23.

Hutchinson, T.C. and Havas, M. (1986) Recovery of previously acidified lakes near Coniston, Ontario, Canada, following reductions in atmospheric sulphur and metal emissions. *Water, Air, and Soil Pollution*, 29, 319–33.

Hvorslev, M.J. (1949) *Subsurface Exploration and Sampling Soils for Civil Engineering Purposes*. Vicksburg, MI: American Society of Civil Engineers, Waterways Experiment Station, Corps of Engineers, U.S. Army.

Ilyashuk, B., Ilyashuk, E., and Dauvalter, V. (2003) Chironomid responses to long-term metal contamination: A paleolimnological study in two bays of Lake Imandra, Kola Peninsula, northern Russia. *Journal of Paleolimnology*, 30, 217–30.

Intergovernmental Panel on Climate Change (IPCC) (2000) *Special Report on Emissions Scenarios*. Cambridge: Cambridge University Press.

Interlandi, S.J. and Kilham, S.S. (1998) Assessing the effects of nitrogen deposition on mountain waters: A study of phytoplankton community dynamics. *Water Science Technology*, 38, 139–46.

Ito, E. (2001) Application of stable isotope techniques to inorganic and biogenic carbonates. In W.M. Last and J.P. Smol (eds.), *Tracking Environmental Change Using Lake Sediments. Volume 2: Physical and Geochemical Methods.* Dordrecht: Kluwer, pp. 351–71.

Jackson, S.T. and Overpeck, J.T. (2000) Responses of plant populations and communities to environmental changes of the late Quaternary. *Paleobiology*, 26 (Suppl. 4), 194–220.

Jankovská, V. and Komárek, J. (2000) Indicative value of Pediastrum and other coccal green algae in palaeoecology. *Folia Geobotanica*, 35, 59–82.

Jenkins, A., Whitehead, P.G., Cosby, B.J., and Birks, H.J.B. (1990) modeling long-term acidification: a comparison with diatom reconstructions and the implications for reversibility. *Philosophical Transactions of the Royal Society of London*, B 327, 435–40.

Jeppesen, E., Madsen, E.A., and Jensen, J.P. (1996) Reconstructing the past density of planktivorous fish and trophic structure from sedimentary zooplankton fossils: a surface sediment calibration data set from shallow lakes. *Freshwater Biology*, 36, 115–27.

Jeppesen, E., Leavitt, P., De Meester, L., and Jensen, J.P. (2001) Functional ecology and palaeolimnology: Using cladoceran remains to reconstruct anthropogenic impacts. *Trends in Ecology and Evolution*, 16, 191–8.

Jeppesen, E., Jensen, J.P., Lauridsen, T.L., et al. (2003) Sub-fossils of cladocerans in the surface sediment of 135 lakes as proxies for community structure of zooplankton, fish abundance and lake temperature. *Hydrobiologia*, 481, 323–30.

Johannsson, O.E., Mills, E.L., and O'Gorman, R. (1991) Changes in the nearshore and offshore zooplankton communities in Lake Ontario: 1981–88. *Canadian Journal of Fisheries and Aquatic Sciences*, 48, 1546–57.

Johnson-Pyrtle, A., Scott, M.R., Laing, T.E., and Smol, J.P. (2000) ^{137}Cs distribution and geochemistry of Lena River (Siberia) drainage basin lake sediments. *The Science of the Total Environment*, 255, 145–59.

Jones, J.M. and Hao, J. (1993) Ombrotrophic peat as a medium for historical monitoring of heavy metal pollution. *Environmental Geochemistry and Health*, 15, 67–74.

Jones, R. (1984) Heavy metals in the sediments of Llangorse Lake, Wales, since Celtic–Roman times. *Verhandlungen der Internationalen Vereinigung von Limnologen*, 22, 1377–82.

Jones, R., Benson-Evans, K., and Chambers, F.M. (1985) Human influence upon sedimentation in Llangorse Lake, Wales. *Earth Surface Processes and Landforms*, 10, 227–35.

Jones, R., Chambers, F.M., and Benson-Evans, K. (1991) Heavy metals (Cu and Zn) in recent sediments of Llangorse Lake, Wales: non-ferrous smelting, Napoleon and the price of wheat – a palaeoecological study. *Hydrobiologia*, 214, 149–54.

Jones, R., Benson-Evans, K., Chambers, F.M., Seddon, A.B., and Tai, Y.C. (1978) Biological and chemical studies of sediments from Llangorse Lake, Wales. *Verhandlungen der Internationalen Vereinigung von Limnologen*, 20, 642–8.

Jones, V.J. and Juggins, S. (1995) The construction of a diatom-based chlorophyll *a* transfer function and its application at three lakes on Signy Island (maritime Antarctic) subject to differing degrees of enrichment. *Freshwater Biology*, 34, 433–45.

Jones, V.J., Hodgson, D.A., and Chepstow-Lusty, A. (2000) Palaeolimnological evidence for marked Holocene environmental changes on Signy Island, Antarctica. *The Holocene*, 10, 43–60.

Jones, V.J., Stevenson, A.C., and Battarbee, R.W. (1989) Acidification of lakes in Galloway, south west Scotland: A diatom and pollen study of the post-glacial history of Round Loch of Glenhead. *Journal of Ecology*, 77, 1–23.

Jones, V.L., Leng, M.J., Solovieva, N., Sloane, H.J., and Tarasov, P. (2004) Holocene climate of the Kola Peninsula; evidence from the oxygen isotope record of diatom silica. *Quaternary Science Reviews*, 23, 833–9.

Jongman, R.H.G., ter Braak, C.J.F., and van Tongeren, O.F.R. (eds.) (1995) *Data Analysis in Community and Landscape Ecology.* Cambridge: Cambridge University Press.

Jordan, P., Rippey, B., and Anderson, N.J. (2001) Modeling diffuse phosphorus loads from land to freshwater using the sedimentary record. *Environmental Science & Technology*, 35, 815–19.

Jørgensen, S.E. (1980) *Lake Management*. Oxford: Pergamon.

Joshi, S.R. and McNeely, R. (1988) Detection and fallout of ^{155}Eu and ^{207}Bi in a ^{210}Pb-dated lake sediment core. *Journal of Radioanalytical and Nuclear Chemistry*, 122, 183–91.

Julius, M.L., Stoermer, E.F., Taylor, C.M., and Schelske, C.L. (1998) Local extirpation of *Stephanodiscus niagarae* (Bacillariophyceae) in the recent limnological record of Lake Ontario. *Journal of Phycology*, 34, 766–71.

Kajak, Z., Kacprzak, K., and Polkowski, R. (1965) Chwytacz rurowy do pobierania prób dna. *Ekologia Polska Seria B*, 11, 159–65.

Kalff, J. (2001) *Limnology*. Upper Saddle River, NJ: Prentice Hall.

Kamenik, C. and Schmidt, R. (2005) Chrysophyte resting stages: a tool for reconstructing winter/spring climate from Alpine lake sediments. *Boreas*, 34, 477–98.

Kansanen, P.H., Jaakola, T., Kulmala, S., and Suutarinen, R. (1991) Sedimentation and distribution of gamma-emitting radionuclides in bottom sediments of southern Lake Päijänne, Finland, after the Chernobyl accident. *Hydrobiologia*, 222, 121–40.

Karst, T.L. and Smol, J.P. (2000) Paleolimnological evidence of a clear-water trophic state equilibrium in a shallow, macrophyte-dominated lake. *Aquatic Sciences*, 62, 20–38.

Keilty, T.J. (1988) A new biological marker layer in the sediments of the Great Lakes: *Bythotrephes cederstroemi* (Schödler) spines. *Journal of Great Lakes Research*, 14, 369–71.

Keller, W. and Gunn, J.M. (1995) Lake water quality improvements and recovering aquatic communities. In J.M. Gunn (ed.), *Restoration and Recovery of an Industrial Region*. New York: Springer-Verlag, pp. 67–80.

Kelley, D., Brachfeld, S., Nater, E., and Wright, H. (2006) Sources of sediment in Lake Pepin on the upper Mississippi River in response to Holocene climatic changes. *Journal of Paleolimnology*, 35, 193–206.

Kemp, A.E.S., Dean, J., Pearce, R.B., and Pike, J. (2001) Recognition and analysis of bedding and sediment fabric features. In W.M. Last and J.P. Smol (eds.), *Tracking Environmental Change Using Lake Sediments*. Volume 2: *Physical and Geochemical Methods*. Dordrecht: Kluwer, pp. 7–22.

Kendall, C. (1998) Tracing nitrogen sources and cycles in catchments. In C. Kendall and J.J. McDonnell (eds.), *Isotope Tracers in Catchment Hydrology*. New York: Elsevier, pp. 519–76.

Kerfoot, W.C. (1974) Net accumulation rates and history of cladoceran communities. *Ecology*, 55, 51–61.

Kerfoot, W.C. and Weider, L.J. (2004) Experimental paleoecology (resurrection ecology): Chasing Van Valen's Red Queen hypothesis. *Limnology and Oceanography*, 49 (4, part 2, suppl.), 1300–16.

Kerfoot, W.C., Lauster, G., and Robbins, J.A. (1994) Paleolimnological study of copper mining around Lake Superior: Artificial varves from Portage Lake provide a high resolution record. *Limnology and Oceanography*, 39, 649–69.

Kerfoot, W.C., Robbins, J.A., and Weider, L.J. (1999) A new approach to historical reconstruction: Combining descriptive and experimental paleolimnology. *Limnology and Oceanography*, 44, 1232–47.

Kilby, G.W. and Batley, G.E. (1993) Chemical indicators of sediment chronology. Australian *Journal of Marine and Freshwater Research*, 44, 635–47.

King, J. and Peck, J. (2001) Use of paleomagnetism in studies of lake sediments. In W.M. Last and J.P. Smol (eds.), *Tracking Environmental Change Using Lake Sediments*. Volume 1: *Basin Analysis, Coring, and Chronological Techniques*. Dordrecht: Kluwer, pp. 371–89.

Kingston, J.C., Birks, H.J.B., Uutala, A.J., Cumming, B.F., and Smol, J.P. (1992) Assessing trends in fishery resources and lake water aluminum from paleolimnological analyses of siliceous algae. *Canadian Journal of Fisheries and Aquatic Sciences*, 49, 116–27.

Kingston, J.C., Cook, R.B., Kreis, Jr., R.G., et al. (1990) Paleoecological investigation of recent acidification in the northern Great Lakes states. *Journal of Paleolimnology*, 4, 153–201.

Kling, H.J., Mugidde, R., and Hecky, R.E. (2001) Recent changes in the phytoplankton community of Lake Victoria in response to eutrophication. In M. Munawar and R.E. Hecky (eds.), *Great Lakes of the World: Food Webs, Health and Integrity*. Leiden: Backhuys, pp. 47–66.

Klink, A. (1989) The Lower Rhine: Palaeoecological analysis. In G.E. Petts (ed.), *Historical Change of Large Alluvial Rivers: Western Europe.* Chichester: John Wiley, pp. 183–201.

Knapp, R.A., Garton, J.A., and Sarnelle, O. (2001) The use of egg shells to infer the historical presence of copepods in alpine lakes. *Journal of Paleolimnology*, 25, 539–43.

Koenig, R. (1999) Digging in the mire for the air of ancient times. *Science*, 284, 900–1.

Koinig, K., Schmidt, R., Sommaruga-Wögrath, S., Tessadri, R., and Psenner, R. (1998a) Climate change as the primary cause of pH shifts in a high alpine lake. *Water, Air, and Soil Pollution*, 104, 167–80.

Koinig, K.A., Sommaruga-Wögrath, S., Schmidt, R., Tessadri, R., and Psenner, R. (1998b) Acidification processes in high alpine lakes. In M.J. Haigh, J. Křeček, G.S. Raijwar, and M.P. Kilmartin (eds.), *Headwaters: Water Resources and Soil Conservation.* Rotterdam: A.A. Balkema, pp. 45–54.

Komárek, J. and Jankovská, V. (2001) *Review of the Green Algal Genus Pediastrum; Implication for Pollen-Analytical Research.* Berlin: J. Cremer.

Korhola, A. and Rautio, M. (2001) Cladocera and other branchiopod crustaceans. In J.P. Smol, H.J.B. Birks, and W.M. Last (eds.), *Tracking Environmental Change Using Lake Sediments.* Volume 4: *Zoological Indicators.* Dordrecht: Kluwer, pp. 5–41.

Korhola, A. and Smol, J.P. (2001) Ebridians. In J.P. Smol, H.J.B. Birks, and W.M. Last (eds.), *Tracking Environmental Change Using Lake Sediments.* Volume 3: *Terrestrial, Algal, and Siliceous Indicators.* Dordrecht: Kluwer, pp. 225–34.

Korhola, A.A. and Tikkanen, M. (1991) Holocene development and early extreme acidification in a small hilltop lake in southern Finland. *Boreas*, 20, 333–56.

Korhola, A., Weckström, J., and Nyman, M. (1999) Predicting the long-term acidification trends in small subarctic lakes using diatoms. *Journal of Applied Ecology*, 36, 1021–34.

Korhola, A., Tikkanen, M., and Weckström, J. (2005) Quantification of the Holocene lake-level changes in Finnish Lapland by means of a cladocera – lake depth transfer model. *Journal of Paleolimnology*, 34, 175–90.

Korsman, T. (1999) Temporal and spatial trends of lake acidity in northern Sweden. *Journal of Paleolimnology*, 22, 1–15.

Korsman, T., Renberg, I., and Anderson, N.J. (1994) A palaeolimnological test of the influence of Norway spruce (*Picea abies*) immigration on lake-water acidity. *The Holocene*, 4, 132–40.

Korsman, T., Dåbakk, E., Nilsson, M.B., and Renbert, I. (2001) Near-Infrared spectrometry (NIRS) in paleolimnology. In W.M. Last and J.P. Smol (eds.), *Tracking Environmental Change Using Lake Sediments.* Volume 2: *Physical and Geochemical Methods.* Dordrecht: Kluwer, pp. 299–317.

Köster, D., Racca, J.M.J., and Pienitz R. (2004) Diatom-based inference models and reconstructions revisited: methods and transformations. *Journal of Paleolimnology*, 32, 233–46.

Kowalewski, M., Avila Serranao, G.E., Flessa, K.W., and Goodfriend, G.A. (2000) Dead delta's former productivity: Two trillion shells at the mouth of the Colorado River. *Geology*, 28, 1059–62.

Krebs, C.J. (1994) *Ecology.* Cambridge, MA: Harper & Row.

Kreiser, A.M., Appleby, P.G., Natkanski, J., Rippey, B., and Battarbee, R.W. (1990) Afforestation and lake acidification: a comparison of four sites in Scotland. *Philosophical Transactions of the Royal Society of London*, B 327, 377–83.

Krisnaswami, S., Lal, D., Martin, J.M., and Meybeck, M. (1971) Geochronology of lake sediments. *Earth and Planetary Science Letters*, 11, 407–14.

Krug, E.C. and Frink, C.R. (1983) Acid rain on acid soil: A new perspective. *Science*, 221, 520–5.

Krümmel, E.M., Gregory-Eaves, I., Macdonald, R., et al. (2005) Concentrations and fluxes of salmon derived PCBs in lake sediments. *Environmental Science & Technology*, 39, 7020–6.

Krümmel, E.M., Macdonald, R., Kimpe, L.E., et al. (2003) Delivery of pollutants by spawning salmon. *Nature*, 425, 255–6.

Kullenberg, B. (1947) The piston core sampler. *Svenska Hydrografisk-Biologiska Kommissionens Skrifter III. Hydrograpfi* 1, pp. 1–46.

Kumke, T., Kienel, U., Weckström, J., Korhola, A., and Hubberten, H.-W. (2004) Inferred Holocene palaeotemperatures from diatoms at Lake Lama, Central Siberia. *Arctic, Antarctic, and Alpine Research*, 36, 624–34.

Kunilov, V.Y. (1994) Geology of the Noril'sk region: The history of the discovery, prospecting, exploration and mining of the Noril'sk deposits. In P.C. Lightfoot and A.J. Naldrett (eds.), *Proceedings of the Sudbury–Noril'sk Symposium*, Volume 5. Ontario: Ministry of Northern Development and Mines, pp. 203–16.

Kuylenstierna, J.C.I., Rodhe, H., Cinderby, S., and Hicks, K. (2001) Acidification in developing countries: Ecosystem sensitivity and the critical load approach on a global scale. *Ambio*, 30, 20–8.

Laird, K. and Cumming, B.F. (2001) A regional paleolimnological assessment of the impact of clear-cutting on lakes from the central interior of British Columbia. *Canadian Journal of Fisheries and Aquatic Sciences*, 58, 492–505.

Laird, K., Cumming, B., and Nordin, R. (2001) A regional paleolimnological assessment of the impact of clear-cutting on lakes from the west coast of Vancouver Island, British Columbia. *Canadian Journal of Fisheries and Aquatic Sciences*, 58, 479–91.

Lamontagne, S. and Schindler, D.W. (1994) Historical status of fish populations in Canadian Rocky Mountain lakes inferred from subfossil *Chaoborus* (Diptera: Chaoboridae) mandibles. *Canadian Journal of Fisheries and Aquatic Sciences*, 51, 1376–83.

Lamoureux, S. (2001) Varve chronology techniques. In W.M. Last and J.P. Smol (eds.), *Tracking Environmental Change Using Lake Sediments. Volume 1: Basin Analysis, Coring, and Chronological Techniques*. Dordrecht: Kluwer, pp. 247–60.

Lamoureux, S. and Gilbert, R. (2004) Physical and chemical properties and proxies of high latitude lake sediments. In J. Pienitz, M.S.V. Douglas, and J.P. Smol (eds.), *Long-Term Environmental Change in Arctic and Antarctic Lakes*. Dordrecht: Springer, pp. 53–87.

Lamoureux, S.F., Stewart, K.A., Forbes, A.C., and Fortin, D. (2006) Multidecadal variations and decline in spring discharge in the Canadian middle Arctic since 1550 AD. *Geophysical Research Letters*, 33; doi:10.1029/2005GL024942.

Landers, D., Gubala, C., Verta, M., et al. (1998) Using lake sediment mercury flux ratios to evaluate the regional and continental dimensions of mercury deposition in Arctic and boreal ecosystems. *Atmospheric Environment*, 32, 919–28.

Last, W.M. (2001a) Textural analysis of lake sediments. In W.M. Last and J.P. Smol (eds.), *Tracking Environmental Change Using Lake Sediments. Volume 2: Physical and Geochemical Methods*. Dordrecht: Kluwer, pp. 41–81.

Last, W.M. (2001b) Mineralogical analysis of lake sediments. In W.M. Last and J.P. Smol (eds.), *Tracking Environmental Change Using Lake Sediments. Volume 2: Physical and Geochemical Methods*. Dordrecht: Kluwer, pp. 143–87.

Last, W.M. and Ginn, F.M. (2005) Saline systems of the Great Plains of western Canada: an overview of the limnogeology and paleolimnology. *Saline Systems*, 1; doi:10.1186/1746-1448-1-10.

Last, W.M. and Smol, J.P. (eds.) (2001a) *Tracking Environmental Change Using Lake Sediments. Volume 1: Basin Analysis, Coring, and Chronological Techniques*. Dordrecht: Kluwer.

Last, W.M. and Smol, J.P. (eds.) (2001b) *Tracking Environmental Change Using Lake Sediments. Volume 2: Physical and Geochemical Methods*. Dordrecht: Kluwer.

Laurion, I., Vincent, W.F., and Lean, D.R.S. (1997) Underwater ultraviolet radiation: development of spectral models for northern high latitude lakes. *Photochemistry and Photobiology*, 65, 107–14.

Leavitt, P.R. and Hodgson, D.A. (2001) Sedimentary pigments. In J.P. Smol, H.J.B. Birks, and W.M. Last (eds.), *Tracking Environmental Change Using Lake Sediments. Volume 3: Terrestrial, Algal, and Siliceous Indicators*. Dordrecht: Kluwer, pp. 295–325.

Leavitt, P.R., Carpenter, S.R., and Kitchell, J.F. (1989) Whole-lake experiments: The annual record of fossil pigments and zooplankton. *Limnology and Oceanography*, 34, 700–17.

Leavitt, P.R., Hodgson, D.A., and Pienitz, R. (2003) Past UV radiation environments and impacts on lakes. In E.W. Helbling and H. Zagarese (eds.), *UV Effects in Aquatic Organisms and Ecosystems*. Comprehensive Series in Photochemistry and Photobiology. Cambridge: The Royal Society of Chemistry, pp. 509–45.

Leavitt, P.R., Brock, C.S., Ebel, C., and Pantoine, A. (2006) Landscape-scale effects of urban nitrogen on a chain of freshwater lakes in central North America. *Limnology and Oceanography*, 51, 2262–77.

Leavitt, P.R., Donald, D.B., Vinebrooke, R., Smol, J.P., and Schindler, D.W. (1997) Past ultraviolet radiation environments revealed using fossil pigments in lakes. *Nature*, 388, 457–9.

Leavitt, P.R., Findlay, D.L., Hall, R.I., Schindler, D.W., and Smol, J.P. (1999) Algal responses to dissolved organic carbon loss and pH decline during whole-lake acidification: Evidence from paleolimnology. *Limnology and Oceanography*, 44, 757–73.

Leavitt, P.R., Hann, B.J., Smol, J.P., et al. (1994) Analysis of whole-lake experiments with paleolimnology: An overview of results from Lake 227, Experimental Lakes Area, Ontario. *Canadian Journal of Fisheries and Aquatic Sciences*, 51, 2322–32.

Leira, M., Jordan, P., Taylor, D., et al. (2006) Assessing the ecological status of candidate reference lakes in Ireland using palaeolimnology. *Journal of Applied Ecology*, 43, 816–27.

Leng M.J. (ed.) (2006) *Isotopes in Palaeoenvironmental Research.* Dordrecht: Springer.

Leng, M.L. and Marshall, J.D. (2004) Palaeoclimate interpretation of stable isotope data from lake sediment archives. *Quaternary Science Reviews*, 23, 811–31.

Leng, M.J., Metcalfe, S., and Davies, S. (2005) Investigating late Holocene climate variability in central Mexico using carbon isotope ratios in organic materials and oxygen isotope ratios from diatom silica within lacustrine sediments. *Journal of Paleolimnology*, 34, 413–31.

Leng, M.J., Lamb, A.L., Heaton, T.H.E., et al. (2006) Isotopes in lake sediments. In M.J. Leng (ed.), *Isotopes in Palaeoenvironmental Research.* Dordrecht: Springer, pp. 148–76.

Leppänen, H. and Oikari, A. (2001) Retene and resin acid concentrations in sediment profiles of a lake recovering from exposure to pulp mill effluents. *Journal of Paleolimnology*, 25, 367–74.

Leroy, S.A.G. and Colman, S.M. (2001) Coring and drilling equipment and procedures for recovery of long lacustrine sequences. In W.M. Last and J.P. Smol (eds.), *Tracking Environmental Change Using Lake Sediments.* Volume 1: *Basin Analysis, Coring, and Chronological Techniques.* Dordrecht: Kluwer, pp. 107–35.

Lesack, L., Marsh, P., and Hecky, R. (1998) Spatial and temporal dynamics of major solute chemistry among Mackenzie Delta lakes. *Limnology and Oceanography*, 43, 1530–43.

Li, Y.F. and Macdonald, R.W. (2005) Sources and pathways of selected organochlorine pesticides to the Arctic and the effect of pathway divergence of HCH trends in biota. *Science of the Total Environment*, 342, 87–106.

Likens, G.E. (ed.) (1985) *An Ecosystem Approach to Aquatic Ecology: Mirror Lake and Its Environment.* New York: Springer-Verlag.

Line, J.M., ter Braak, C.J.F., and Birks, H.J.B. (1994) WACALIB version 3.3: a computer intensive program to reconstruct environmental variables from fossil assemblages by weighted averaging and to derive sample-specific errors of prediction. *Journal of Paleolimnology*, 10, 147–52.

Lipiatou, E., Hecky, R.E., Eisenreich, S.J., Lockhart, L., Muir, D., and Wilkinson, P. (1996) Recent ecosystem changes in Lake Victoria reflected in sedimentary natural and anthropogenic compounds. In T.C. Johnson and E. Odada (eds.), *The Limnology, Climatology and Paleoclimatology of the East African Lakes.* Toronto: Gordon and Breach, pp. 523–42.

Little, J. and Smol, J.P. (2000) Changes in fossil midge (Chironomidae) assemblages in response to cultural activities in a shallow polymictic lake. *Journal of Paleolimnology*, 23, 207–12.

Little, J. and Smol, J.P. (2001) A chironomid-based model for inferring late-summer hypolimnetic oxygen in Southeastern Ontario lakes. *Journal of Paleolimnology*, 26, 259–79.

Little, J., Hall, R., Quinlan, R., and Smol, J.P. (2000) Past trophic status and hypolimnetic anoxia during eutrophication and remediation of Gravenhurst Bay, Ontario: Comparison of diatoms, chironomids, and historical records. *Canadian Journal of Fisheries and Aquatic Sciences*, 57, 333–41.

Liu, X., Sun, L., Xie, Z., Yin, X., and Wang, Y. (2005) A 1,300 year record of penguin populations at Ardley Island in the Antarctic, as deduced from the geochemical data in the ornithogenic lake sediments. *Arctic, Antarctic, and Alpine Research*, 37, 490–8.

Liu, Z., Henderson, A.C.G., and Huang, Y. (2006) Alkenone-based reconstruction of late-Holocene surface temperature and salinity changes in Lake Qinghai, China. *Geophysical Research Letters*, 33; doi:10.1029/2006GL026151.

Livingstone, D.A. (1955) A lightweight piston sampler for lake deposits. *Ecology*, 36, 137–9.

Lockhart, W.L., Macdonald, R.W., Outridge, P.M., et al. (2000) Tests of the fidelity of lake sediment core records of mercury deposition to known histories of mercury contamination. *The Science of the Total Environment*, 260, 171–80.

Lodge, D.M. (1993) Species invasions and deletions: Community effects and response to climate and habitat change. In P.M. Karieva, J.G. Kingsolver, and R.B. Huey (eds.), *Biotic Interactions and Global Change*. Sunderland: Sinauer, pp. 367–87.

Lotter, A.F. (1998) The recent eutrophication of Baldeggersee (Switzerland) as assessed by fossil diatom assemblages. *The Holocene*, 8, 395–405.

Lotter, A.F. and Bigler, C. (2000) Do diatoms in a high mountain lake in the Swiss Alps reflect the length of ice-cover? *Aquatic Sciences*, 62, 125–41.

Lotter, A.F., Pienitz, R., and Schmidt, R. (1999) Diatoms as indicators of environmental change near Arctic and alpine treeline. In E.F. Stoermer and J.P. Smol (eds.), *The Diatoms: Applications for the Environmental and Earth Sciences*. Cambridge: Cambridge University Press, pp. 205–26.

Lotter, A.F., Birks, H.J.B., Hofmann, W., and Marchetto, A. (1997a) Modern diatom, cladocera, chironomids, and chrysophyte cyst assemblages as quantitative indicators for the reconstruction of past environmental conditions in the Alps. I. Climate. *Journal of Paleolimnology*, 18, 395–420.

Lotter, A.F., Birks, H.J.B., Hofmann, W., and Marchetto, A. (1998) Modern diatom, cladocera, chironomids, and chrysophyte cyst assemblages as quantitative indicators for the reconstruction of past environmental conditions in the Alps. II. Nutrients. *Journal of Paleolimnology*, 19, 443–63.

Lotter, A., Sturm, M., Teranes, J., and Wehrli, B. (1997b) Varve formation since 1885 and high-resolution varve analyses in hypereutrophic Baldeggersee (Switzerland). *Aquatic Sciences*, 59, 304–26.

Lozon, J.D. and MacIsaac, H.J. (1997) Biological invasions: Are they dependent on disturbance? *Environmental Reviews*, 5, 131–44.

Ludlam, S., Feeny, S., and Douglas, M.S.V. (1996) Changes in the importance of lotic and littoral diatoms in a High Arctic lake over the last 191 years. *Journal of Paleolimnology*, 16, 184–204.

Ludwig, J.P., Kurita-Matsuba, H., Auman, H.J., et al. (1996) Deformities, PCBs, and TCDD-equivalents in double-crested cormorants (*Phalacrocorax auritus*) and Caspian terns (*Hydroprogne caspia*) of the upper Great Lakes 1986–1991: Testing a cause–effect hypothesis. *Journal of Great Lakes Research*, 22, 172–97.

MacDonald, G.M. (2001) Conifer stomata. In J.P. Smol, H.J.B. Birks, and W.M. Last (eds.), *Tracking Environmental Change Using Lake Sediments. Volume 3: Terrestrial, Algal, and Siliceous Indicators*. Dordrecht: Kluwer, pp. 33–47.

MacDonald, G.M., Beukens, R.P., Kieser, W.E., and Vitt, D.H. (1991) Comparative radiocarbon dating of terrestrial plant macrofossils and aquatic moss from the "ice-free corridor" of western Canada. *Geology*, 15, 837–40.

MacDonald, G.M., Edwards, T., Moser, K., Pienitz, R., and Smol, J.P. (1993) Rapid response of treeline vegetation and lakes to past climate warming. *Nature*, 361, 243–6.

Macdonald, R.W., Harner, T., and Fyfe, J. (2005) Recent climate change in the Canadian Arctic and its impact on contaminant pathways and interpretation of temporal trend data. *Science of the Total Environment*, 342, 5–86.

Macdonald, R.W., Ikonomou, M.G., and Paton, D.W. (1998) Historical inputs of PCDDs, PCDFs and PCBs to a British Columbia interior lake: the effect of environmental controls on pulp mill emissions. *Environmental Science & Technology*, 32, 331–7.

Mackay, A.W., Ryves, D.B., Morley, D.W., Jewson, D.J., and Rioual, P. (2006) Assessing the vulnerability of endemic diatom species in Lake Baikal to predicted future climate change: a multivariate approach. *Global Change Biology*, 12, 2297–315.

Mackay, A.W., Flower, R., Kuzima, A., et al. (1998) Diatom succession trends in recent sediments from Lake Baikal and their relation to atmospheric pollution and to climate change. *Philosophical Transactions of the Royal Society of London*, 335, 1011–55.

Mackereth, F.J.H. (1958) A portable core sampler for lake deposits. *Limnology and Oceanography*, 3, 181–91.

Mackereth, F.J.H. (1966) Some chemical observations on post-glacial lake sediments. *Philosophical Transactions of the Royal Society of London*, B 250, 165–213.

Mackereth, F.J.H. (1969) A short core sampler for sub-aqueous deposits. *Limnology and Oceanography*, 14, 145–51.

Magnuson, J. (1990) Long-term ecological research and the invisible present. *BioScience*, 40, 495–501.

Maher, B.A. and Thompson, R. (eds.) (1999) *Quaternary Climates, Environments and Magnetism*. Cambridge: Cambridge University Press.

Maher, B.A., Thompson, R., and Hounslow, M.W. (1999) Introduction. In B.A. Maher and R. Thompson (eds.), *Quaternary Climates, Environments and Magnetism*. Cambridge: Cambridge University Press, pp. 1–48.

Manca, M. and Comoli, P. (1995) Temporal variations of fossil Cladocera in the sediments of Lake Orta (N. Italy) over the last 400 years. *Journal of Paleolimnology*, 14, 113–22.

Marsicano, L.J., Hartranft, J.L., Siver, P.A., and Hamer, J.S. (1995) An historical account of water quality changes in Candlewood Lake, Connecticut, over a sixty year period using paleolimnology and ten years of monitoring data. *Lake and Reservoir Management*, 11, 15–28.

Martínez-Cortizas, A., Pontevedra-Pombal, X., García-Rodeja, E., Nóvoa-Muñoz, J.C., and Shotyk, W. (1999) Mercury in a Spanish peat bog: Archive of climate change and atmospheric deposition. *Science*, 284, 939–42.

Mathewes, R.W. and D'Auria, J.M. (1982) Historic changes in an urban watershed determined by pollen and geochemical analyses of lake sediments. *Canadian Journal of Botany*, 19, 2114–25.

Mattson, M.D. and Godfrey, P.J. (1994) Identification of road salt contamination using multiple regression and GIS. *Environmental Management*, 18, 767–73.

McCarthy, J.J., Canziani, O.F., Leary, N.A., Dokken, D.J., and White, K.S. (eds.) (2001) *Climate Change 2001: Impacts, Adaptation & Vulnerability*. Cambridge: Cambridge University Press.

McCrea, J.M. (1950) On the isotopic chemistry of carbonates and a paleotemperature scale. *Journal of Chemical Physics*, 18, 849–57.

McGowan, S., Britton, G., Haworth, E., and Moss, B. (1999) Ancient blue-green blooms. *Limnology and Oceanography*, 44, 436–9.

McKee, P.M., Snodgrass, W.J., Hart, D.R., Duthie, H.C., McAndrews, J.H., and Keller, W. (1987) Sedimentation rates and sediment core profiles of ^{238}U and ^{232}Th decay chain radionuclides in a lake affected by uranium mining and milling. *Canadian Journal of Fisheries and Aquatic Sciences*, 44, 390–8.

Mergeay, J., Verschuren, D., and De Meester, L. (2005) *Daphnia* species diversity in Kenya, and a key to the identification of their ephippia. *Hydrobiologia*, 542, 261–74.

Meriläinen, J. (1967) The diatom flora and the hydrogen ion concentration of water. *Annales Botanici Fennici*, 4, 51–8.

Meriläinen, J.J., Hynynen, J., Palomäki, A., et al. (2001) Pulp and paper mill pollution and subsequent ecosystem recovery of a large boreal lake in Finland: A palaeolimnological analysis. *Journal of Paleolimnology*, 26, 11–35.

Messerli, B., Grosjean, M., Hofer, T., Núñez, L., and Pfister, C. (2000) From nature-dominated to human-dominated environmental change. *Quaternary Science Reviews*, 19, 459–79.

Meyers, P.A. and Ishiwatari, R. (1993) Lacustrine organic geochemistry – an overview of indicators of organic matter sources and digenesis in lake sediments. *Organic Geochemistry*, 20, 867–900.

Meyers, P.A. and Lallier-Vergès, E. (1999) Lacustrine sedimentary organic matter records of Late Quaternary paleoclimates. *Journal of Paleolimnology*, 21, 345–72.

Meyers, P.A. and Teranes, J.L. (2001) Sediment organic matter. In W.M. Last and J.P. Smol (eds.), *Tracking Environmental Change Using Lake Sediments*. Volume 2: *Physical and Geochemical Methods*. Dordrecht: Kluwer, pp. 239–69.

Mezquita, F., Hernández, R., and Rueda, J. (1999) Ecology and distribution of ostracods in a polluted Mediterranean river. *Palaeogeography, Palaeoclimatology, Palaeoecology*, 148, 87–103.

Michelutti, N., Laing, T., and Smol, J.P. (2001a) Diatom assessment of past environmental changes in lakes located near Noril'sk (Siberia) smelters. *Water, Air, and Soil Pollution*, 125, 231–41.

Michelutti, N., Douglas, M.S.V., and Smol, J.P. (2002a) Tracking recent recovery from eutrophication in a High Arctic lake (Meretta Lake, Cornwallis Island, Nunavut, Canada) using fossil diatom assemblages. *Journal of Paleolimnology*, 28, 377–81.

Michelutti, N., Douglas, M.S.V., and Smol, J.P. (2002b) Tracking recovery in a eutrophied, High Arctic lake (Meretta Lake, Cornwallis Island, Canadian Arctic) using periphytic diatoms. *Verhandlungen der Internationale Vereinigung von Limnologie*, 28, 1533–7.

Michelutti, N., Hay, M., Marsh, P., Lesack, L., and Smol, J.P. (2001b) Diatom changes in lake sediments from the Mackenzie Delta, N.W.T., Canada: Paleohydrological applications. *Arctic, Antarctic, and Alpine Research*, 33, 1–12.

Millennium Ecosystem Assessment (2005) *Millennium Ecosystem Assessment, Ecosystems and Human Well-Being*. Washington, DC: Island Press.

Miller, B.B. and Tevesz, M.J.S. (2001) Freshwater molluscs. In J.P. Smol, H.J.B. Birks, and W.M. Last (eds.), *Tracking Environmental Change Using Lake Sediments*. Volume 4: *Zoological Indicators*. Dordrecht: Kluwer, pp. 153–71.

Miller, L.B., McQueen, D.J., and Chapman, L. (1997) *Impacts of Forest Harvesting on Lake Ecosystems: A Preliminary Literature Review*. British Columbia Ministry of Environment, Lands and Park Wildlife Branch, Wildlife Bulletin No. B-84.

Mills, E.L., Leach, J.H., Carlton, J.T., and Secor, C.L. (1993) Exotic species in the Great Lakes: A history of biotic crises and anthropogenic introductions. *Journal of Great Lakes Research*, 19, 1–54.

Mills, E.L., Leach, J.H., Carlton, J.T., and Secor, C.L. (1994) Exotic species and the integrity of the Great Lakes. *BioScience*, 44, 666–76.

Mischke, S., Herzschuh, U., Massmann, G., and Zhang, C. (2007) An ostracod-conductivity transfer function for Tibetan lakes. *Journal of Paleolimnology*, doi:10.1007/S10933-006-9087-5.

Miskimmin, B.M. and Schindler, D.W. (1994) Long-term invertebrate community response to toxaphene treatment in two lakes: 50-yr records reconstructed from lake sediments. *Canadian Journal of Fisheries and Aquatic Sciences*, 51, 923–32.

Moiseenko, T.I., Dauvalter, V.A., and Kagan, L.Ya. (1997) Mountain lakes as indicators of air pollution. *Water Resources*, 24, 556–64.

Moore, J.W. and Ramamoorthay, S. (1984) *Heavy Metals in Natural Waters*. New York: Springer-Verlag.

Morel, F.M.M., Kraepiel, A.M.L., and Amyot, M. (1998) The chemical cycle and bioaccumulation of mercury. *Annual Review of Ecology and Systematics*, 29, 543–66.

Moser, K.A., Korhola, A., Wekström, J., et al. (2000) Paleohydrology inferred from diatoms in northern latitude regions. *Journal of Paleolimnology*, 24, 93–107.

Moss, B. (1978) The ecological history of a medieval man-made lake, Hickling Broad, Norfolk, United Kingdom. *Hydrobiologia*, 60, 23–32.

Moss, B. (1990) Engineering and biological approaches to the restoration from eutrophication of shallow lakes in which aquatic plant communities are important components. *Hydrobiologia*, 200/201, 367–77.

Moss, B. (2001) *The Broads*. London: HarperCollins.

Muir, D., Omelchenko, A., Grift, N., et al. (1996) Spatial trends and historical deposition of polychlorinated biphenyls in Canadian midlatitudes and Arctic lake sediments. *Environmental Science & Technology*, 30, 3609–17.

Munch, C.S. (1980) Fossil diatoms and scales of Chrysophyceae in the recent history of Hall Lake, Washington. *Freshwater Biology*, 10, 61–6.

Naiman, R.J., Magnuson, J.J., McKnight, D.M., and Stanford, J.A. (eds.) (1995) *The Freshwater Imperative: A Research Agenda*. Washington, DC: Island Press.

NAPAP Aquatic Effects Working Group (1991) *National Acidic Precipitation Assessment Program 1990 Integrated Assessment Report*. Washington, DC: National Acid Precipitation Program.

Neff, J.M. (1985) Polycyclic aromatic hydrocarbons. In G.M. Rand and S.R. Petrocelli (eds.), *Fundamentals of Aquatic Toxicology*. Washington, DC: Hemisphere, pp. 416–54.

Nicholls, K.E. (1997) Planktonic green algae in western Lake Erie: The importance of temporal scale in the interpretation of change. *Freshwater Biology*, 38, 419–25.

Nicholls, K.H. and Gerrath, J.F. (1985) The taxonomy of *Synura* (Chrysophyceae) in Ontario with special reference to taste and odour in water supplies. *Canadian Journal of Botany*, 63, 1482–93.

Nilsson, J. (ed.) (1986) *Critical Loads for Nitrogen and Sulphur.* Report 1986: 11, Copenhagen: The Nordic Council of Ministers.

Nilsson, M. and Renberg, I. (1990) Viable endospores of *Thermoactinomycetes vulgaris* in lake sediments as indicators of agricultural history. *Applied and Environmental Microbiology,* 56, 2025–8.

Norris, G. and McAndrews, J.H. (1970) Dinoflagellate cysts from post-glacial lake muds, Minnesota (U.S.A.). *Review of Palaeobotany and Palynology,* 10, 131–56.

Nowaczyk, N.R. (2001) Logging of magnetic susceptibility. In W.M. Last and J.P. Smol (eds.), *Tracking Environmental Change Using Lake Sediments.* Volume 1: *Basin Analysis, Coring, and Chronological Techniques.* Dordrecht: Kluwer, pp. 155–70.

Nriagu, J.O. (1983) Arsenic enrichment in lakes near smelters at Sudbury, Ontario. *Geochimica et Cosmochimica Acta,* 47, 1523–6.

Nriagu, J.O. (1988) A silent epidemic of environmental metal pollution? *Environmental Pollution,* 50, 139–61.

Nriagu, J.O. (1996) A history of global metal pollution. *Science,* 272, 223–4.

Nriagu, J.O. (1998) Paleoenvironmental research: Tales told in lead. *Science,* 281, 1622–3.

Nriagu, J.O. and Coker, R.D. (1983) Sulphur in sediments chronicles past changes in lake acidification. *Nature,* 303, 692–4.

Nriagu, J.O. and Pacyna, J.M. (1988) Quantitative assessment of worldwide contamination of air, water and soils by trace metals. *Nature,* 333, 134–9.

Nriagu, J.O. and Rao, S.S. (1987) Response of lake sediments to changes in trace metal emission from the smelters at Sudbury, Ontario. *Environmental Pollution,* 44, 211–18.

Nriagu, J.O. and Wong, H.K. (1983) Selenium pollution in lakes near the smelters at Sudbury, Ontario. *Nature,* 301, 55–7.

Nriagu, J.O. and Wong, H.K.T. (1986) What fraction of the total metal flux into lakes is retained in the sediments? *Water, Air, and Soil Pollution,* 31, 999–1006.

Nriagu, J.O., Wong, H.K.T., and Coker, R.D. (1982) Deposition and chemistry of pollutant metals in lakes around smelters at Sudbury, Ontario. *Environmental Science & Technology,* 16, 551–60.

Nygaard, G. (1956) Ancient and recent flora of diatoms and chrysophycea in Lake Gribsö. Studies on the humic acid lake Gribsö. *Folia Limnologica Scandinavica,* 8, 32–94.

O'Connell, J., Reavie, E.D., and Smol, J.P. (1997) Diatom epiphytes on *Cladophora* in the St. Lawrence River (Canada). *Diatom Research,* 12, 55–70.

O'Hara, S.L., Street-Perrott, F.A., and Burt, T.P. (1993) Accelerated soil erosion around a Mexican highland lake caused by prehispanic agriculture. *Nature,* 362, 48–51.

O'Sullivan, P.E. (1992) The eutrophication of shallow coastal lakes in Southwest England – understanding and recommendations for restoration, based on palaeolimnology, historical records, and the modeling of changing phosphorus loads. *Hydrobiologia,* 243/244, 421–34.

Odén, S. (1968) *The Acidification of Air Precipitation and its Consequences in the Natural Environment.* Ecology Committee Bulletin No. 1. Stockholm: Swedish National Research Council.

Odgaard, B.V. and Rasmussen, P. (2001) The occurrence of egg-cocoons of the leech *Piscicola geometra* (L.) in recent lake sediments and their relationship with remains of submerged macrophytes. *Archiv für Hydrobiologie,* 152, 671–86.

Odum, E.P. (1985) Trends expected in stressed ecosystems. *BioScience,* 35, 419–22.

Ogden, R.W. (2000) Modern and historical variation in aquatic macrophyte cover of billabongs associated with catchment development. *Regulated Rivers: Research & Management,* 16, 497–512.

Oldfield, F. (1991) Environmental magnetism – a personal perspective. *Quaternary Science Reviews,* 10, 73–85.

Oldfield, F. and Appleby, P.G. (1984) Empirical testing of ^{210}Pb-dating models for lake sediments. In E.Y. Haworth and J.W.G. Lund (eds.), *Lake Sediments and Environmental History.* Minneapolis: University of Minnesota Press, pp. 93–124.

Ostrofsky, M.L. and Duthie, H.C. (1980) Trophic upsurge and the relationship between phytoplankton biomass and productivity in Smallwood Reservoir, Canada. *Canadian Journal of Botany,* 58, 1174–80.

Pacyna, J.M., Scholtz, T., and Li, Y.-F. (1995) Global budgets of trace metal sources. *Environmental Reviews*, 3, 145–59.

Palm, F., Stenson, J.A.E., and Lagergren, R. (2005) Which paleolimnoloical zooplankton records can indicate changes in planktivorous fish predation? *Verhandlungen der Internationale Vereinigung von Limnologie*, 29, 661–6.

Panfili, J., Pontual, H. (de), Troadec, H., and Wright, P.J. (eds.) (2002) *Manual of Fish Sclerochronology*. Brest: Ifremer–IRD coedition.

Parker, D. (2004) Climate: large-scale warming is not urban. *Nature*, 432, 290.

Paterson, A.M., Cumming, B.F., Smol, J.P., and Hall, R.I. (2004) Marked recent increases of colonial scaled chrysophytes in boreal lakes: Implications for the management of taste and odour events. *Freshwater Biology*, 49, 199–207.

Paterson, A.M., Cumming, B.F., Smol, J.P., Blais, J.M., and France, R. (1998) Assessment of the effects of logging and forest fires on lakes in northwestern Ontario: A 30-year paleolimnological perspective. *Canadian Journal of Forest Research*, 28, 1546–56.

Paterson, A.M., Cumming, B.F., Smol, J.P., Blais, J.M., and France, R. (2000) A paleolimnological assessment of the effects of logging, forest fires, and drought on lakes in Northwestern Ontario, Canada. *Verhandlungen der Internationalen Vereinigung von Limnologen*, 27, 1214–19.

Paterson, A.M., Cumming, B.F., Smol, J.P., Morimoto, D., and Szeicz, J. (2002) A paleolimnological investigation of the effects of fire on lake water quality in northwestern Ontario over the past ca. 150 years. *Canadian Journal of Botany*, 80, 1329–36.

Patterson, R.T. and Kumar, A. (2000) Use of Arcellacea (Thecamoebians) to gauge levels of contamination and remediation in industrially polluted lakes. In R.E. Martin (ed.), *Environmental Micropalaeontology*. Dordrecht: Kluwer, pp. 257–78.

Patterson, W.P. and Smith, G.R. (2001) Fish. In J.P. Smol, H.J.B. Birks, and W.M. Last (eds.), *Tracking Environmental Change Using Lake Sediments*. Volume 4: *Zoological Indicators*. Dordrecht: Kluwer, pp. 173–87.

Pearson, D.A.B., Lock, A.S., Belzile, N., and Bowins, R.J. (2002) Assessing Kelly Lake's history as a natural trap for mining industry and municipal effluent during the growth of Sudbury. In D. Rousell and K. Jansons (eds.), *The Physical Environment of the City of Greater Sudbury*. Special Volume 6. Sudbury, Ontario: Ontario Geological Survey, pp. 174–92.

Peck, A.M., Linebaugh, E.K., and Hornbuckle, K. (2006) Synthetic musk fragrances in Lake Erie and Lake Ontario sediment cores. *Environmental Science & Technology*, 40, 5630–35.

Pelley, J. (2006) Synthetic fragrances perfume lake sediments. *Environmental Science & Technology*, 40, 5588.

Pennington, W. (1984) Long-term natural acidification of upland sites in Cumbria: Evidence from post-glacial sediments. In *Freshwater Biological Association Report* 52, pp. 28–46.

Pennington, W., Cambray, R.S., and Fisher, E.M. (1973) Observations on lake sediments using fallout ^{137}Cs as a tracer. *Nature*, 242, 324–6.

Peters, A.J., Jones, K.C., Flower, R.J., et al. (2001) Recent environmental change in North African wetland lakes: a baseline study of organochlorine contaminant residues in sediments from nine sites in the CASSARINA Project. *Aquatic Ecology*, 35, 449–59.

Petrovský, E. and Ellwood, B.B. (1999) Magnetic monitoring of air-, land- and water-pollution. In B.A. Maher and R. Thompson (eds.), *Quaternary Climates, Environments and Magnetism*. Cambridge: Cambridge University Press, pp. 279–322.

Pienitz, R. and Vincent, W. (2000) Effect of climate change relative to ozone depletion on UV exposure in subarctic lakes. *Nature*, 404, 484–7.

Pienitz, R., Douglas, M.S.V., and Smol, J.P. (eds.) (2004) *Long-Term Environmental Change in Arctic and Antarctic Lakes*. Dordrecht: Springer.

Pienitz, R., Smol, J.P., and Birks, H.J.B. (1995) Assessment of freshwater diatoms as quantitative indicators of past climatic change in the Yukon and Northwest Territories, Canada. *Journal of Paleolimnology*, 13, 21–49.

Pienitz, R., Smol, J.P., and Lean, D.R.S. (1997a) Physical and chemical limnology of 24 lakes located between Yellowknife and Contwoyto Lake, Northwest Territories (Canada). *Canadian Journal of Fisheries and Aquatic Sciences*, 54, 347–58.

Pienitz, R., Smol, J.P., and Lean, D.R.S. (1997b) Physical and chemical limnology of 59 lakes located between the Southern Yukon and the Tuktoyaktuk Peninsula, Northwest Territories (Canada). *Canadian Journal of Fisheries and Aquatic Sciences*, 54, 330–46.

Pienitz, R., Smol, J.P., and MacDonald, G.M. (1999) Paleolimnological reconstruction of Holocene climatic trends from two boreal treeline lakes, Northwest Territories, Canada. *Arctic, Antarctic, and Alpine Research*, 31, 82–93.

Pimentel, D., Lach, L., Zuniga, R., and Morrison, D. (2000) Environmental and economic costs of nonindigenous species in the United States. *BioScience*, 50, 53–65.

Pimentel, D., Harvey, C., Resosudarmo, P., et al. (1995) Environmental and economic costs of soil erosion and conservation benefits. *Science*, 267, 1117–23.

Piperno, D.R. (2001) Phytoliths. In J.P. Smol, H.J.B. Birks, and W.M. Last (eds.), *Tracking Environmental Change Using Lake Sediments. Volume 3: Terrestrial, Algal, and Siliceous Indicators*. Dordrecht: Kluwer, pp. 235–51.

Pla, S. (2001) Chrysophycean cysts from the Pyrenees. *Bibliotheca Phycologia*, 109, 1–179.

Pontual, H. (de), Panfili, J., Wright, P.J., and Troadec, H. (2002) Introduction. In J. Panfili, H. de Pontual, H. Troadec, and P.J. Wright (eds.), *Manual of Fish Sclerochronology*. Brest: Ifremer–IRD coedition, pp. 17–28.

Postel, S. and Carpenter, S. (1997) Freshwater ecosystem services. In G.C. Daily (ed.), *Nature's Services*. Washngton, DC: Island Press, pp. 195–214.

Potapova, M. (1996) Epilithic algal communities in rivers of the Kloyma Mountains, NE Siberia, Russia. *Nova Hedwigia*, 63, 309–34.

Potvin, R.R. and Negusanti, J.J. (1995) Declining industrial emissions, improving air quality, and reduced damage to vegetation. In J.M. Gunn (ed.), *Restoration and Recovery of an Industrial Region*. New York: Springer-Verlag, pp. 51–65.

Prowse, C.W. (1987) The impacts of urbanization on major ion flux through catchments: A case study in southern England. *Water, Air, and Soil Pollution*, 32, 277–92.

Psenner, R. and Schmidt, R. (1992) Climate-driven pH control of remote alpine lakes and effects of acid deposition. *Nature*, 356, 781–3.

Punt, W., Hoen, P.P., Blackmore, S., Nilsson, S., and Le Thomas, A. (2007) Glossary of pollen and spore terminology. *Review of Palaeobotany and Palynology*, 143, 1–81.

Quinlan, R. and Smol, J.P. (2001) Chironomid-based inference models for estimating end-of-summer hypolimnetic oxygen from south-central Ontario lakes. *Freshwater Biology*, 46, 1529–51.

Quinlan, R., Smol, J.P., and Hall, R.I. (1998) Quantitative inferences of past hypolimnetic anoxia in south-central Ontario lakes using fossil midges (Diptera: Chironomidae). *Canadian Journal of Fisheries and Aquatic Sciences*, 55, 587–96.

Quinlan, R., Douglas, M.S.V., and Smol, J.P. (2005) Food web changes in Arctic ecosystems related to climate warming. *Global Change Biology*, 11, 1381–6.

Raask, E. (1984) Creation, capture and coalescence of mineral species in coal flames. *Journal of Institutional Energy*, 57, 231–9.

Racca, J.M.J., Gregory-Eaves, I., Pienitz, R., and Prairie, Y.T. (2004) Tailoring palaeolimnological diatom-based transfer functions. *Canadian Journal of Fisheries and Aquatic Sciences*, 61, 2440–54.

Racca, J.M.J., Philibert, A., Racca, R., and Prairie, Y.T. (2001) A comparison between diatom-based pH inference models using Artificial Neural Networks (ANN), Weighted Averaging (WA) and Weighted Averaging Partial Least Squares (WA-PLS) regressions. *Journal of Paleolimnology*, 26, 411–22.

Racca, J.M.J., Wild, M., Birks, H.J.B., and Prairie, Y.T. (2003) Separating wheat from chaff: Diatom taxon selection using an artificial neural network pruning algorithm. *Journal of Paleolimnology*, 29, 123–33.

Rada, R.G., Wiener, J.G., Winfrey, M.R., and Powell, D.E. (1989) Recent increases in atmospheric deposition of mercury to north-central Wisconsin lakes inferred from sediment analyses. *Archives of Environmental Contamination and Toxicology*, 18, 175–81.

Raiswell, R., Buckley, F., Berner, R., and Anderson, F. (1988) Degree of pyritization of iron as a paleoenvironmental indicator of bottom-water oxygenation. *Journal of Sedimentary Petrology*, 58, 812–19.

Ramstack, J.M., Fritz, S.C., and Engstrom, D.R. (2004) Twentieth century water quality trends in Minnesota lakes compared with pre-settlement variability. *Canadian Journal of Fisheries and Aquatic Sciences*, 61, 561–76.

Ramstack, J.M., Fritz, S.C., Engstrom, D.R., and Heiskary, S.A. (2003) The application of a diatom-based transfer function to evaluate regional water-quality trends in Minnesota since 1970. *Journal of Paleolimnology*, 29, 79–84.

Rasmussen, P.E. (1994) Current methods of estimating atmospheric mercury fluxes in remote areas. *Environmental Science & Technology*, 28, 2233–41.

Rasmussen, P. and Anderson, N.J. (2005) Natural and anthropogenic forcing of aquatic macrophyte development in a shallow Danish lake during the last 7000 years. *Journal of Biogeography*, 32, 1993–2005.

Ravera, O., Trincherini, P.R., Beone G.M., and Maiolini, B. (2005) The trend from 1934 to 2001 of metal concentrations in bivalve shells (*Unio pictorum*) from two small lakes: Lake Levico and Lake Caldonazzo (Trento Province, northern Italy). *Journal of Limnology*, 64, 113–18.

Reavie, E.D. and Smol, J.P. (1997) Diatom-based model to infer past littoral habitat characteristics in the St. Lawrence River. *Journal of Great Lakes Research*, 23, 339–48.

Reavie, E.D. and Smol, J.P. (1998) Epilithic diatoms of the St. Lawrence River and their relationships to water quality. *Canadian Journal of Botany*, 76, 251–7.

Reavie, E.D. and Smol, J.P. (1999) Diatom epiphytes in the St. Lawrence River (Canada): Characterization and relation to environmental conditions. In S. Mayama, M. Ideai, and I. Koizumi (eds.), *Proceedings of the Fourteenth International Diatom Symposium*. Koenigstein: Koeltz Scientific Books, pp. 489–500.

Reavie, E.D., Douglas, M.S.V., and Williams, N.E. (2001) Paleoecology of a groundwater outflow using siliceous microfossils. *Ecoscience*, 8, 239–46.

Reavie, E.D., Smol, J.P., and Dillon, P.J. (2002) Inferring long-term nutrient changes in southeastern Ontario lakes: Comparing paleolimnological and mass-balance models. *Hydrobiologia*, 481, 61–74.

Reavie, E.D., Smol, J.P., Carignan, R., and Lorrain, S. (1998) Diatom paleolimnology of two fluvial lakes in the St. Lawrence River: A reconstruction of environmental changes during the last century. *Journal of Phycology*, 34, 446–56.

Reid, M.A. and Ogden, R.W. (2006) Trend, variability or extreme event? The importance of long-term perspectives in river ecology. *River Research and Applications*, 22, 167–77.

Reid, M.A., Sayer, C.D., Kershaw, A.P., and Heijnis, H. (2007) Palaeolimnological evidence for submerged plant loss in a floodplain lake associated with accelerated catchment soil erosion (Murray River, Australia). *Journal of Paleolimnology*, 38, 191–208.

Reid, M., Fluin, J., Ogden, R., Tibby, J., and Kershaw, P. (2002) Long-term perspectives on human impacts on floodplain–river ecosystems, Murray–Darling basin, Australia. *Verhandlungen der Internationale Vereinigung von Limnologie*, 28, 710–16.

Renberg, I. (1981) Improved methods for sampling, photographing and varve-counting of varved lake sediments. *Boreas*, 10, 255–58.

Renberg, I. (1986) A sedimentary record of severe acidification in Lake Blåmissusjön, N. Sweden, through natural soil processes. In J.P. Smol, R.W. Battarbee, R.B. Davis, and J. Meriläinen (eds.), *Diatoms and Lake Acidity*. Dordrecht: Kluwer, pp. 213–19.

Renberg, I. (1990) A 12,600 year perspective of the acidification of Lilla Öresjön, southwest Sweden. *Philosophical Transactions of the Royal Society of London*, B 327, 3357–61.

Renberg, I. (1991) The HON-Kajak sediment corer. *Journal of Paleolimnology*, 6, 167–70.

Renberg, I. and Battarbee, R.W. (1990) The SWAP Palaeolimnology Programme: A synthesis. In B.J. Mason (ed.), *The Surface Waters Acidification Programme*. Cambridge: Cambridge University Press, pp. 281–300.

Renberg, I. and Hultberg, H. (1992) A paleolimnological assessment of acidification and liming effects on diatom assemblages in a Swedish lake. *Canadian Journal of Fisheries and Aquatic Sciences*, 49, 65–72.

Renberg, I. and Nilsson, M. (1991) Dormant bacteria in lake sediments as palaeoecological indicators. *Journal of Paleolimnology*, 7, 127–35.

Renberg, I. and Wik, M. (1984) Dating of recent lake sediments by soot article counting. *Verhandlungen der Internationalen Vereinigung von Limnologen*, 22, 712–18.

Renberg, I., Korsman, T., and Birks, H.J.B. (1993a) Prehistoric increases in the pH of acid-sensitive Swedish lakes caused by land-use changes. *Nature*, 362, 824–6.

Renberg, I., Korsman, T., and Anderson, N.J. (1993b) A temporal perspective of lake acidification in Sweden. *Ambio*, 22, 264–71.

Renberg, I., Wik Persson, M., and Emteryd, O. (1994) Pre-industrial atmospheric lead contamination detected in Swedish lake sediments. *Nature*, 368, 323–6.

Renberg, I., Brodin, Y.W., El-Daoushy, F., et al. (1990) Recent acidification and biological changes in Lilla Öresjön, southwest Sweden, and the relation to atmospheric pollution and land-use history. *Philosophical Transactions of the Royal Society of London*, B 327, 391–6.

Ricciardi, A. and MacIsaac, H.J. (2000) Recent mass invasion of the North American Great Lakes by Ponto-Caspian species. *Trends in Ecology and Evolution*, 15, 62–5.

Ricciardi, A. and Rasmussen, J. (1999) Extinction rates of North American freshwater fauna. *Conservation Biology*, 13, 1220–2.

Rigler, F.H. (1974) *Char Lake Project PF-2, Final Report 1974*. Toronto: Canadian Committee for the International Biological Programme.

Rippey, B., Anderson, N.J., and Foy, R.H. (1997) Accuracy of diatom-inferred total phosphorus concentrations, and the accelerated eutrophication of a lake due to reduced flushing and increased internal loading. *Canadian Journal of Fisheries and Aquatic Sciences*, 54, 2637–46.

Robbins, J.A. (1978) Geochemical and geophysical applications of radioactive lead. In J.O. Nriagu (ed.), *Biogeochemistry of Lead in the Environment*. Amsterdam: Elsevier, pp. 285–393.

Rognerud, S. and Fjeld, E. (2001) Trace element contamination of Norwegian lake sediments. *Ambio*, 30, 11–19.

Rose, N.L. (2001) Fly-ash particles. In W.M. Last and J.P. Smol (eds.), *Tracking Environmental Change Using Lake Sediments*. Volume 2: *Physical and Geochemical Methods*. Dordrecht: Kluwer, pp. 319–49.

Rose, N.L. and Appleby, P.G. (2005) Regional applications of lake sediment dating by spheroidal carbonaceous particle analysis I: United Kingdom. *Journal of Paleolimnology*, 34, 349–61.

Rosenberg, D.M. (1998) A national ecosystem health program for Canada: We should go against the flow. *Bulletin of the Entomological Society of Canada*, 30, 144–53.

Rosenqvist, I.T. (1977) *Acid Soil – Acid Water*. Oslo: Ingeniörforlaget.

Rosenqvist, I.T. (1978) Alternative sources of acidification of river waters in Norway. *The Science of the Total Environment*, 10, 39–49.

Rosqvist, G., Jonsson, C., Yam, R., Karlén, W., and Shemesh, A. (2004) Diatom oxygen isotopes in pro-glacial lake sediments from northern Sweden: A 5000 year record of atmospheric circulation. *Quaternary Science Reviews*, 23, 851–9.

Rowell, H.C. (1996) Paleolimnology of Onondaga Lake: The history of anthropogenic impacts on water quality. *Lake and Reservoir Management*, 12, 35–45.

Ruddiman, W.F. (2003) The Anthropocene greenhouse era began thousands of years ago. *Climatic Change*, 61, 261–93.

Ruggiu, D., Luglié, A., Cattaneo, A., and Panzani, P. (1998) Paleoecological evidence for diatom responses to metal pollution in Lake Orta (N. Italy). *Journal of Paleolimnology*, 20, 333–45.

Rühland, K., Priesnitz, A., and Smol, J.P. (2003) Evidence for recent environmental changes in 50 lakes across the Canadian Arctic treeline. *Arctic, Antarctic, and Alpine Research*, 35, 110–23.

Saarinen, T. and Petterson, G. (2001) Image analysis techniques. In W.M. Last and J.P. Smol (eds.), *Tracking Environmental Change Using Lake Sediments*. Volume 2: *Physical and Geochemical Methods*. Dordrecht: Kluwer, pp. 23–39.

Sabater, S. and Roca, J.R. (1990) Some factors affecting the distribution of diatom assemblages in Pyrenean springs. *Freshwater Biology*, 24, 493–508.

Sakamoto, M. (1966) Primary production by phytoplankton community in some Japanese lakes and its dependence on lake depth. *Archiv für Hydrobiologie*, 62, 1–28.

Sala, O.E., Chapin, III, F.S., Armesto, J., et al. (2000) Global biodiversity scenarios for the year 2100. *Science*, 287, 1770–4.

Salonen, V., Tuovinen, N., and Valpola, S. (2006) History of mine drainage impact on Lake Orijärvi algal communities, SW Finland. *Journal of Paleolimnology*, 35, 289–303.

Sandgren, P. and Fredskild, B. (1991) Magnetic measurements recording Late Holocene man-induced erosion, S. Greenland. *Boreas*, 20, 315–31.

Sandgren, P. and Snowball, I. (2001) Application of mineral magnetic techniques to paleolimnology. In W.M. Last and J.P. Smol (eds.), *Tracking Environmental Change Using Lake Sediments. Volume 2: Physical and Geochemical Methods*. Dordrecht: Kluwer, pp. 217–37.

Sandgren, C., Smol, J.P., and Kristiansen, J. (eds.) (1995) *Chrysophyte Algae: Ecology, Phylogeny and Development*. Cambridge: Cambridge University Press.

Sandman, A., Meriläinen, J.J., Simola, H., et al. (2000) Short-core paleolimnological investigation of Lake Pihlajavesi in the Saimaa Lake complex, eastern Finland: assessment of habitat quality for an endemic and endangered seal population. *Journal of Paleolimnology*, 24, 317–29.

Sayer, C.D., Roberts, N., Sadler, J., David, C., and Wade, P.M. (1999) Biodiversity changes in a shallow lake ecosystem: A multi-proxy palaeolimnological analysis. *Journal of Biogeography*, 26, 97–114.

Sayer, C.D., Jackson, M.J., Hoare, D., et al. (2006) TBT causes regime shift in shallow lakes. *Environmental Science & Technology*, 40, 5269–75.

Scheffer, M. (1998) *Ecology of Shallow Lakes*. London: Chapman & Hall.

Scheffer, M., Hosper, S.H., Meijer, M.-L., Moss, B., and Jeppesen, E. (1993) Alternate equilibria in shallow lakes. *Trends in Ecology and Evolution*, 8, 275–9.

Schelske, C.L. (1999) Diatoms as mediators of biogeochemical silica depletion in the Laurentian Great Lakes. In E.F. Stoermer and J.P. Smol (eds.), *The Diatoms: Applications for the Environmental and Earth Sciences*. Cambridge: Cambridge University Press, pp. 73–84.

Schelske, C.L. and Hodell, D. (1991) Recent changes in productivity and climate of Lake Ontario detected in isotopic analysis of sediments. *Limnology and Oceanography*, 36, 961–75.

Schelske, C.L., Stoermer, E.F., Conley, D.J., Robbins, J.A., and Glover, R.M. (1983) Early eutrophication in the lower Great Lakes: New evidence from biogenic silica in sediments. *Science*, 222, 320–2.

Schelske, C.L., Donar, C.M., and Stoermer, E.F. (1999) A test of paleolimnological proxies for the planktonic/benthic ratio of microfossil diatoms in Lake Apopka. In I. Mayana, I. Masahiko, and I. Koizumi (eds.), *Proceedings of the Fourteenth International Diatom Symposium*. Koenigstein: Koeltz Scientific Books, pp. 367–82.

Schiermeier, Q. (2006) Putting the carbon back: The hundred billion tonne challenge. *Nature*, 442, 620–3.

Schindler, D.E., Leavitt, P.R., Johnson, S.P., and Brock, C.S. (2006) A 500-year context for the recent surge in sockeye salmon (*Oncorhynchus nerka*) abundance in the Alagnak River, Alaska. *Canadian Journal of Fisheries and Aquatic Sciences*, 63, 1439–44.

Schindler, D.E., Leavitt, P.R., Brock, C., Johnson, S.P., and Quay, P.D. (2005) Marine-derived nutrients, commercial fisheries, and the production of salmon and lake algae in Alaska. *Ecology*, 86, 3225–31.

Schindler, D.W. (1974) Eutrophication and recovery in experimental lakes: Implications for lake management. *Science*, 184, 897–9.

Schindler, D.W. (1988) Effects of acid rain on freshwater ecosystems. *Science*, 239, 149–56.

Schindler, D.W. (1990) Experimental perturbations of whole lake as test of hypotheses concerning ecosystem structure and function. *Oikos*, 57, 25–41.

Schindler, D.W. (1998) A dim future for boreal waters and landscapes. *BioScience*, 48, 157–64.

Schindler, D.W. (2001) The cumulative effects of climate warming and other human stresses on Canadian freshwater in the new millennium. *Canadian Journal of Fisheries and Aquatic Sciences*, 58, 18–29.

Schindler, D.W. (2006) Recent advances in the understanding and management of eutrophication. *Limnology and Oceanography*, 51, 356–63.

Schindler, D.W. and Donahue, W.F. (2006) An impending water crisis in Canada's western prairie provinces. *Proceedings of the National Academy of Sciences*, 103, 7210–16.

Schindler, D.W., Curtis, P.J., Parker, B., and Stainton, M. (1996) Consequences of climate warming and lake acidification for UV-B penetration in North American boreal lakes. *Nature*, 379, 705–8.

Schindler, D.W., Kalff, J., Welch, H.E., Brunskill, G.J., Kling, H., and Kritsch, N. (1974) Eutrophication in the High Arctic – Meretta Lake, Cornwallis Island (75°N lat.). *Journal of the Fisheries Research Board of Canada*, 31, 647–62.

Schindler, D.W., Mills, K.H., Malley, D.F., et al. (1985) Long-term ecosystem stress: The effects of years of experimental acidification of a small lake. *Science, 228*, 1395–401.

Schmidt, R., Kamenik, C., Tessadri, R., and Koinig, K. (2006) Climatic changes from 12,000 to 4000 years ago in the Austrian central Alps tracked by sedimentological and biological proxies of a lake sediment core. *Journal of Paleolimnology*, 35, 491–505.

Schnurrenberger, D., James Russell, J., and Kelts, K. (2003) Classification of lacustrine sediments based on sedimentary components. *Journal of Paleolimnology*, 29, 141–54.

Scholz, C.A. (2001) Applications of seismic sequence stratigraphy in lacustrine basins. In W.M. Last and J.P. Smol (eds.), *Tracking Environmental Change Using Lake Sediments*. Volume 1: *Basin Analysis, Coring, and Chronological Techniques*. Dordrecht: Kluwer, pp. 7–22.

Schustera, M., Roquin, C., Duringer, P., et al. (2005) Holocene Lake Mega-Chad palaeoshorelines from space. *Quaternary Science Reviews*, 24, 1821–7.

Schwalb, A. (2003) Lacustrine ostracodes as stable isotope recorders of late-glacial and Holocene environmental dynamics and climate. *Journal of Paleolimnology*, 29, 265–351.

Scully, N.M., Leavitt, P.R., and Carpenter, S.R. (2000) Century-long effects of forest harvest on the physical structure and autotrophic community of a small temperate lake. *Canadian Journal of Fisheries and Aquatic Sciences*, 57 (Suppl. 2), 50–9.

Seehausen, O., Van Alphen, J.J.M., and Witte, F. (1997) Cichlid fish diversity threatened by eutrophication that curbs sexual selection. *Science*, 277, 1808–11.

Selbie, D.T., Lewis, B.A., Smol, J.P., and Finney, B.P. (2007) Long-term population dynamics of the endangered Snake River sockeye salmon: Evidence of past influences on stock decline and impediments to recovery. *Transactions of the American Fisheries Society*, 136, 800–21.

Shapiro, J. (1995) Lake restoration by biomanipulation – a personal view. *Environmental Reviews*, 3, 83–93.

Shotyk, W., Norton, S.A., and Farmer, J.G. (eds.) (1997) *Peat Bog Archives of Atmospheric Metal Deposition*. Dordrecht: Kluwer.

Shotyk, W., Weiss, D., Applebly, P.G., et al. (1998) History of atmospheric lead deposition since 12,370 ^{14}C yr BP from a peat bog, Jura Mountains, Switzerland. *Science*, 281, 1635–40.

Shotyk, W., Weiss, D., Kramers, J.D., et al. (2001) Geochemistry of the peat bog at Etang de la Gruère, Jura Mountains, Switzerland, and its record of atmospheric Pb and lithogenic trace elements (Sc, Ti, Y, Zr, Hf and REE) since 12,370 ^{14}C yr BP. *Geochimica et Cosmochimica Acta*, 65, 2337–60.

Siegel, F., Kravitz, J., and Galasso, J. (2000) Arsenic in Arctic sediment cores: Pathfinder to chemical weapons dump sites? *Environmental Geology*, 39, 705–6.

Siegenthaler, U., Stocker, T.F., Monnin, E., et al. (2005) Stable carbon cycle–climate relationship during the late Pleistocene. *Science*, 310, 1313–17.

Simberloff, D. and Van Holle, B. (1999) Positive interactions of nonindigenous species: Invasional meltdown? *Biological Invasions*, 1, 21–32.

Simcik, M., Eisenreich, S., Golden, K., et al. (1996) Atmospheric loading of polycyclic aromatic hydrocarbons to Lake Michigan as recorded in sediments. *Environmental Science & Technology*, 30, 3039–46.

Simola, H. (1977) Diatom succession in the formation of annually laminated sediment in Lovojärvi, a small eutrophicated lake. *Annals Botanica Fennici*, 18, 160–8.

Simola, H. (1983) Limnological effects of peatland drainage and fertilization as reflected in the varved sediments of a deep lake. *Hydrobiologia*, 106, 43–57.

Simola, H., Hanski, I., and Liukkonen, M. (1990) Stratigraphy, species richness and seasonal dynamics of plank-tonic diatoms during 418 years in Lake Lovojärvi, south Finland. *Annals Botanica Fennici*, 27, 241–59.

Simola, H., Kukkonen, M., Lahtinen, J., and Tossavainen, T. (1995) Effects of intensive forestry and peatland management on forest ecosystems in Finland: Sedimentary records of diatom floral changes. In D. Marino and M. Montresor (eds.), *Proceedings of the Thirteenth International Diatom Symposium 1994*, Bristol: Biopress Ltd, pp. 121–8.

Siver, P.A. (1991) *The Biology of Mallomonas*. Dordrecht: Kluwer.

Siver, P.A. (1993) Inferring the specific conductivity of lake water with scaled chrysophytes. *Limnology and Oceanography*, 38, 1480–92.

Siver, P.A. and Hamer, J.S. (1992) Seasonal periodicity of Chrysophyceae and Synurophyceae in a small New England lake: Implications for paleolimnological research. *Journal of Phycology*, 28, 186–98.

Siver, P.A., Lott, A.M., Cash, E., Moss, J., and Marsicano, J. (1999) Century changes in Connecticut, U.S.A., lakes as inferred from siliceous algal remains and their relationships to land-use changes. *Limnology and Oceanography*, 44, 1928–35.

Smith, A.J., Davis, J.W., Palmer, D.F., Forester, R.M., and Curry, B.B. (2003) Ostracodes as hydrologic indicators in springs, streams and wetlands: A tool for environmental and paleoenvironmental assessment. In L.E. Park and A.J. Smith (eds.), *Bridging the Gap: Trends in the Ostracode Biological and Geological Sciences*. The Paleontological Society Special Papers. Volume. 9. New Haven, CT: Yale University Press, pp. 203–22.

Smith, D.G. (1998) Vibracoring: a new method for coring deep lakes. *Palaeogeography, Palaeolimatology, Palaeoecology*, 140, 433–40.

Smith, J.N. and Levy, E.M. (1990) Geochronology for polycyclic aromatic hydrocarbon contamination in sediments of the Saguenay Fjord. *Environmental Science & Technology*, 24, 874–9.

Smith, S.V., Bradley, R.S., and Abbott, M.B. (2004) A 300 year record of environmental change from Lake Tuborg, Ellesmere Island, Nunavut, Canada. *Journal of Paleolimnology*, 32, 137–48.

Smol, J.P. (1980) Fossil synuracean (Chrysophyceae) scales in lake sediments: A new group of paleoindicators. *Canadian Journal of Botany*, 58, 458–65.

Smol, J.P. (1981) Problems associated with the use of "species diversity" in paleolimnological studies. *Quaternary Research*, 15, 209–12.

Smol, J.P. (1985) The ratio of diatom frustules to chrysophycean statospores: A useful paleolimnological index. *Hydrobiologia*, 123, 199–208.

Smol, J.P. (1986) Chrysophycean microfossils as indicators of lakewater pH. In J.P. Smol, R.W. Battarbee, R.B. Davis, and J. Meriläinen (eds.), *Diatoms and Lake Acidity*. Dordrecht: Dr W. Junk Publishers, pp. 275–87.

Smol, J.P. (1988) Paleoclimate proxy data from freshwater Arctic diatoms. *Verhandlungen der Internationale Vereinigung von Limnologie*, 23, 837–44.

Smol, J.P. (1990a) Are we building enough bridges between paleolimnology and aquatic ecology? *Hydrobiologia*, 214, 201–6.

Smol, J.P. (1990b) Paleolimnology – Recent advances and future challenges. *Memoire dell'Istituto Italiano di Idrobiologia*, 47, 253–76.

Smol, J.P. (1992) Paleolimnology: An important tool for effective ecosystem management. *Journal of Aquatic Ecosystem Health*, 1, 49–58.

Smol, J.P. (1995a) Paleolimnological approaches to the evaluation and monitoring of ecosystem health: Providing a history for environmental damage and recovery. In D. Rapport, C. Gaudet, and P. Calow (eds.), *Evaluating and Monitoring the Health of Large-Scale Ecosystems*. NATO ASI Series, Volume 128. Stuttgart, Springer-Verlag, pp. 301–18.

Smol, J.P. (1995b) Application of chrysophytes to problems in paleoecology. In C. Sandgren, J.P. Smol, and J. Kristiansen (eds.), *Chrysophyte Algae: Ecology, Phylogeny and Development*. Cambridge: Cambridge University Press, pp. 303–29.

Smol, J.P. (2002) *Pollution of Lakes and Rivers: A Paleoenvironmental Perspective*, 1st edn. London: Arnold/New York: Oxford University Press.

Smol, J.P. and Cumming, B.F. (2000) Tracking long-term changes in climate using algal indicators in lake sediments. *Journal of Phycology*, 36, 986–1011.

Smol, J.P. and Dickman, M.D. (1981) The recent histories of three Canadian Shield lakes: a paleolimnological experiment. *Archiv für Hydrobiologie*, 93, 83–108.

Smol, J.P. and Douglas, M.S.V. (2007) From controversy to consensus: Making the case for recent climatic change in the Arctic using lake sediments. *Frontiers in Ecology and the Environment* (in press).

Smol, J.P., Birks, H.J.B., and Last, W.M. (eds.) (2001a) *Tracking Environmental Change Using Lake Sediments*. Volume 3: *Terrestrial, Algal, and Siliceous Indicators*. Dordrecht: Kluwer.

Smol, J.P., Birks, H.J.B., and Last, W.M. (eds.) (2001b) *Tracking Environmental Change Using Lake Sediments*. Volume 4: *Zoological Indicators*. Dordrecht: Kluwer.

Smol, J.P., Walker, I.R., and Leavitt, P.R. (1991) Paleolimnology and hindcasting climatic trends. *Verhandlungen der Internationalen Vereinigung von Limnologen*, 24, 1240–6.

Smol, J.P., Battarbee, R.W., Davis, R.B., and Meriläinen, J. (eds.) (1986) *Diatoms and Lake Acidity: Reconstructing pH from Siliceous Algal Remains in Lake Sediments*. Dordrecht: Dr W. Junk Publishers.

Smol, J.P., Cumming, B.F., Dixit, A.S., and Dixit, S.S. (1998) Tracking recovery patterns in acidified lakes: A paleolimnological perspective. *Restoration Ecology*, 6, 318–26.

Smol, J.P., Wolfe, A.P., Birks, H.J.B., et al. (2005) Climate-driven regime shifts in the biological communities of Arctic lakes. *Proceedings of the National Academy of Sciences*, 102, 4397–402.

Solhøy, T. (2001) Oribatid mites. In J.P. Smol, H.J.B. Birks, and W.M. Last (eds.), *Tracking Environmental Change Using Lake Sediments*. Volume 4: *Zoological Indicators*. Dordrecht: Kluwer, pp. 81–104.

Sommaruga, R., Psenner, R., Schafferer, E., Koinig, K., and Sommaruga-Wögrath, S. (1999) Dissolved organic carbon concentration and phytoplankton biomass in high-mountain lakes of the Austrian Alps: Potential effect of climatic warming on UV underwater attenuation. *Arctic, Antarctic, and Alpine Research*, 31, 247–53.

Sommaruga-Wögrath, R., Koinig, K., Schmidt, R., Sommaruga, R., Tessadi, R., and Psenner, R. (1997) Temperature effects on the acidity of remote alpine lakes. *Nature*, 387, 64–7.

Sörenson, S.P.L. (1909) Enzyme studies II. The measurement and meaning of hydrogen ion concentration in enzymatic processes. *Biochemische Zeitschrift*, 21, 131–200.

Sorvari, S., Korhola, A., and Thompson, R. (2002) Lake diatom response to recent Arctic warming in Finnish Lapland. *Global Change Biology*, 8, 171–81.

Spahni, R., Chappellaz, J., Stocker, T.F., et al. (2005) Atmospheric methane and nitrous oxide of the late Pleistocene from Antarctic ice cores. *Science*, 310, 1317–21.

Spaulding, S. and McKnight, D. (1999) Diatoms as indicators of environmental change in Antarctic waters. In E.F. Stoermer and J.P. Smol (eds.), *The Diatoms: Applications for the Environmental and Earth Sciences*. Cambridge: Cambridge University Press, pp. 245–63.

Speidel, D.H. and Agnew A.F. (eds.) (1988) *Perspectives on Water: Uses and Abuses*. New York: Oxford University Press.

St. Jacques, J.-M., Douglas, M.S.V., Price, N., Drakulic, N., and Gubala, C.P. (2005) The effect of fish introductions on the diatom and cladoceran communities of Lake Opeongo, Ontario Canada. *Hydrobiologia*, 549, 99–113.

Stanley, D.J. and Warne, A.G. (1993) Nile Delta: Recent geological evolution and human impact. *Science*, 260, 628–34.

Stephenson, M., Klaverkamp, J., Motycka, M., Baron, C., and Schwartz, W. (1996) Coring artefacts and contaminant inventories in lake sediment. *Journal of Paleolimnology*, 15, 99–106.

Stern, N. (2007) *The Economics of Climate Change. The Stern Review*. Cambridge: Cambridge University Press.

Stevenson, A.C., Jones, V.J., and Battarbee, R.W. (1990) The cause of peat erosion: a palaeolimnological approach. *New Phytologist*, 114, 727–35.

Stine, S. (1994) Extreme and persistent drought in California and Patagonia during mediaeval time. *Nature*, 369, 546–9.

Stoermer, E.F. (1998) Thirty years of diatom studies on the Great Lakes at the University of Michigan. *Journal of Great Lakes Research*, 24, 518–30.

Stoermer, E.F. and Smol, J.P. (eds.) (1999) *The Diatoms: Applications for the Environmental and Earth Sciences.* Cambridge: Cambridge University Press.

Stoermer, E.F., Emmert, G., and Schelske, C.L. (1989) Morphological variation of *Stephanodiscus niagarae* (Bacillariophyta) in a Lake Ontario sediment core. *Journal of Paleolimnology,* 2, 227–36.

Stoermer, E.F., Wolin, J., and Schelske, C.L. (1993) Paleolimnological comparison of the Laurentian Great Lakes based on diatoms. *Limnology and Oceanography,* 38, 1311–16.

Stoermer, E.F., Emmert, G., Julius, M.L., and Schelske, C.L. (1996) Paleolimnologic evidence of rapid recent change in Lake Erie's trophic status. *Canadian Journal of Fisheries and Aquatic Sciences,* 53, 1451–8.

Stoermer, E.F., Wolin, J., Schelske, C.L., and Conley, D. (1985a) An assessment of ecological changes during the recent history of Lake Ontario based on siliceous algal microfossils preserved in the sediments. *Journal of Phycology,* 21, 257–76.

Stoermer, E.F., Wolin, J., Schelske, C.L., and Conley, D. (1985b) Variations in *Melosira islandica* valve morphology in Lake Ontario sediments related to eutrophication and silica depletion. *Limnology and Oceanography,* 30, 414–18.

Stoermer, E.F., Wolin, J., Schelske, C., and Conley, D. (1985c) Postsettlement diatom succession in the Bay of Quinte, Lake Ontario. *Canadian Journal of Fisheries and Aquatic Sciences,* 42, 754–67.

Stuiver, M., Reimer, P.J., Bard, E., et al. (1998) INTCAL98 radiocarbon age calibration, 24,000–0 cal BP. *Radiocarbon,* 40, 1041–83.

Suarez, A.V. and Tsutsui, N.D. (2004) The value of museum collections for research and society. *BioScience,* 54, 66–77.

Sullivan, T.J, Cosby, B.J., Driscoll, C.T., Charles, D.F., and Hemond, H.F. (1996) Influence of organic acids on model projections of lake acidification. *Water, Air, and Soil Pollution,* 91, 271–82.

Sun, L.G., Yin, X.B., Pan, C.P., and Wang, Y.H. (2005) A 50-years record of dichloro-diphenyl-trichloroethanes and hexachlorocyclohexanes in lake sediments and penguin droppings on King George Island, Maritime Antarctic. *Journal of Environmental Science – China,* 17, 899–905.

Swadling, K.M., Dartnall, H.J.G., and Gibson, J.A.E. (2001) Fossil rotifers and the early colonization of an Antarctic lake. *Quaternary Research,* 55, 380–4.

Swain, E., Engstrom, D., Gringham, M., and Brezonik, P. (1992) Increasing rates of atmospheric mercury deposition in midcontinental North America. *Science,* 257, 784–7.

Swedish NGO Secretariat on Acid Rain (1995) *Critical Loads.* Environmental Factsheet No. 6, Göteborg.

Sweetman, J.N. and Smol, J.P. (2006) Reconstructing past shifts in fish populations using subfossil *Chaoborus* (Diptera: Chaoboridae) remains. *Quaternary Science Reviews,* 25, 2013–23.

Sweets, P.R. (1992) Diatom paleolimnological evidence for lake acidification in the trail ridge region of Florida. *Water, Air, and Soil Pollution,* 65, 43–57.

Sweets, P.R., Bienert, R.W., Crisman, T.L., and Binford, M.W. (1990) Paleoecological investigations of recent lake acidification in northern Florida. *Journal of Paleolimnology,* 4, 103–37.

Talbot, M.R. (2001) Nitrogen isotopes in palaeolimnology. In W.M. Last and J.P. Smol (eds.), *Tracking Environmental Change Using Lake Sediments.* Volume 2: *Physical and Geochemical Methods.* Dordrecht: Kluwer, pp. 401–39.

Telford, R.J. and Birks, H.J.B. (2005) The secret assumption of transfer functions: problems with spatial autocorrelation in evaluating model performance. *Quaternary Science Reviews,* 24, 2173–9.

ter Braak, C.J.F. (1986) Canonical correspondence analysis: a new eigenvector technique for multivariate direct gradient analysis. *Ecology,* 67, 1167–79.

Thienemann, A. (1921) Seetypen. *Die Naturwissenschaften,* 18, 643–6.

Thompson, L.G., Mosley-Thompson, E., Bolzan, J.F., and Koci, B.R. (1985) A 1500-year record of tropical precipitation in ice cores from the Quelccaya ice cap, Peru. *Science,* 229, 971–3.

Thompson, R., Battarbee, R., O'Sullivan, P., and Oldfield, F. (1975) Magnetic susceptibility of lake sediments. *Limnology and Oceanography,* 20, 687–98.

Thompson, R., Price, D., Cameron, N., et al. (2005) Quantitative calibration of remote mountain lake sediments as climate recorders of ice-cover duration. *Arctic, Antarctic, and Alpine Research,* 37, 626–35.

Thoms, M.C., Ogden, R.W., and Reid, M.A. (1999) Establishing the condition of lowland floodplain rivers: a palaeo-ecological approach. *Freshwater Biology*, 41, 407–23.

Thornton, K.W. (1990) Perspectives on reservoir limnology. In K.W. Thornton, B.L. Kimmel, and F.E. Payne (eds.), *Reservoir Limnology: Ecological Perspectives*. New York: John Wiley, pp. 1–13.

Thornton, K.W., Kimmel, B.L., and Payne, F.E. (eds.) (1990) *Reservoir Limnology: Ecological Perspectives*. New York: John Wiley.

Tilman, D. (1989) Ecological experimentation: Strengths and conceptual problems. In G.E. Likens (ed.), *Long Term Studies in Ecology: Approaches and Alternatives*. New York: Springer-Verlag, pp. 136–57.

Tolkien, J.R.R. (1954) *The Lord of the Rings*. London: George Allen & Unwin.

Tomiyasu, T., Matsuyama, A., Eguchi, T., et al. (2006) Spatial variations of mercury in sediment of Minamata Bay, Japan. *Science of the Total Environment*, 368, 283–90.

Trefry, J.H., Metz, S., and Trocine, R.P. (1985) A decline in lead transport by the Mississippi River. *Science*, 230, 439–41.

Troels-Smith, J. (1955) Characterization of unconsolidated sediments. *Danmarks Geologiske Undersogelse IV*, 3, 38–73.

Tuchman, M.L., Stoermer, E.F., and Carney, H.J. (1984) Effects of increased salinity on the diatom assemblage in Fonda Lake, Michigan. *Hydrobiologia*, 109, 179–88.

Turkia, J., Sandman, O., and Huttunen, P. (1998) Palaeolimnological evidence of forestry practices disturbing small lakes in Finland. *Boreal Environment Research*, 3, 45–61.

Turner, E.R. and Rabalais, N.N. (1994) Coastal eutrophication near the Mississippi River delta. *Nature*, 368, 619–21.

Turney, C.S.M. and Lowe, J.J. (2001) Tephrochronology. In W.M. Last and J.P. Smol (eds.), *Tracking Environmental Change Using Lake Sediments. Volume 1: Basin Analysis, Coring, and Chronological Techniques*. Dordrecht: Kluwer, pp. 451–71.

Turton, C.L. and McAndrews, J.H. (2006) Rotifer loricas in second millennium sediment of Crawford Lake, Ontario, Canada. *Review of Palaeobotany and Palynology*, 141, 1–6.

Tyler, P.A. and Bowling, L.C. (1990) The wax and wane of meromixis in estuarine lakes in Tasmania. *Verhandlungen der Internationalen Vereinigung von Limnologen*, 24, 117–21.

Tyler, P.A. and Vyverman, W. (1995) The microbial market place – trade offs at the chemocline of meromictic lakes. *Progress in Phycology*, 11, 325–70.

United Nations (1995) *World Population Prospectus: The 1994 Revision*. New York: United Nations.

United States Environmental Protection Agency (EPA) (2000) *Western Ecology Division Research Update*, May 2000 edn. Corvallis, OR: USEPA.

Uutala, A.J. (1990) *Chaoborus* (Diptera: Chaoboridae) mandibles – paleolimnological indicators of the historical status of fish populations in acid-sensitive lakes. *Journal of Paleolimnology*, 4, 139–51.

Uutala, A.J. and Smol, J.P. (1996) Paleolimnological reconstructions of long-term changes in fisheries status in Sudbury area lakes. *Canadian Journal of Fisheries and Aquatic Sciences*, 53, 174–80.

Uutala, A.J., Yan, N., Dixit, A.S., Dixit, S.S., and Smol, J.P. (1994) Paleolimnological assessment of declines in fish communities in three acidic, Canadian Shield lakes. *Fisheries Research*, 19, 157–77.

Vallentyne, J.R. (1974) The algal bowl: lakes and man. *Fisheries Research Board of Canada, Miscellaneous Special Publication*, 22, 1–186.

van Geel, B. (2001) Non-pollen palynomorphs. In J.P. Smol, H.J.B. Birks, and W.M. Last (eds.), *Tracking Environmental Change Using Lake Sediments. Volume 3: Terrestrial, Algal, and Siliceous Indicators*. Dordrecht: Kluwer, pp. 99–119.

van Geel, B. and Aptroot, A. (2006) Fossil ascomycetes in Quaternary deposits. *Nova Hedwigia*, 82, 313–29.

van Geel, B., Mur, L.R., Ralska-Jasiewiczowa, M., and Goslar, T. (1994) Fossil akinetes of *Aphanizomenon* and *Anabaena* as indicators for medieval phosphate-eutrophication of Lake Gosciaz (central Poland). *Review of Palaeobotany and Palynology*, 83, 97–105.

Van Metre, P.C., Callendar, E., and Fuller, C.C. (1997) Historic trends in organochlorine compounds in river basins identified using sediment cores from reservoirs. *Environmental Science & Technology*, 31, 2339–44.

Vandekerkhove, J., Declerck, S., Brendonck, L., et al. (2005) Uncovering hidden species: Hatching diapausing eggs for the analysis of cladoceran species richness. *Limnology and Oceanography: Methods*, 3, 399–407.

Vasko, K., Toivonen, H.T.T., and Korhola, A. (2000) A Bayesian multinomial Gaussian response model for organism-based environmental reconstruction. *Journal of Paleolimnology*, 24, 243–50.

Vaughn, J.C. (1961) Coagulation difficulties of the south district filtration plant. *Pure Water*, 13, 45–9.

Verleyen, E., Hodgson, D.A., Sabbe, K., and Vyverman, W. (2005) Late Holocene changes in ultraviolet radiation penetration recorded in an east Antarctic Lake. *Journal of Paleolimnology*, 34, 191–202.

Verschuren, D. (2000) Freeze coring soft sediments in tropical lakes. *Journal of Paleolimnology*, 24, 361–5.

Verschuren, D. and Marnell, L.F. (1997) Fossil zooplankton and the historical status of westlslope cutthroat trout in a headwater lake of Glacier National Park, Montana. *Transactions of the American Fisheries Society*, 126, 21–34.

Verschuren, D., Edgington, D.N., Kling, H.J., and Johnson, T.C. (1998) Silica depletion in Lake Victoria: Sedimentary signals at offshore station. *Journal of Great Lakes Research*, 24, 118–30.

Verschuren, D., Tibby, J., Leavitt, P.R., and Roberts, C.N. (1999) The environmental history of a climate-sensitive lake in the former "white Highlands" of Central Kenya. *Ambio*, 28, 494–501.

Verschuren, D., Johnson, T.C., Kling, H.J., et al. (2002) The chronology of human impact on Lake Victoria, East Africa. *Proceedings of the Royal Society of London*, B 269, 289–94.

Vincent, W. and Pienitz, R. (1996) Sensitivity of high-latitude freshwater ecosystems to global change: Temperature and solar radiation. *Geoscience Canada*, 23, 231–6.

Vincent, W.F. and Roy, S. (1993) Solar ultraviolet-B radiation and aquatic primary production: damage, protection and recovery. *Environmental Reviews*, 1, 1–12.

Vinebrooke, R.D., Hall, R.I., Leavitt, P.R., and Cumming, B.F. (1998) Fossil pigments as indicators of phototrophic response to salinity and climatic changes in lakes of western Canada. *Canadian Journal of Fisheries and Aquatic Sciences*, 55, 668–81.

Vitousek, P.M., Aber, J.D., Howarth, R.W., et al. (1997) Human alteration of the global nitrogen cycle: Sources and consequences. *Ecological Applications*, 7, 737–50.

Vollenweider, R.A. (1968) *Scientific Fundamentals of Lake and Stream Eutrophication, with Particular Reference to Phosphorus and Nitrogen as Eutrophication Factors*. Technical Report DAS/DSI/68.27. Paris: OECD.

Vollenweider, R.A. (1975) Input–output models with special reference to the phosphorus loading concept. *Schweizer Zeitschrift Hydrobiologie*, 37, 58–83.

Vyverman, W. and Sabbe, K. (1995) Diatom-temperature transfer functions based on the altitudinal zonation of diatom assemblages in Papua New Guinea: A possible tool in the reconstruction of regional palaeoclimatic changes. *Journal of Paleolimnology*, 13, 65–77.

Walker, I.R. (2001) Midges: Chironomidae and related Diptera. In J.P. Smol, H.J.B. Birks, and W.M. Last (eds.), *Tracking Environmental Change Using Lake Sediments*. Volume 4: *Zoological Indicators*. Dordrecht: Kluwer, pp. 43–66.

Walker, I.R. and Cwynar, L.C. (2006) Midges and palaeotemperature reconstruction – the North American experience. *Quaternary Science Reviews*, 25, 1911–25.

Walker, I.R., Smol, J.P., Engstrom, D.R., and Birks, H.J.B. (1991) An assessment of Chironomidae as quantitative indicators of past climatic change. *Canadian Journal of Fisheries and Aquatic Sciences*, 48, 975–87.

Wania, F. and Mackay, D. (1993) Global fractionation and cold condensation of low volatility organochlorine compounds in polar regions. *Ambio*, 22, 10–18.

Wania, F. and Mackay, D. (1996) Tracking the distribution of persistent organic pollutants. *Environmental Science & Technology*, 30, 390A–396A.

Wansard, G. and Mezquita, F. (2001) The response of ostracod shell chemistry to seasonal change in a Mediterranean freshwater spring environment. *Journal of Paleolimnology*, 25, 9–16.

Warwick, W.F. (1980) Palaeolimnology of the Bay of Quinte, Lake Ontario: 2800 years of cultural influence. *Canadian Bulletin of Fisheries and Aquatic Sciences*, 206, 1–117.

Weatherhead, P.J. (1986) How unusual are unusual events? *American Naturalist*, 128, 150–4.

Webber, J.A., Jackson, K.W., Parekh, P.P., and Bopp, R.F. (2004) Reconstruction of a century of airborne asbestos concentrations. *Environmental Science & Technology*, 38, 707–14.

Weckström, J., Korhola, A., Erästö, P., and Holmström, L. (2006) Temperature patterns over the past eight centuries in Northern Fennoscandia inferred from sedimentary diatoms. *Quaternary Research*, 66, 78–86.

Weider, L.J., Lampert, W., Wessels, M., Colbourne, J., and Lamburg, P. (1997) Long-term genetic shifts in a microcrustacean egg bank associated with anthropogenic changes in the Lake Constance ecosystem. *Proceedings of the Royal Society of London*, B 264, 1613–18.

Weiss, D., Shotyk, W., Cheburkin, A.K., Gloor, M., and Reese, S. (1997) Atmospheric lead deposition from 12,400 to ca. 2000 yrs BP in a peat bog profile, Jura Mountains, Switzerland. *Water, Air, and Soil Pollution*, 100, 311–24.

Wetzel, R.G. (2001) *Limnology: Lake and River Ecosystems*. San Diego: Academic Press.

Whitehead, D.R., Charles, D.F., and Goldstein, R.A. (1990) The PIRLA project (Paleoecological Investigation of Recent Lake Acidification): An introduction to the synthesis of the project. *Journal of Paleolimnology*, 3, 187–94.

Whiting, M.C., Whitehead, D.R., Holmes, R.W., and Norton, S.A. (1989) Paleolimnological reconstruction of recent acidity changes in four Sierra Nevada lakes. *Journal of Paleolimnology*, 2, 285–304.

Whitlock, C. and Larsen, C.P.S. (2001) Charcoal as a fire proxy. In J.P. Smol, H.J.B. Birks, and W.M. Last (eds.), *Tracking Environmental Change Using Lake Sediments*. Volume 3: *Terrestrial, Algal, and Siliceous Indicators*. Dordrecht: Kluwer, pp. 75–97.

Wik, M. and Renberg, I. (1996) Environmental records of carbonaceous fly-ash particles from fossil-fuel combustion. A summary. *Journal of Paleolimnology*, 15, 192–206.

Wilkinson, A.N., Hall, R.I., and Smol, J.P. (1999) Chrysophyte cysts as paleolimnological indicators of environmental change due to cottage development and acidic deposition in the Muskoka–Haliburton region, Ontario, Canada. *Journal of Paleolimnology*, 22, 17–39.

Wilkinson, A.N., Zeeb, B., and Smol, J.P. (2001) *Atlas of Chrysophycean Cysts*, Volume II. Dordrecht: Kluwer.

Willersev, E., Hansen, A.J., Binladen, J., et al. (2003) Diverse plant and animal records from Holocene and Pleistocene sediments. *Science*, 300, 791–5.

Williams, N.E. and Williams, D.D. (1997) Palaeoecological reconstruction of natural and human influences on groundwater outflows. In P.J. Boon and D.L. Howell (eds.), *Freshwater Quality: Defining the Indefinable?* Edinburgh: Scottish Natural Heritage, pp. 172–80.

Williams, W.D. (1987) Salinization of rivers and streams: An important environmental hazard. *Ambio*, 16, 180–5.

Williams, W.D. (1999) Salinisation: A major threat to water resources in the arid and semi-arid regions of the world. *Lakes & Reservoirs: Research and Management*, 4, 85–91.

Winkler, M.G. (1988) Paleolimnology of a Cape Cod kettle pond: Diatoms and reconstructed pH. *Ecological Monographs*, 58, 197–214.

Wolfe, A.P. (2002) Climate modulates the acidity of Arctic lakes on millennial timescales. *Geology*, 30, 215–18.

Wolfe, A.P., Baron, J.S., and Cornett, R.J. (2001) Anthropogenic nitrogen deposition induces rapid ecological change in alpine lakes of the Colorado Front Range (U.S.A.). *Journal of Paleolimnology*, 25, 1–7.

Wolfe, A.P., Vinebrooke, R.D., Michelutti, N., Rivard, B., and Das, B. (2006) Experimental calibration of lake-sediment spectral reflectance to chlorophyll *a* concentrations: Methodology and paleolimnological validation. *Journal of Paleolimnology*, 36, 91–100.

Wolfe, B.B., Edwards, T.W.D., Elgood, R.J., and Beuning, K.R.M. (2001) Carbon and oxygen isotope analysis of lake sediment cellulose: methods and applications. In W.M. Last and J.P. Smol (eds.), *Tracking Environmental Change Using Lake Sediments*. Volume 2: *Physical and Geochemical Methods*. Dordrecht: Kluwer, pp. 373–400.

Wolfe, B.B., Karst-Riddoch, T.L., Vardy, S.R., Falcone, M.D., Hall, R.I., and Edwards, T.W.D. (2005) Impacts of climate and river flooding on the hydro-ecology of a floodplain basin, Peace–Athabasca Delta, Canada since A.D. 1700. *Quaternary Research*, 64, 147–62.

Wolfe, B.B., Hall, R.I., Last, W.M., et al. (2006) Reconstruction of multi-century flood histories from oxbow lake sediments, Peace–Athabasca Delta, Canada. *Hydrological Processes*, 20, 4131–53.

Wolin, J.A. and Duthie, H.C. (1999) Diatoms as indicators of water level change in freshwater lakes. In E.F. Stoermer and J.P. Smol (eds.), *The Diatoms: Applications for the Environmental and Earth Sciences*. Cambridge: Cambridge University Press, pp. 183–204.

Wolin, J.A., Stoermer, E.F., and Schelske, C.L. (1991) Regional changes in Lake Ontario 1981–1987: Microfossil evidence of phosphorus reduction. *Journal of Great Lakes Research*, 17, 229–40.

Wong, C.S., Sanders, G., Engstrom, D., Long, D., Swackhamer, D., and Eisenreich, S. (1995) Accumulation, inventory, and diagenesis of chlorinated hydrocarbons in Lake Ontario sediments. *Environmental Science & Technology*, 29, 2661–72.

Wooller, M.J., Francis, D., Fogel, M.L., Miller, G.H., Walker, I.R., and Wolfe, A.P. (2004) Quantitative paleotemperature estimates from $\delta^{18}O$ of chironomid head capsules preserved in Arctic lake sediments. *Journal of Paleolimnology*, 31, 267–74.

Wright, H.E., Jr. (1990) An improved Hongve sampler for surface sediments. *Journal of Paleolimnology*, 4, 91–2.

Wright, H.E., Jr. (1991) Coring tips. *Journal of Paleolimnology*, 6, 37–49.

Wrona, F.J., Prowse, T.D., Reist, J.D., et al. (2006) Effects of ultraviolet radiation and contaminant-related stressors on Arctic freshwater ecosystems. *Ambio*, 35, 388–401.

Wyn, B., Sweetman, J., Leavitt, P.R., and Donald, D.B. (2007) Historical metal concentrations in lacustrine food webs revealed using fossil ephippia from *Daphnia* spp. *Ecological Applications*, 17, 754–64.

Yan, N.D. and Pawson, T.W. (1997) Changes in the crustacean zooplankton community of Harp Lake, Canada, following invasion by *Bythotrephes cederstroemi*. *Freshwater Biology*, 37, 409–25.

Yan, N.D., Keller, W., Scully, N., Lean, D., and Dillon, P. (1996) Increased UV-B penetration in a lake owing to drought-induced acidification. *Nature*, 381, 141–3.

Yang, J.-R. and Duthie, H.C. (1993) Morphology and ultrastructure or teratological forms of the diatoms *Stephanodiscus niagarae* and *S. parvus* from Hamilton Harbour (Lake Ontario, Canada). *Hydrobiologia*, 269/270, 57–66.

Yang, J.-R., Duthie, H., and Delorme, L.D. (1993) Reconstruction of the recent environmental history of Hamilton Harbour (Lake Ontario, Canada) from analysis of siliceous microfossils. *Journal of Great Lakes Research*, 19, 55–71.

Yin, X., Liu, X., Sun, L., Zhu, R., Xie, Z., and Wang, Y. (2006) A 1500-year record of lead, copper, arsenic, cadmium, zinc level in Antarctic seal hairs and sediments. *Science of the Total Environment*, 371, 252–357.

Yunker, M.B., Macdonald, R.W., Vingarzan, R., Mitchell, R.H., Goyette, D., and Sylvestre, S. (2002) PAHs in the Fraser River basin: a critical appraisal of PAH ratios as indicators of PAH source and composition. *Organic Geochemistry*, 33, 489–515.

Zeeb, B.A. and Smol, J.P. (1995) A weighted-averaging regression and calibration model for inferring lake-water salinity using chrysophycean stomatocysts from western Canadian lakes. *International Journal of Salt Lake Research*, 4, 1–23.

Zeeb, B.A. and Smol, J.P. (1991) Paleolimnological investigation of the effects of road salt seepage on scaled chrysophytes in Fonda Lake, Michigan. *Journal of Paleolimnology*, 5, 263–6.

Zeeb, B.A. and Smol, J.P. (2001) Chrysophyte scales and cysts. In J.P. Smol, H.J.B. Birks, and W.M. Last (eds.), *Tracking Environmental Change Using Lake Sediments. Volume 3: Terrestrial, Algal, and Siliceous Indicators*. Dordrecht: Kluwer, pp. 203–23.

Zhang, G., Parker, A., House, A., et al. (2002) Sedimentary records of DDT and HCH in the Pearl River Delta, South China. *Environmental Science & Technology*, 36, 3671–7.

Zhao, Y., Sayer, C., Birks, H., Hughes, M., and Peglar, S. (2006) Spatial representation of aquatic vegetation by macrofossils and pollen in a small and shallow lake. *Journal of Paleolimnology*, 35, 335–50.

Zolitschka, B., Mingram, J., van der Gaast, S., Jansen, J.H.F., and Naumann, R. (2001) Sediment logging techniques. In W.M. Last and J.P. Smol (eds.), *Tracking Environmental Change Using Lake Sediments. Volume 1: Basin Analysis, Coring, and Chronological Techniques*. Dordrecht: Kluwer, pp. 137–53.

Züllig, H. (1989) Role of carotenoids in lake sediments for reconstructing trophic history during the late Quaternary. *Journal of Paleolimnology*, 2, 23–40.

Index